MAPS

NAVTEQ
The map driving a mobile world

Sponsor Statement

The need to understand where we are and where we are going is as old as the human race itself. Printed maps have addressed this need over time. But in today's world where location, direction, and guidance are essential components of our daily lifestyle, digital maps are transforming the way we use such information and helping us find our way to people, places, and opportunities more easily and safely than ever before.

NAVTEQ—a world leader in digital maps—is pleased to sponsor this unique exhibition about the story of our timeless fascination with maps. NAVTEQ is a Chicago-based company with employees around the globe creating digital maps of the world. Our people are passionate about leveraging technologies in the science of digital mapmaking. NAVTEQ maps are revolutionizing the way people think about and interact with maps, and they are used billions of times each year on the Internet, in portable devices and cell phones, in navigation systems, and in a myriad of business and government applications. We are pleased to be a part of this unique project that celebrates humanity's passion for and interaction with maps.

On behalf of the employees of NAVTEQ, we hope you enjoy Maps: Finding Our Place in the World and this companion book.

Judson C. Green
President and
Chief Executive Officer
NAVTEQ Corporation

The Field Museum · THE NEWBERRY LIBRARY · CHICAGO

COPUBLISHED WITH *The Field Museum*, IN ASSOCIATION WITH *The Newberry Library*

MAPS

FINDING OUR PLACE IN THE WORLD

Edited by James R. Akerman and Robert W. Karrow Jr.

THE

UNIVERSITY OF

CHICAGO PRESS

CHICAGO AND

LONDON

The University of Chicago Press, Chicago 60637
The University of Chicago Press, Ltd., London
© 2007 by The Field Museum
All rights reserved. Published 2007
Printed in the United States of America

16 15 14 13 12 11 10 09 08 07   2 3 4 5

ISBN-13: 978-0-226-01075-5 (cloth)
ISBN-10: 0-226-01075-9 (cloth)

Library of Congress Cataloging-in-Publication Data

Maps : finding our place in the world / edited by James R. Akerman
and Robert W. Karrow, Jr.
       p.   cm.
  "Copublished with The Field Museum, in association with
The Newberry Library."
Includes bibliographical references and index.
ISBN-13: 978-0-226-01075-5 (hardcover : alk. paper)
ISBN-10: 0-226-01075-9 (hardcover : alk. paper)
1. Maps. 2. Maps--History. I. Akerman, James R. II. Karrow,
Robert W. III. Field Museum of Natural History. IV. Newberry Library.
GA108.M34 2007
912—dc22   2007002819

This book is printed on acid-free paper.

CONTENTS

FOREWORD

The Field Museum and the Newberry Library are pleased to present Maps: Finding Our Place in the World, a far-reaching examination of the human endeavor of mapmaking featuring more than one hundred of the world's most important maps, globes, and other artifacts. Several years ago, I posed a question to Kenneth Nebenzahl, author and map expert: when was the last, great comprehensive maps exhibition in the United States? His answer: The World Encompassed at the Baltimore Art Museum in 1952. And when I followed with, "Is the world ready for another one?" Ken's answer was a resounding "Yes!"

The Maps exhibition shows us that the content of a given map is as much determined by culture, historical circumstances, and the interests of mapmakers and map users as it is by the geography that it attempts to depict. From the earliest maps on clay tablets to today's in-car navigation systems, maps tell us not just where we are but who we are. They are artifacts of—and witnesses to—history. And they continue to inspire us to wonder about our place in the world, and mark it for others to see.

Ken Nebenzahl became the guiding spirit of the project, leading The Field Museum to engage David A. Woodward, professor emeritus of geography at the University of Wisconsin–Madison and coeditor of *The History of Cartogra-*

phy, as exhibition curator. With David's untimely death in 2004—an all-too-early loss for the cartography community—Ken rightly suggested that we turn to a heralded map institution for guidance: the Newberry Library.

The Field Museum and the Newberry Library collaborated previously when the Field exhibited The Dead Sea Scrolls; The *Endurance*: Shackleton's Legendary Antarctic Expedition; and Cleopatra: From History to Myth. Building on this Chicago crosstown relationship, the Field accepted with enthusiasm the offer by Newberry's then president and librarian, Charles T. Cullen, of the library's two leading historians of cartography, James Akerman and Robert Karrow, to carry on David Woodward's leadership. As curators of the exhibition—and editors of this volume—Jim and Bob cast their nets wide in an effort to bring the most significant maps to museum visitors. The exhibition and this companion volume reflect this team's deep knowledge of the history of maps as well as their commitment to sharing the subject with a broad audience. Jim and Bob have been assisted in this endeavor through the tireless efforts of Diane Dillon, Todd Tubutis, and Gretchen Baker, whose work has been invaluable in keeping the project moving forward smoothly. Finally, I am happy to note that our institutional partnership continues with the great support of David Spadafora, current president and librarian of the Newberry, who joins me in the sentiments expressed in this foreword.

We are grateful to NAVTEQ Corporation, a leading provider of digital map data worldwide based here in Chicago, and the sponsor of Maps. NAVTEQ's support has allowed us to undertake this complex project and present cutting-edge, twenty-first-century technology to the public. I want to thank especially Judson Green, chief executive officer of NAVTEQ, for his support of this major endeavor.

FOREWORD

This project has also received generous support from the steering committee of the citywide Festival of Maps, under the enthusiastic leadership of Barry MacLean: Roger S. Baskes, Harry Chandler, John A. Edwardson, Arthur Holzheimer, Arthur L. Kelly, Frederick J. Manning, Andrew McNally IV, and Kenneth Nebenzahl. Coordinated by Anna Siegler, this dedicated group galvanized a number of institutions and organizations in Chicago to celebrate the multifarious aspects of maps and mapmaking through related programs and exhibitions. Five of the men mentioned are trustees of the Newberry Library, and the other two have long involvement with Newberry maps and programs. I am especially pleased that Mayor Richard M. Daley has agreed to serve as honorary chairman of this citywide festival.

And it is only fitting that Maps will travel to Baltimore—the site of The World Encompassed over fifty years ago—where it will appear at the Walters Art Museum as the exhibition's sole other venue in spring 2008. We are proud to acknowledge the inspiration for this project by returning another marvelous maps exhibition to Baltimore. Director Gary Vikan and Curator of Manuscripts and Rare Books William Noel have been central to this effort. I want to thank especially the anonymous donor who has been so generous and helped the Field and the Walters present this important exhibition.

Finally, The Field Museum and the Newberry Library are indebted to all of the institutions and private lenders from around the world who graciously lent their one-of-a-kind treasures to this effort. Without their generosity, this exhibition would not be possible.

John W. McCarter Jr.
PRESIDENT AND CHIEF EXECUTIVE OFFICER :: *The Field Museum*

INTRODUCTION

Robert W. Karrow Jr.

There are two kinds of people in the world: those who love maps, and those who can leave them alone. If you are reading this, chances are that you fall into the first category. You're one of those people who can happily pass the time just looking at a map, even—or maybe especially—if it shows a place you've never been. When you were a kid, you were the one who held the road map on car trips. As an adult visiting a new city, you have to beg, buy, or borrow a map before you take your first walking tour. Reading a travel book or a history, you're flipping through it to find the appropriate map and cursing if it's not there.

Maybe you've never really thought about *why* you love maps—is it a visceral reaction to the splash of colors and lines, the kind of reaction you might have to a Jackson Pollock painting? Is it joy at the ease with which a map can help you imagine a future trip or re-create a treasured journey? Is it admiration for the hard labor behind the map, the explorers mushing through Arctic ice or the engineers measuring their precise angles? Maybe it's all of these things, and maybe it's something else entirely. We've designed this book to help you think about maps and why you like them; about how they work and why they're made; about the many kinds of maps and the many kinds of people who made them.

This is not a history of cartography. There are lots of histories out there, and we'll give you some good advice about where to find additional information if you want it. Rather, this is an *introduction* to maps and to their history, a book that we hope will make you want to see more maps and to learn more about what you see.

Should we start by defining a "map"? This will not be easy, and in fact has become a rather contentious issue in the field. Perhaps we should take the tack of Potter Stewart, the associate justice of the Supreme Court who famously declined to define *obscenity*, but maintained that he knew it when he saw it. In fact, *map* conjures up so many powerful images in the popular mind that the word has long had figurative connotations far beyond those we consider in this book. Administrators and politicians "map strategy," teachers use an "English curriculum map," and diplomats follow a "road map" toward peace. Our conception of the noun *map* and the verb *mapping* is not that broad. Still, if you have already glanced at some of the illustrations in this book, you may have thought, "That's not a map!" or "Is *that* a map?" One of the things we hope to do in this book is to stretch your idea of "mapness," to show you a number of objects that help people in other cultures understand their environments graphically. If we can begin to see how some initially unfamiliar constructions can function in maplike ways, perhaps we can also begin to see how culturally and historically conditioned our notions of "mapness" are.

Some of the types of objects you will see illustrated here are maps of the common *orthographic* type—that is, maps in which the viewpoint is straight down, from points directly above the surface being mapped. Some objects are more narrowly defined as *views*; these show an area in *profile*, as though the mapmaker (or, in such a case, should we call him an artist?) were drawing a city from the opposite shore of a river, or *obliquely*, as if the mapmaker had assumed a bird's-eye view. Some objects combine orthographic and oblique elements, as when a city map shows streets as they would appear from above and buildings in profile or from a bird's-eye view. Some objects are as much about representing concepts as places, as in a *cosmographical diagram* purporting to depict the universe, or an *Aboriginal dreaming* in which topographical features, personal and tribal histories, and the supernatural are combined in a complex depiction. What all of these objects have in common is that they are expressed *visually*, in a picture, drawing, carving, or sculpture that conveys a sense of a real or imagined environment.

What would such an operational definition *eliminate* as a "map"? For one thing, it would leave out purely textual or oral accounts, such as medieval itineraries or in-car navigation systems that tell us to "turn left at the next intersection and go seven-tenths of a mile to destination." It would also leave out purely numerical accounts, such as Ptolemy's tables of coordinates, or the masses of

INTRODUCTION

ones and zeros that go into a digital image. While such compilations of letters and numbers are sometimes essential to the *making* of maps, we take the view that until they are somehow reconstituted *visually*, they are not maps.

HOW DO MAPS WORK?

Before we get into the *history* of maps, it may be helpful to look at some common characteristics of maps that help determine how a map, how *any* map, works. These structural features present the mapmaker, *any* mapmaker, with a number of choices. The most basic of these is scale.

Scale

The *scale* of a map is the ratio between the size of the map and the size of the piece of the environment it is trying to show. In modern maps, this is expressed by a fraction, like 1/50,000 or 1:50,000 (both are read as "one to fifty thousand"). This particular "representative fraction," as mapmakers call it, means that one unit on the map represents fifty thousand units on the ground. The fraction holds true no matter what unit of measurement we want to use: one inch represents fifty thousand inches, one centimeter represents fifty thousand centimeters, and so on. It is the representative fraction, and not the physical size of a map, that lies behind the terms *large-scale* and *small-scale* maps. *Large* and *small* in these expressions refer to the relative size of the representative fraction. 1/500,000 is a much *smaller fraction* than 1/50,000, so a map at 1:500,000 scale is at a *smaller scale* than a map at 1:50,000. On a given sheet of paper, a smaller-scale map will show more area than a larger-scale map.

Mapmakers before the nineteenth century seldom used representative fractions on their maps. Western mapmakers after the year 1500 often expressed the scales of their maps in terms of statements of equivalency, such as "one inch [on the map] = one mile [on the ground]." To convert the scale statement "1 inch = 1 mile" to a representative fraction, we need to express the second term (the ground measurement) in the same units as the first term (the map measurement). There are 63,360 inches (160,934.4 cm) in a mile (1.6 km), so "1 inch = 1 mile" becomes the representative fraction 1:63,360, a relatively large-scale map. "1 inch = 100 miles" would be 1:6,336,000, a relatively small-scale map. Awkward representative fractions like 1:633,600 are today most often found in maps made in the United States, where we obstinately continue to reckon distances in feet and inches. In most of the rest of the world, use of the metric system makes scales like 1:25,000 and 1:100,000 much more natural.

With many maps, the question of scale becomes problematic or even irrelevant. A map may not have any kind of scale statement, and may not have been drawn with any attention to mathematical scale at all. We can often calculate some kind of rough scale if we know the size of the area shown and the size of the map. If you draw the route to grandmother's house on the back of an envelope for a friend, you've made a map and some kind of rough scale statement could be derived from it, but there's probably no need for you to express the scale—except, perhaps, to tell your friend that the trip will take about an hour. If a sixteenth-century cartographer drew a hemisphere of the earth on a sheet of foolscap paper, we *could* calculate the scale as roughly 1:42,000,000, since we know that the diameter of the earth is about 8,000 miles and a sheet of foolscap is a little more than a foot wide. Some maps have variable scales: a larger scale for one part, a smaller for another, with the scale constantly changing in between. On the other hand, a Buddhist monk who decides to paint a mandala on the wall of a temple is not concerned about its scale, nor should we be; such a calculation would necessitate knowing just how big, in light-years, say, the monk's cosmos was. The calculation would be meaningless.

But relative scale, as a general characteristic of maps, is *highly* meaningful, since it determines so much of what is put in and what is left out. The distinction between chapters 2 and 3 of this book is primarily one of scale.

Selection

Cartographers refer to the question of what is put in and what is left out by the term *selection*. Any place we might wish to map is filled with myriad details, and very early on we need to decide which details we want to show and which we want to leave out. One thing is certain: we will be able to show *very few* details; we will need to be *very* selective. Just as a memoirist does not feel compelled to report the time of a conversation to the minute, or to chronicle his every waking thought or action, so too the cartographer—*every* cartographer—must leave out a great deal of the truth as she attempts to represent the real world.

If she is mapping a city, she will probably want to show streets, but should she try to show all streets, or only the arterials? What about alleys? Is it important to show which way the traffic runs on one-way streets? Is she obligated to indicate which areas of the city might be considered dangerous, or would that be viewed as scare tactics and politically incorrect? Should she try to show all buildings, or only commercial buildings? Or only commercial buildings in certain areas? Or only major public buildings in certain areas? And just what is a "major" public building, or a "public building," for that matter? Someone is paying her to make the map—what would they think? Decisions, decisions, decisions. But finally, her map is finished. There it is, in all its crisp newness, attractively colored, sharply printed, exuding trustworthiness, as if it were showing everything that mattered—just the truth and nothing but the truth.

Maps are not usually labeled as *nonfiction*. A map, unless titled "imaginary" or employing such gross exaggerations that most people recognize it as inherently false, tends to carry an invisible *nonfiction* label, an implied certification that it is factual and trustworthy. This is especially true when the map meets high standards of design and production, and really "looks the part." And many maps *are* "true" and "trustworthy"—relatively so, for a limited number of purposes and with many qualifications.

One of the great contributions of modern students of maps has been to cast a critical eye on selection. The late J. B. Harley, borrowing a term from literary criticism, began to speak of the "silences" on maps, by which he meant the things maps *didn't* show (Harley 1988). He realized that by looking at what had been *selected out*, we could learn more about why the map was made and what function it served. When a map of colonial America makes no attempt to indicate the presence of large armed Indian nations in and around the seaboard settlements, that silence is not accidental—it represents a choice and *means something*. While you study the illustrations in this book, we invite you to ask yourself what the maps are leaving out, and why.

Generalization

Generalization is a characteristic of maps that is closely related to both selection and scale. It refers to the need to modify the depiction of a feature for the sake of clarity and legibility. Imagine that our hypothetical city mapper wants to put Chicago and its suburbs on a sheet of paper 8 1/2 × 11 inches (21.6 × 27.9 cm). She'll have to use a scale of about 1:275,000, or 1 inch = about 4.3 miles (2.54 cm = 6.9 km). She knows a city street is 46 feet (13.8 m) wide, and, having been urged to be very accurate, calculates that at that scale, the street on her map should be about 1/500 of an inch wide. But that's only half the thickness of a human hair! She doesn't have a pen fine enough to even draw such a line. And even if she did, could the printer print it? And even if he could, would anyone be able to see it? She finally decides that the street has to be at least 1/64 of an inch wide to be reasonably visible. But that would, in effect, be saying that the actual street was more than 350 feet (105 m) wide! She'd be lying, and outrageously at that. Oh, well. Maybe no one will notice. Clearly, there are limits to this accuracy thing.

Even if she possesses the strongest will in the world, our mapmaker is forced to *exaggerate* many of the things she does depict (by drawing city streets as though they are as wide as a football field is long, for example), to *obscure* things (by placing textual labels, such as street names, over them), and even to *lie* outright about much that she knows (like depicting the Chicago River as sky blue when in fact it is a dirty brown, or insisting that the Newberry Library is shaped like a square).

Indeed, there are very real limits to accuracy and precision, and every mapmaker *must* adjust, and fudge, and lie a little here and simplify a little there and put down something that isn't literally true in order to make the map work. Rivers tend to weave back and forth, and so we indicate them with wavy lines that can only be a generalization of their actual course. The makers of portolan charts in the fourteenth century developed a characteristic way of depicting coastlines that may strike us as quaint, but is a perfectly reasonable way of generalizing the ins and outs of an irregular coast. As you ponder a map in this book, we invite you to ask yourself how the depiction of something might have been simplified, and what difference (if any) that makes to the map.

Sign language

All communication requires the use of *signs*. In this technical sense, a sign may be defined as something that someone interprets as standing for something else. All maps consist of signs of various kinds; in fact, any map itself is a sign. We agree to interpret a map as standing for, or representing, something else. Maps convey meaning through sets of *conventional signs* or *symbols*, and the cartographer's choice of signs is inextricably related to the choices made about scale, selection, and generalization. Some signs are *iconic*, meaning that they actually look somewhat like the thing or concept they are depicting. Our city mapper might draw a beach umbrella to indicate a recreational beach, or a cross inside a rectangle to indicate a Christian cemetery. Some signs are more *arbitrary* in that they represent what they do, not because a map reader sees something that reminds him of an object in the "real world," but because of a more or less arbitrary agreement between mapmaker and map user. For instance, a mapmaker may decide to color the British Empire red. He indicates this choice in a legend or key, and the map reader agrees to the proposition that "Red = British Empire."

But although maps may be dense with icons, lines, colors, and tonal variations—all signs conveying meaning—it's hard to imagine a modern one that could rely on graphics alone. Today's maps need texts in the form of names and other labels. Textual signs on maps are also arbitrary, because they can function only if the mapmaker and the map user share a common language. The word *forest* is a conventional sign; speakers of English have agreed to use that combination of letters to mean "a dense concentration of trees." Other language communities would use the sign *bosque*, or *Wald*, or пуща to indicate that concept.

Mapmakers and historians of cartography have paid a lot of attention lately to the ways that maps deploy signs to communicate, and to how maps mediate between an outside environment and an individual map reader. Alfred Korzybski, the founder of a field he called "general semantics," observed that "a map is not the territory it represents" (Korzybski 1933). This seems self-evident, and yet we very often put a faith in maps that is misplaced. Thinking about such a basic, philosophical aspect of mapping as its use of signs (semiotics) has helped us understand just how arbitrary and contingent maps are. In your study of this book, we invite you to think about the ways in which the maps you see might "stand for something" to someone.

We are used to thinking of books as having an author, a flesh-and-blood person who is ultimately responsible for the words that appear between the covers. Maps, however, often have a studied anonymity that can make us mistake the map for the territory. The maker of a medieval world map would never have signed his work; if pressed, he might have attributed it to divine inspiration. Today the name of a map's author might be unstated; wrapped in a corporate mantle like "Rand McNally & Co." or "U. S. Coast and Geodetic Survey"; or only implied, as in "from the best authorities." To be critical map readers, we need to ask of any map, "Who made this map (or paid for it), and why?" Our hypothetical city mapper is unlikely to depict neighborhoods she may consider dangerous if her sponsor is trying to stimulate tourism, but she might adopt representational techniques that would exaggerate the amount of green space if her employer were the park commission.

As with their contemporary counterparts, maps in the past were made by real people, profit-minded businesses, and politically sensitive entities. Some old maps, like some modern ones, were blatantly propagandistic; their makers had axes to grind and deployed all the cartographic tools at their disposal to try to make their maps "stand for" a specific something in the minds of their readers. But propaganda aside, *all* maps, old and new, inevitably reflect points of view, personal and political perspectives, and authorial biases. Recent scholarship in the history of cartography has tried to peel back the layers of anonymity and objectivity with which many maps have been clothed. In your perusal of *Maps: Finding Our Place in the World*, we encourage you to think about the authors, personal and corporate, who stand, sometimes closely and sometimes distantly, behind their maps.

Power

Finally, as a way to summarize all of these internal characteristics of maps, we need to acknowledge the simple *power* of maps. Maps have an undeniable, if sometimes elusive, way of expressing knowledge of, mastery of, and control over the environments they depict; knowledge, mastery, and control are undoubtedly kinds of power. This power of maps rather subtly commands respect, deference, and subordination. Maps charm, intimidate, beguile, and browbeat—by their authority, their signs, what they show and how they show it, even by their scale. The huge format and commanding bird's-eye viewpoint of Jacopo de' Barbari's 1500 view of Venice (fig. 63) seems to say, with Carl Sandburg, "We are the greatest city, the greatest nation; nothing like us ever was." The phrase "the power of maps" has taken on a special significance since the appearance of a wonderful exhibition (and an accompanying book) with that title (Wood and Fels 1992).

Because maps and mapping are such ancient endeavors, we might imagine that the history of cartography has also been around for a long time, but that would be a mistake. The historical study of old maps has only been practiced for about 150 years. Had this book been written anytime in the first 130 of those years, it would have been very different. "Well, of course," you say, "that's because new maps have been made, and we need to update our history." Fifty years ago, mapping from satellites, computer cartography, and geographic information systems (GIS) didn't exist. But it's not just that *cartography* has changed; the *history*, the stories we tell about it, has changed as well.

Few people besides professional historians devote much time to thinking about historiography, or the history of writing history. Nevertheless, the stories that are told about the past are changing all the time, because the people who tell them change. Every generation faces the past with a different perspective, and brings everything it has learned to that task. Inevitably, each generation will view that past differently—in a very real sense, each views a *different* past—and their history will reflect those differences.

Sometimes the stories we tell about the past can change fairly rapidly. In totalitarian states, history can be rewritten virtually overnight, changing heroes to arch criminals and exonerating enemies of the state (probably posthumously). More often, especially in open, democratic societies, our histories change at a leisurely pace as we discover a new fact here, find a new letter there, or learn about the actions of someone we never knew was involved. Gradually, we begin to build up a slightly different picture of the past, a past in which there are more actors and more varied motivations, a more nuanced past in which it is sometimes harder to characterize events as illustrating either glorious righteousness or unalloyed villainy.

But did anything really change? History happened, in all its sloppy, complicated, and random richness. Millions of people made it happen, helped it happen, or had it happen to them. Some of them acted venally, some nobly; some testified truthfully, some lied. Most of what happened is unknown and unknowable: a large portion of the written and graphic record has been destroyed, and historians are left to pick over the detritus, hunting for previously unexamined evidence and poring over the old evidence with new eyes and new questions. Thus do our stories about the past change.

In our review of the external characteristics of maps, we have already seen some of the innovations that modern scholarship has introduced. In fact, even the kind of analysis that identified and elaborated on those characteristics has grown out of what we might call the "new cartographic history." How has the history of cartography changed over the past couple of decades? The sections that follow describe the most crucial changes, in no particular order of importance.

Cross-cultural viewpoint

In the last thirty years, the history of cartography as a field has made a conscious effort to embrace a more cross-cultural perspective. Traditionally, it had emphasized mapping by Europeans, and by the European-trained cartographers in that continent's overseas possessions. This Eurocentric viewpoint paid scant attention to the contributions of Asian and Middle Eastern mapmakers and completely ignored the cartographic achievements of the indigenous peoples of Africa, Australia, Oceania, and the Americas. It is important to recognize that in making this point, the scholarship, value, or legitimacy of any individual contribution to history is not being criticized. Not every study can be, or should try to be, culturally comprehensive; European cartography deserves intensive study, as do Italian and Welsh cartography. We simply point out that scholarship as a whole, and most accounts that professed to be "general" histories of cartography, were heavily tilted toward the experience of Western cultures, and more recent scholarship has tried to redress that balance. As have we, in choosing the objects for the exhibition and in crafting the essays for this volume.

Broader chronological perspective

Traditional histories tended to find most interesting those maps made by Europeans during the early modern and baroque periods—from, say, 1400 to 1700. An immensely creative and productive time for cartography, these centuries include the age of discoveries, when new continents were added to European maps, and the golden era of Dutch cartography, when publishers achieved standards of map engraving and decoration that continue to captivate collectors and scholars. In his influential one-volume *History of Cartography*, finished in 1939, Leo Bagrow ended his story in the second half of the eighteenth century, "at the point where maps ceased to be works of art, the products of individual minds, and where craftsmanship was finally superseded by specialized science and the machine" (Bagrow 1964, 22). Again, there is nothing "wrong" with this emphasis in and of itself; but as a steady diet, it misses the huge and fascinating histories of Enlightenment and nineteenth- and twentieth-century mapping. In this volume, the history of cartography begins in the mists of time and ends yesterday.

More generous definition of the map and mapping behavior

Bagrow approved as "perfectly adequate" the definition written by the mathematician Joseph-Louis, Comte Lagrange, in 1770: "A geographical map is a plane figure representing the surface of the earth, or a part of it." No one would dispute that objects covered by this definition are maps, but many *would* argue that such a definition leaves out a great many objects that undeniably function as maps, including maps that show extraterrestrial or teleological phenomena, and maps depicting completely imaginary environments. A map of the Land of Oz or Middle Earth is not any less a map because Oz and Middle Earth are not real places. An artist who uses maps or maplike imagery to engage his viewers is manipulating the power of maps every bit as much as the most accomplished cartographer. To ignore such representations would be to lose much that enriches our experience of maps and strengthens the history of cartography.

Readers of this book will see illustrated a number of objects that do not fit Lagrange's definition, and in fact, an entire section (chapter 6) is devoted to maps that refuse to represent "the earth or a part of it." We believe that contemplating such objects can only enhance readers' appreciation of the power and ubiquity of the mapping impulse.

More interdisciplinary

Traditionally, the study of old maps has been the province of geographers, historians, librarians, and archivists. One of the most striking developments of the past thirty years has been the flowering of interest in the history of maps evinced by scholars from a wide variety of disciplines. This disciplinary expansion of the history of cartography is evident in the contributions to scholarly journals, which have included articles by anthropologists, philosophers, linguists, and studio artists, among others. Their unique points of view have enormously enriched and stimulated the whole field. The authors of this book are a good indicator of this tendency, for they include three geographers, a historian, a statistician-psychologist, a literary scholar, and an art historian.

Less concern with the "beauty" of old maps

Many treatments of the history of cartography have been excessively concerned with the appearance of the maps as works of art—as splendid monuments of engraving, printing, and coloring. All of us, for obvious reasons, are drawn to maps that have an immediate and arresting visual impact, and graphic brilliance is an important component of the power of maps. We trust that our readers will find maps that engage them in this way, but we also hope to intrigue them with maps that may seem crude, or naïve, or graphically bland, but that are important exemplars of the science, technology, and cultural or personal predilections that lie behind their production and use.

More attention to thematic maps

Thematic maps are maps whose primary function is not to show the obvious geographic features of an environment, but to depict the distribution of phenomena that would be invisible to the average observer. Maps showing the distribution of temperature or languages, the subsurface geology, the variations in magnetic declination, or the density of population are examples of thematic maps. Such maps only became widely used in the nineteenth century, and have been largely passed over by private collectors. And yet, as the authors of our fifth chapter demonstrate, many of them demand our attention and admiration, not only for their contributions to the physical and social sciences, but also for their brilliant and often beautiful solutions to complicated problems of graphic design.

Less concern with "great men"

Queens, kings, statesmen, and great inventors and artists are, of course, intrinsically interesting people, and we can all be inspired and instructed by reading their stories. But one of the achievements of twentieth-century historiography has been to bring more everyday, common, anonymous people into the picture. While not denying the importance and influence of great men and women, such "grassroots" history or "history from the bottom up" has given us more realistic and democratic stories about our past. In cartography, the stories of the Ptolemies, Mercators, and Cassinis loom large, and continue to inspire scholars and popular writers alike. But modern histories of cartography are likely to look beyond these luminaries to try to appreciate the more ordinary men and women who have made and used maps. Nor are the contemporary accounts exclusively concerned with "the products of individual minds," in Bagrow's phrase; modern cartography, like modern science, is likely to be a group effort in which it is difficult to assign personal responsibility for particular innovations.

Less attention to notions of "progress"

Traditionally, historians of cartography have often cast their stories as chronicles of progress, with maps being arranged chronologically as a record of more or less inevitable movement toward some qualitative goal of "accuracy" or "precision": map D is more accurate (or "better") than map C, which is "better" than map B, and so on. Such comparisons can be useful and meaningful when the maps being compared were created at a time and in a place where geometric accuracy and mathematical precision were, in fact, normative goals of mapmaking, as in the Western world since about 1800. But to say that a map drawn by an Ioway Indian in 1836 (fig. 28), for example, is an "inaccurate" or a "crude" map is to hold it to a completely inappropriate standard. One might as well call the classic map of the London Underground (fig. 24) a "crude" map or a "bad" map. True, the location of stations is not geometrically correct, compared to their positions in the real world. But clearly, the important point is that *for the purposes for which they were made*, the Underground map and the Ioway map are effective, even elegant, solutions. Especially in a survey that tries to cover as much chronological and cultural ground as this book does, comparisons such as accurate/inaccurate, good/bad, worse/better are not likely to be very helpful. It was our goal to try, as much as possible, to take each map on its merits, considering the circumstances of its production and the functions it served in the society that produced it.

Greater concern with map use

The story of a map does not end with its creation, any more than the story of a book ends with its publication. Indeed, there is a school of modern criticism that holds that the real story only *starts* with a work's reception by readers. Map readers will rely on their imagination to fill in real and perceived gaps; they will read between the lines and make judgments based on their personal knowledge, prejudices, or ignorance.

Those who acquire, organize, read, and use maps contribute enormously to the cartographic enterprise, and all of our authors have tried to look at the ways individual maps functioned in their time and place. The last chapter of the book, "Consuming Maps," is particularly concerned with the relatively unstudied theme of map use.

These important shifts of emphasis in the historiography of mapmaking have, as I've said, evolved over the past thirty years or so, and are now exemplified in many current monographs and articles in the field—some of which you will find cited in the list of references at the back of this book. One work, however, has been especially crucial in this evolution, and that is the multivolume *History of Cartography*, conceived by David Woodward and J. B. Harley (Harley and Woodward 1987–). Both men were, alas, taken too early by death, and this book is dedicated to their memory. The *History*, among its many other achievements, pioneered the serious examination of maps made by indigenous peoples; its treatment of the cartography of Asian peoples and of the Islamic world is the most comprehensive to appear in any language. The available volumes of the *History* were constant companions as we worked on this book and the exhibition that inspired it; indeed, neither could have been conceived in its present form without the *History*.

THE ORGANIZATION OF THE BOOK

Maps: Finding Our Place in the World is an attempt to introduce readers to the widest possible range of maps of different types, from all times and a variety of different cultures, made for divergent purposes, and depicting a range of environments, without neglecting the famous, the important, the beautiful, the groundbreaking, or the amusing. An impossible assignment? Probably, but we accepted the challenge anyway.

Arrangement was our first concern. If we were to try to reflect the best that has been written about the history of cartography in the past thirty years, it was clear that a purely chronological or national approach would not serve. We decided instead to envision a framework built around the *functions* of maps: the kinds of things maps do, and have done, in a variety of cultures at a variety of times. Despite their contextual differences, maps through time and space have been made to address certain core questions:

→ Where am I and how do I get where I want to be?
→ What does the world look like, and what is my place in it?
→ What does my part of the world look like, and how do I belong there?
→ What happened here, what will happen here, and how are these events important to me?
→ How can maps help me comprehend things that I can't even see?
→ How can maps enhance my literary and artistic experiences?
→ How can I get access to maps, acquire and use them?

These core questions framed our choice of objects for the seven sections of the exhibition and guided our authors in writing the seven chapters of this book. For a starting point, we selected a function of maps that we thought would already be more or less familiar to the largest group of our readers: the use of maps to help us get from here to there.

Chapter One: Finding Our Way

The book begins by examining the popular assumption that maps are primarily concerned with *wayfinding*, the ways in which we orient ourselves and navigate in physical space. We assume that all of our readers will have relied on such maps to navigate around their neighborhood, their country, and their world. But at other times and places, various kinds of wayfinding maps guided Egyptian officials to a part of the Nubian Desert where building stones and gold could be mined; helped Roman legions plan their travels in the Empire; guided medieval pilgrims (or crusaders) to the Middle East; taught Pacific Islanders about wave patterns; made it easier for traders to ply the oceans; and guided bombers and airlines in wartime and peacetime. Today, many motorists can use a GPS (global positioning system) display to find the best route to a restaurant and avoid traffic jams. While all of these maps have essentially the same purpose, the way that they accomplish it depends on geographic and historical circumstances and cultural differences. It turns out that each of these wayfinding maps tells us a great deal about the peoples that made and used them.

In this volume, our pilot through the world of wayfinding maps is James Akerman, the co-curator of the exhibition, a scholar with an abiding interest in how maps help us get around, and editor of the volume of essays *Cartographies of Travel and Navigation* (2006).

Chapter Two: Mapping the World

In chapter 2, we look at maps of the smallest scale; that is, maps that helped their makers to envision the broadest possible human stage—the world, the heavens, even the act of creation itself. Here is a dazzling array of cartographic artifacts that show how different cultures throughout history (and prehistory) have depicted and understood their world cosmologically, religiously, geographically, and politically. To comprehend, to visualize such enormous spaces and abstract concepts, different cultures have employed theological speculation, geometric reasoning, and the most advanced space imagery.

In this volume, we've asked Denis Cosgrove to help us understand how people use maps to structure the wider world and to imagine their place in it. Professor Cosgrove, a cultural geographer, is the author of *Apollo's Eye: A Cartographic Genealogy of the Earth in the Western Imagination* (2001).

Chapter Three: Mapping Parts of the World

The next chapter looks at maps that are, literally, at the opposite end of the scale from cosmological and world maps. These relatively *large-scale maps* represent specific parts of the world and provide a close-up view of areas in which their makers, lived, worked, and moved. They were areas that could be directly seen—at least a small piece at a time—and, if the culture required it, measured. Being able to create and manipulate a model of even a very familiar place aids our understanding of the place, helps solve some practical problems of living there, and expresses the importance of the place to us. Like world maps, these regional and local maps are ancient, come from many cultural traditions, and illustrate a wide variety of representational styles (including true plans, pictorial maps, bird's-eye views, and relief models) and subjects (including river courses, farmsteads, and cities and towns). Though often more detailed than the world and cosmological maps discussed in chapter 2, the appearance and content of these maps is no less influenced by culture, social and political matters, and even the sentimental attachments of their makers and users.

Our introduction to the world of regional and local mapping is led by Matthew Edney, who is the director of the University of Wisconsin project producing the multivolume *History of Cartography*. Professor Edney is also the author of *Mapping an Empire: The Geographical Construction of British India, 1765–1843* (1997).

Chapter Four: Mapping American History

Having shown how maps help us get around and make sense of our greater and lesser worlds, we wanted to show how certain maps can be linked to particular historical events. Maps from all cultures embody elements of their history, of their standard narrative, and readers will notice more or less historical references in maps discussed in all seven chapters. But in this chapter, we wished to concentrate on maps that helped to interpret, or even make, history. To give this broad topic more coherence, we decided to restrict our inquiry to maps illustrative of American history (making it the only chapter of the book with an explicitly national viewpoint). Chapter 4 shows how maps have helped Americans to understand their past, cope with current events, and plan their national future.

Our guide through some five centuries of American history in maps is Professor Susan Schulten, author of *The Geographical Imagination in America, 1880–1950* (2001).

Chapter Five: Visualizing Nature and Society

So far, we have been concerned with maps depicting environments that we can see with our eyes, whether at the scale of the cosmos or a city block, whether depicting a road or a revolution. In this chapter, we pass on to maps whose subject matter is invisible and requires other kinds of observation to be apprehended. The observation may use scientific instruments, such as compasses or magnetometers. It may rely on social surveys, population censuses, or historical records. Its production may involve statistical analysis and require innovative graphic solutions. But the end result is the representation of otherwise invisible data in the form of a map. Particularly since the advent of the scientific revolution, such maps have been employed increasingly often, both to pose questions and to answer them. Among these are such questions as, What is the character and origin of geologic formations? How is it that diseases spread? Why doesn't the compass point to true north? What does the ocean floor look like? and What will the weather be like tomorrow?

To conduct our tour of these *thematic maps* in chapter 5, we welcome Michael Friendly, a psychologist specializing in statistical graphics; and Gilles Palsky, a geographer. Professor Friendly is coauthor of the comprehensive Web site Milestones in the History of Thematic Cartography, Statistical Graphics, and Data Visualization, and Professor Palsky has written *Des chiffres et des cartes: Naissance et développement de la cartographie quantitative français au XIXe siècle* (1996).

In maps, as in fiction, there are "alternate worlds," environments that are not real in any physical sense, but which it inspires or amuses us to treat as real. All the worlds and places that appear on maps are in some sense the creation of the cartographers' imagination. It requires imagination to "see" geologic formations, to conceive of the world as round (before the advent of space flight), to identify oneself as part of a nation, even to see the Yangtze River as a single landscape feature. But the objects discussed in this chapter capitalize on the imaginative power of maps to create entire fictional worlds, from Hell to Utopia, to Middle Earth, to the fantasy game World of Warcraft. Here are maps that somehow complement and complete the texts they often accompany, as well as examples of the imaginative use of cartographic themes in modern art, where they are their own text. This chapter may serve to remind us that all maps pretend to be what they represent, not to deceive us—although they can do that—but to help us tell stories about the world around us.

Ricardo Padrón, a literary scholar, will conduct us on our tour of maps and the imagination. Professor Padrón is the author of *The Spacious Word: Cartography, Literature, and Empire in Early Modern Spain* (2004).

Why and how do people acquire maps? Who uses maps and how do they get them? Virtually all cultures have made maps, but before the age of printing few had been seen by more than a handful of people. Printing expanded cartography's horizons, making it possible for any map to be distributed among hundreds, then thousands, then hundreds of thousands of users. Today's broadcast and electronic technologies make it possible for almost anyone to access millions of maps and even to create and distribute their own. The last chapter traces the origins of map consumption in various cultures and ponders the impact of widely available cartography on modern society.

Our escort through the world of map consumption is Diane Dillon, an art historian and assistant curator of the exhibition that inspired this book. Dr. Dillon is the author of "Mapping Enterprise: Cartography and Commodification at the 1893 World's Columbian Exposition" (2003).

▲

This volume was developed as a companion to the major exhibition of the same name, organized jointly by The Field Museum and the Newberry Library. The relationship between the exhibition and the book is an intimate one, but the experiences of reading the book and seeing the exhibition are meant to complement, rather the duplicate, each other. The themes of the seven chapters in this book were inspired by the seven themes developed in the exhibition, but we asked the contributors to the volume to expand on the themes of the exhibition, drawing in, as they saw fit, maps and ideas that have not been included in the exhibition. The result is not a catalog of the exhibition but a far more wide-ranging excursion into the history and interpretation of cartography than we could have ever imagined ourselves. The book and the exhibition are also united by the title they share. The subtitle, *Finding Our Place in the World*, expresses the simple but powerful idea that maps, in all their diversity, tell us not only where things are but also who *we* are, how we exist and function in relation to the world, how we find our place in the world (on this, see Crampton 2003, 4, 172–79). Maps depict the physical characteristics and spatial organization of our planet. But the content of maps is also determined by, and expresses, the culture, historical circumstances, and ideas and interests of mapmakers and map users. Yet, despite the great diversity of mapmaking reflected in this volume and the exhibition, people across time and space use maps for broadly similar purposes. Each of the chapters that follow is concerned with one of these core ways in which individuals and communities comprehend the world and find their place in it.

Attentive readers will notice maps discussed in one chapter that well could have been taken up in another. After all, the Waldseemüller world map discussed in chapter 4 (fig. 89) is as much a way of looking at the world as it is a document of American history. The fragmentary marble map of ancient Rome (fig. 176) is as much an example of local topographic mapping as it is of map consumption. The earliest known portolan chart (fig. 21) might have been used as an example of regional mapping as easily as it served us to talk about wayfinding maps.

But the use of these overlapping, intertwining, and permeable categories is really one of the points of our book. Many maps happily serve multiple functions, and their richness is partly a measure of the different ways they can be read. We hope that you will be similarly stimulated by further contemplation of the complex and valuable records of human cultural diversity that we call maps.

1234567

FINDING OUR WAY

James R. Akerman

The maps most familiar to Americans today are probably those we use to find our way by car through the nation's highways, back roads, and streets. For the better part of the past century, road maps have been extraordinarily easy to obtain in the United States. Since the mid-1920s, when many service stations adopted the practice of issuing free paper road maps (fig. 1) to their customers, until the present time, when high-quality digital road maps and trip-planning tools are widely available online (fig. 2), Americans have come to view these navigational tools as essential parts of their highly mobile lifestyle—so much so that a road map very likely is what most Americans mean when they use the word *map*. Our own comfort with the idea of using a map to help us navigate by automobile, and indeed with our own geographic mobility, should not color our expectations of wayfinding in other contexts. While the wayfinding maps made across human history share many common traits, whether and how societies used them depended on historical, cultural, and environmental circumstances. Wayfinding maps, it seems, do not just tell us where we are going, they also tell us who we are.

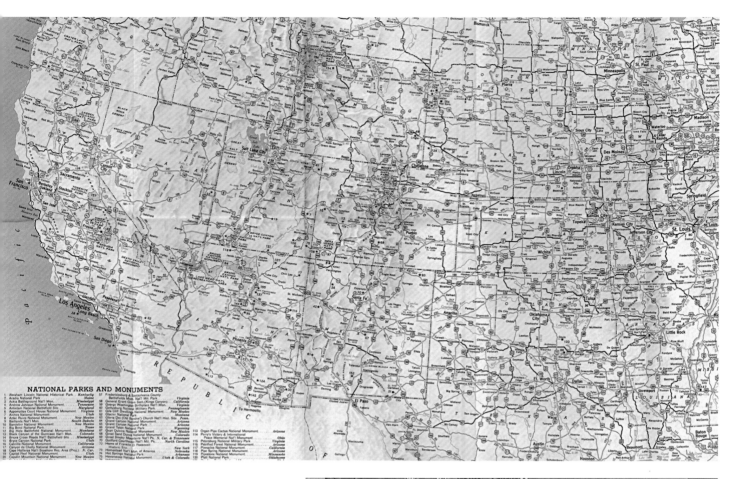

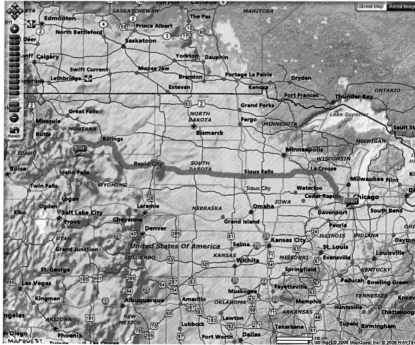

FIGURE 1.

"Road Map of United States" (1941), detail.

FIGURE 2.

Driving directions from Chicago to Old Faithful
Lodge, Yellowstone National Park (2006).

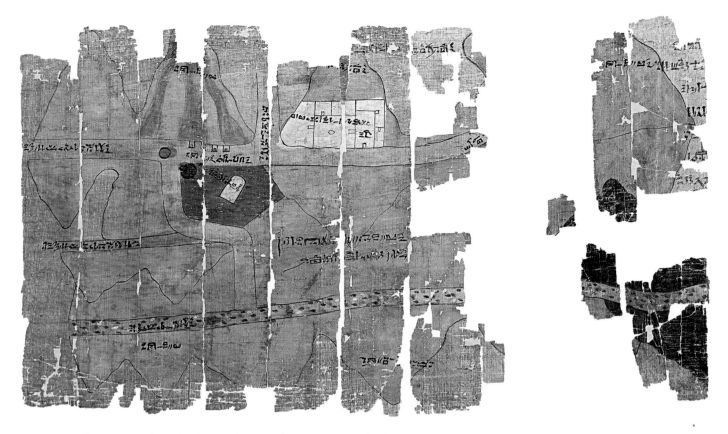

FIGURE 3.

Fragment of the Turin papyrus map
(ca. 1160 BCE).

Maps showing roads and other pathways of movement on land or water are indeed ancient. One of the oldest surviving regional maps of any kind, an Egyptian map drawn on papyrus and dating from about 1160 BCE, is sometimes characterized as the earliest road map. Now preserved in a museum in Turin, Italy, it is in two large fragments, the first of which (fig. 3) shows three routes traversing a mountainous gold- and silver-mining region in the desert east of the Nile. What appear to be roads are actually generalized routes through valleys or along seasonally dry watercourses, or *wadis*. The more important of these is the lower route on this section of the map, Wadi al-Hammamat, which is speckled to represent the rocky character of its dry bed. A smaller valley connects this wadi to a parallel route, where a mining settlement and a well (shown as a red dot) are located. A second fragment, with an uncertain geographic relationship to the first, shows about 9 miles (14.4 km) of the Wadi al-Hammamat, leading to a sandstone formation whose stone was quarried for use in monumental constructions. The map may have been made to help Pharaoh Ramses IV obtain blocks of the sandstone for use in statuary (Harrell and Brown 1992; Shore 1987, 121–25).

There are also ancient Chinese maps showing routes. The oldest surviving regional maps made in China are seven maps drawn on wooden boards that were discovered in a Qin dynasty tomb dated about 300 BCE near Fangmatan in Gansu Province. Some of these depict specific roads, and accompanying inscriptions give the distances to the location of rich resources of timber marked on the map, suggesting that their maker shared one motive for creating a way-finding map—access to natural resources—with the maker of the Turin papyrus (Hsu 1993; Yee 1994a, 37–40). In some instances it is hard to tell whether the thin lines on these maps represent roads on dry land or river courses, and it may be that the cartographer had no need to distinguish between the two.

The greatest road builders of the ancient world, the Romans, left behind one rather spectacular route map, known as the Peutinger map (named for its early sixteenth-century owner). It probably shows geographic information dating from the fourth century CE, but it survives only in a copy dating from the twelfth or thirteenth century (fig. 4A–L). The Peutinger map shows an extensive network of routes leading from Rome to all corners of the known world. Its distinctive notches appear to represent different stages or stops along each route, and it was long thought that the map served as a master map for potential travelers. Most contemporary scholarship agrees, however, that the Peutinger map more likely had a commemorative or ornamental purpose. Nevertheless, it was probably based on practical wayfinding information, including the oral reports of travelers as well as written itineraries, which are verbal (that is, not cartographic) written guides and lists describing particular travel routes (Albu 2005; Delano-Smith 2006, 58–59; Salway 2005; Talbert 2004).

The ancient itineraries of the greater Mediterranean world presumably served a broad range of travelers on military, political, and commercial errands. One of the most complete that survives from those times, the so-called Antonine itinerary (third century CE), includes detailed lists of the land and sea routes of the Roman Empire, staging places, and intervening distances, possibly of interest to the emperors of the Antonine dynasty for military and civil purposes (Dilke 1987b, 234–36). Although Roman civil routes maintained for public communication were marked with milestones listing destinations and distances to them, it is not unlikely that Roman travelers often carried simple itineraries with them on the road as well (Talbert 2006). Ancient and medieval travelers at sea wrote and used sailing directions that indicated distances between harbors and described navigational hazards and currents, prevailing winds, and coastal physical features that would help sailors confirm their location. Known to Greek sailors as *periploi*, to Italians as *portolani*, and to the English as *rutters*, these guides were especially useful to sailors in an era when they preferred to maintain close visual contact with the coast. The need for these guides did not decline after the invention of the sea chart, however. Predominantly verbal sailing directions such as the U.S. Coast and Geodetic Survey's annual *United States Coast Pilot* continue to be published to the present day.

The great significance many cultures placed on religious pilgrimage also spawned itineraries and guides more particular to the needs of travelers on spiritual journeys. Medieval and early modern European pilgrims to Rome and the Holy Land occasionally created detailed accounts of their journey that provided practical travel information (Delano-Smith 2006, 24–29). Buddhists created guides for pilgrims making the circuit of sacred sites in Tibet (Schwartzberg 1994b, 657–59); and guidebooks for Islamic pilgrims, emphasizing the description of shrines, were common by the fourteenth and fifteenth centuries (Rogers 1992, 239). Sanskrit texts that encouraged Hindu faithful to visit the many pilgrimage places, or *tirthas*, scattered about the Indian Subcontinent often included instructions on how to get from site to site (Schwartzberg 1992, 330).

Before later modern times, however, the construction and use of graphic wayfinding aids, of maps designed explicitly to guide travelers, were rare. The Turin papyrus, the Fangmatan maps, and the Peutinger map were made for the use of elites and would have been seen by only a small number of people. Each of these maps helped military, administrative, or political authorities comprehend and control the territories the drawings represented, and were not purely practical wayfinding tools. An understanding of travel routes was apparently important to the readers of these maps, but was not the only, or even their primary, purpose. With few exceptions, maps made explicitly for wayfinding by general populations simply did not exist. When most travel on land was on foot and horseback, travelers could expect to cover no more than 25–30 miles (40–48 km) each day (Delano-Smith 2006, 17). They had no need to form a comprehensive picture of the countryside beyond what they could see with their own eyes or learn from people they met along the way.

The chief mechanism of way-finding has always been oral. Once the traveller was on the road, all maps were redundant so long as he had his itinerary. Regular checks with other travellers and local people at meeting places such as inns, relay and coaching stations, and river crossings ensured that the traveller was following the right road or knew which branch of the track to follow at a bifurcation. There may have been signposts, or markers of some sort . . . [but] the traveller was also told which landmarks to look out for. (Ibid., 45–46)

There could be little additional practical value gained from transforming written itineraries into graphic representations. Until the last two or three centuries, the practical value of maps to wayfinding, as we usually understand it, was largely confined to the planning of trips rather than to use on the road to solve navigational problems of the moment (ibid., 57).

We begin to understand, then, that the automobile road map so familiar to us is the product of, historically speaking, fairly unique circumstances. In the United States roads are mostly paved, well maintained, and well marked. Peaceful conditions prevail over a vast continent-sized space, and most Americans can both afford a car and (at press time!) the fuel required to power it. Our movement on land is not constrained by political or economic conditions like those that prevailed in ancient and medieval times or that prevail in many parts of the world today. When we travel by car today we have so many options and we can travel so far and so quickly that we truly need maps to grasp the

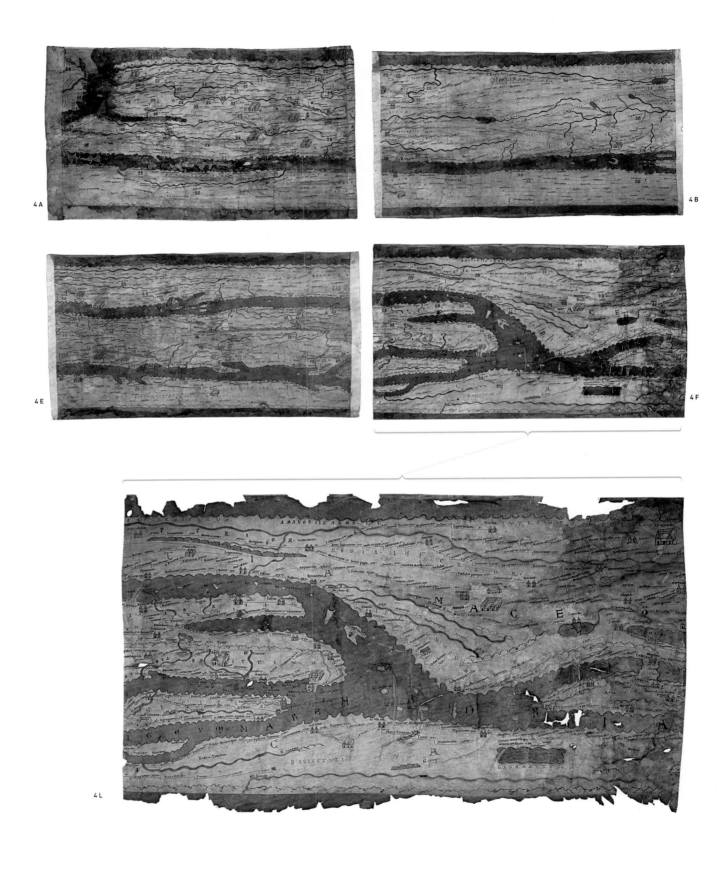

4 A

4 B

4 E

4 F

4 L

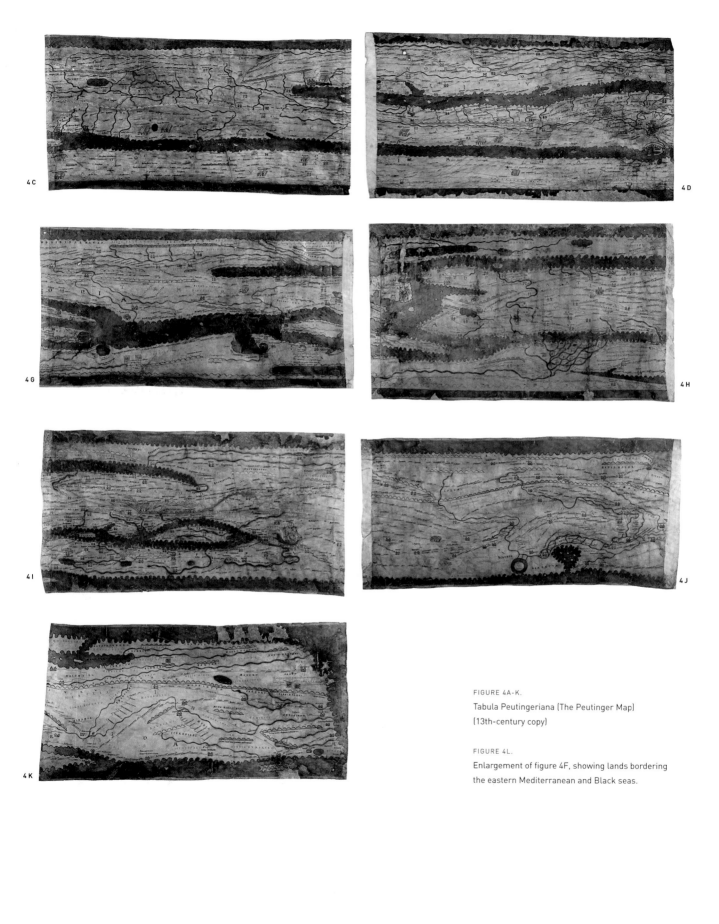

FIGURE 4A-K.

Tabula Peutingeriana (The Peutinger Map)
(13th-century copy)

FIGURE 4L.

Enlargement of figure 4F, showing lands bordering
the eastern Mediterranean and Black seas.

territory into which we venture—and at a level of detail and comprehensiveness that perhaps only an Egyptian pharaoh or a Roman official could dream of in the past. Fortunately, far from being the valuable and exceptional objects like the Turin papyrus and the Peutinger map, modern road maps cost less than a sandwich and can be easily acquired at the nearest convenience store.

This said, we have much in common with travelers of the past. The ready availability of wayfinding maps has not eliminated the need or inclination of modern travelers to ask "the locals" for directions. The popular reality television series *The Amazing Race* pits apparently typical Americans against one another in elaborate worldwide races. Some of the more dramatic moments in each episode revolve around one or more of the teams getting lost, often by misinterpreting a map. Forced to ask the local inhabitants for directions, the teams eventually find their way, but sometimes too late to avoid being eliminated from the competition. Such moments add dramatic tension to the narrative of the television series, but they also appeal to the show's audience, because each of us—even those of us who think we are expert map readers—has had the experience of getting lost in an unfamiliar place and finding our map momentarily indecipherable. Even in our cartographically rich society, there are limits to the utility of maps as wayfinding tools.

PUTTING WAYFINDING MAPS IN CONTEXT

Simply put, *wayfinding* refers to the act of moving along a path or through space from one point to another effectively and successfully. Usually, this requires answering three fundamental questions:

1. Where am I? That is, where am I now, at the beginning of my journey in relation to the larger space I wish to navigate?
2. Where do I want to go? That is, what is my destination and where is it in relation to where I am now?
3. How do I get there? That is, what are the means by which I will get there? What pathways can or should I take? What are the specific instructions that I need to follow that will ensure that I complete this journey successfully?

It may seem that the third question is the most essential one for a wayfinding tool such as a map or written itinerary to answer. We always know where we are in the most immediate sense. I don't need a map to tell me where I am when I'm at home or at my office, but when I walk in the door of an unfamiliar shopping mall in search of a children's shoe store, a map that obligingly tells me "you are here" is most welcome. Likewise, I won't begin my journey toward my destination if the map doesn't tell me which shoe stores are in the mall and where they are.

JAMES R. AKERMAN

We also know from our own experience that we do not always need way-finding tools to find our way. When we move about our homes, workplaces, and schools, we do not refer to maps because we have long ago committed "mental maps" of these environments to memory that we recall instantaneously and unconsciously. Neither do we need maps to execute successfully our routine journeys out in the larger world to well-known destinations, such as the local supermarket, or even to familiar places several hundred miles distant. Three or four times each year, my family drives to suburban Atlanta to visit my parents and extended family. We always have a road atlas handy, just in case, but we hardly need it. We've made the journey so often that even the most minute details—the unusual highway bridge at Columbus, Indiana, the distance in hungry miles from Nashville to the Smoke House restaurant in Monteagle, Tennessee, and the fact that the average speed of drivers on Georgia's Interstate 75 is 85 miles per hour (136 km/hr)—are well known to us.

We usually only refer to wayfinding maps when we must navigate unfamiliar territories and spaces. Large urban hospitals, for example, are often complex spaces that are difficult to navigate. Fortunately, most of us do not visit hospitals often enough to become very familiar with them. Moreover, navigation is further hampered by the tendency of hospital buildings to grow new wings, each of which may have been designed in a different decade, according to a different plan, and to serve a different function or medical technology. Consequently, many hospitals supply visitors with floor plans—like this one showing the plan of Evanston Hospital in Evanston, Illinois—as they enter the building, or else display these prominently at entrances and near elevators (fig. 5). We keep road atlases in our cars or refer to Internet wayfinding services for the same reason: to guide us to and through unfamiliar places. As skilled as I think I am with wayfinding, I would not dream of traveling by car to Yellowstone National Park from my home in Chicago without referring to maps of Wyoming (fig. 6) and several other states, both to plan the trip and to execute my plan.

In each case, these maps help me determine a proper pathway by orienting me in the space the map describes, helping me locate my goal, and describing possible pathways between my initial location and my destination. Their approach, however, is unique to their specific wayfinding contexts. To begin, Evanston Hospital is a confined space surrounded by walls, and so its plan presumes that I will enter the building from relatively few locations. Because of their wayfinding importance, the map marks these entrances with prominent dark rectangles and large labels. Wyoming, in contrast, has open boundaries. Motorists are free to enter the state from any direction, and consequently it would be absurd and counterproductive for a road map to limit the number of possible starting points. (In fact, the map identifies about seventy roads leading into the state.) But note that "where we are" to start on this map is a question not just of geographic location, but also of belonging and identity. If the user of this road map is a resident of the state, their starting point could be anywhere on the map. Here the map helps out by naming most of the populated places in the state and locating them in relation to the state's highway system. Though it is permeable, the boundary drawn around the state on the map reminds all other travelers (residents of Illinois, California, or Ontario) that they may be *in* Wyoming but they are not *of* Wyoming; their origins and a major part of their identity lie somewhere on or beyond the light-green margins of the map.

Both maps do their best to anticipate our most likely destinations. The hospital plan emphasizes areas of particular interest to hospital outsiders, including a cancer care center, a women's hospital, the emergency department, and various labs. Doorways indicate the precise entrance to each of these possible destinations, while several other spaces, presumably not open to the public, remain unnamed and doorless. Many more possible destinations appear on the Wyoming map, but here too there is subtle selectivity at work. Detailed inset maps of Yellowstone National Park and of Casper and Cheyenne, the largest cities in the state, reflect the expectation that these are common destinations. Points of particular interest to travelers appear in red, and national forests, parks, grasslands, and recreation areas are marked in shades of green.

The assumptions made by each cartographer concerning how map users will get from place to place are profoundly different. Evanston Hospital is an interior space and a much smaller space than Wyoming, and it is navigated on foot or by wheelchair. Its map identifies the principal pathways through the building by rendering them in white and the closed interiors of specific

FIGURE 5.
Ground floor, Evanston Hospital, Evanston, IL (2006).

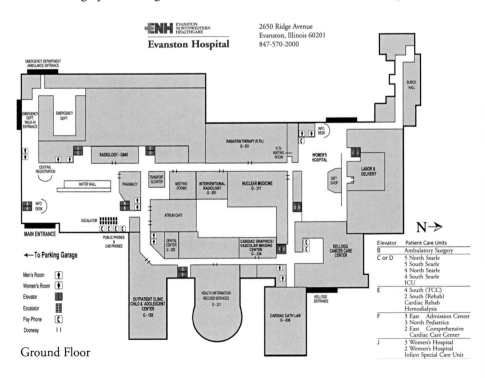

JAMES R. AKERMAN

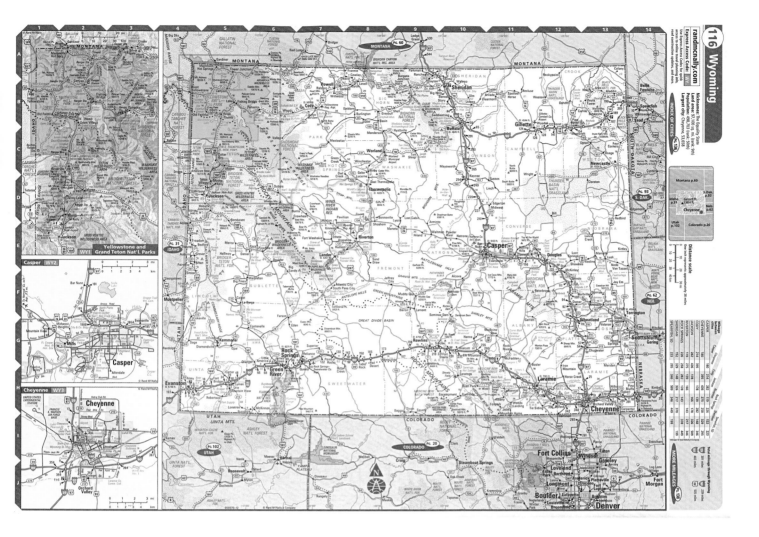

FIGURE 6.

"Wyoming." From *Rand McNally,
The Road Atlas '05* (2005).

rooms in gray, but it also does something more. Every variation in the width of the hallway is represented, so that as we course through the building we may be able to confirm not just the direction of our travel but our current location by the shape and width of the pathway we are following. Though the hospital plan is only two dimensional, we gain from it a fairly good picture of its interior spaces, something that is essential to first-time navigators. Faced with the challenge of representing a space the size of a western state on a small piece of paper, the road map is necessarily more schematic. Even so, we do gain some sense of the width and character of the roads. Fast, limited-access highways (the interstates) appear as wide blue lines. Major state and federal trunk highways appear as red lines of varying widths. Faint gray lines indicate local, but not highly recommended roads; the solid gray indicating paved roads, the double lines indicating unpaved roads. But here the experience of being on any one road, the vision one has of the space around it, is not as important as the quality of the road surface, the speed at which one can travel on it, and its numerical identity.

The landmarks and intermediate destinations the hospital plan identifies also reflect the relatively small scale of this environment and its concern with human comfort. The main "points of interest" here are telephones, rest rooms, restaurants, information desks, and gift shops. Comfort is no less a concern to the automobile traveler, but on the road map it is assumed that the nearest rest room, restaurant, hotel, or service station is in the next town. These are accessed in a timely fashion by correctly calculating distances, reading route numbers correctly, and judging the relative sizes of towns. The road map addresses these wayfinding needs by indicating the distances between towns and crossroads, naming large cities with large letters and small towns with small letters, and clearly marking each major highway with its route number.

As these two examples suggest, how wayfinding maps answer our third question, "How do I get there?" depends on the context of the wayfinding. Part of this context involves the physical setting and the purpose of the wayfinding. But if we hold these factors constant, we might also appreciate the importance of historical, technological, even cultural factors. If, for example, instead of driving to Yellowstone I chose to fly to Jackson, Wyoming, and then rent a car for the final leg of the trip, I would not need a map at all until I was on the ground in Jackson, because I trust the pilot to get me that far safely while I read the in-flight magazine. There was a time, however, when the maps airlines provided passengers were elaborate and even beautiful confections that provided a fairly comprehensive description of the landscape over which their plane was flying. A *United Air Lines Map and Air Log* published in 1931, made for passengers on the "Chicago–San Francisco division of United's coast-to-coast service," helped them "identify many of the most conspicuous geographical divisions of this interesting trip" (fig. 7). Though it couldn't have had any practical navigational value to the passive airline passenger confined to his or her seat, the map enhanced the entertainment value of looking out the window by identifying lakes, railroad lines, rivers, towns, and topographic features visible from the "airway." Beacons and auxiliary fields marked on the map may have assured the nervous passenger in the early days of commercial flight that the plane can never be lost or far from a safe refuge on the ground (Ehrenberg 2003). An accompanying verbal itinerary (or "log") helped the passenger keep track of the number of minutes that passed between each intervening stop at Iowa City, Des Moines, Omaha, Cheyenne, and Rock Springs (cities marked in all capital letters on the map). Frequent stops were necessary, because the number of operating passenger routes was so few, and the distance planes could travel before refueling was relatively limited.

After World War II, radar and the jet engine made it possible for planes to fly both higher and faster. (My theoretical journey to Jackson by air today might offer me little chance to see the ground unless I get a window seat and the day is cloudless, and the opportunity to enjoy the view would be limited, since the flight would be over in three hours or less.) As air routes multiplied, the maps

JAMES R. AKERMAN

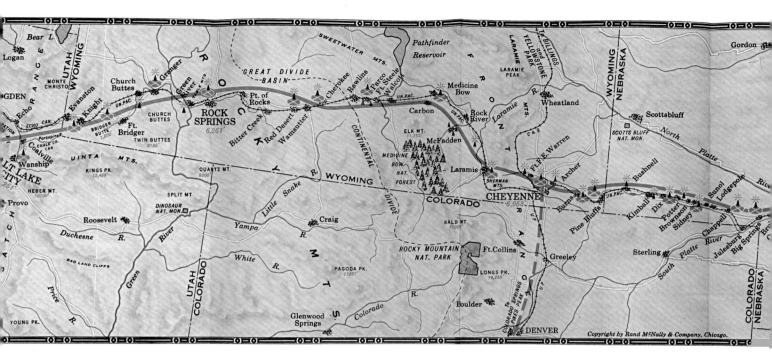

given out to passengers were greatly simplified in order to show an airline's complete route system. These were still fairly elaborate and attractive maps, and were handed out with some ceremony by stewardesses—along with a deck of playing cards and pin-on pilot's wings, as I recall from my first childhood flight—since air travel was still a comparative novelty even in the 1950s and 1960s. So disinterested are today's airline passengers in what is going on below, that most airlines bury their map in the back of in-flight magazines. Some of these magazines dispense with system maps altogether, although they usually include detailed plans of hub airports to guide harried travelers through an airport to connecting flights and ground transportation.

A century ago, virtually all travelers to Yellowstone arrived by one of several train routes serving the various gateways to the park. Almost certainly they would have consulted a map *before* beginning their trip, perhaps one attached to a railroad timetable or in a brochure promoting rail service to the park, like this map (fig. 8) printed on the back of a brochure published by the Chicago, Burlington, and Quincy Railroad (Burlington Route) in 1906. This inexpensively printed but carefully designed map provided only a sketch of the rail network of the western United States, while emphasizing the lines operated by the CB&Q (shown in bold) radiating from its eastern terminus in Chicago. Rail passengers may not have realized that the map offered a highly selective and manipulative view of the western rail network. The lines of connecting railroads that operated trains originating on CB&Q rails are thinner on the map, so as to distin-

FIGURE 7.

Portion of *United Air Lines Map and Air Log Chicago to San Francisco* (1931).

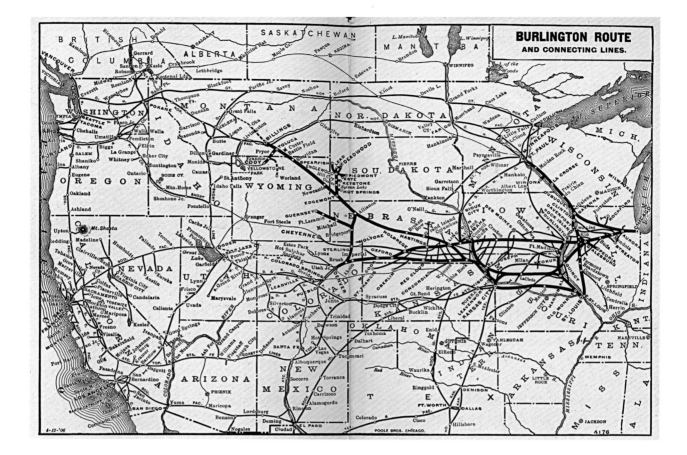

"Burlington Route and Connecting Lines" (1906).

guish them from Burlington lines. Railroads that competed more directly with the Burlington Route, such as the Chicago and Northwestern, do not appear at all on the map. Following a practice common to railroad map design at the time, the scale of distances was also manipulated by the cartographer (Musich 2006, 115). The far western states were compressed in the east–west direction, while the Great Plains states were elongated east to west, adding to the impression that the Burlington occupied a central and dominant geographic position among railroads heading west from Chicago. The omission of most rail lines emanating from Chicago's midwestern urban rivals may also have had the effect of channeling unsuspecting passengers to that city in order to increase the chances that they would choose the Burlington for their travel to Yellowstone.

Once on the trip, however, the map would be of little navigational use to the traveler, and did not need to be. Rail passengers do not control the speed or the direction of the train; neither, in most cases, do they determine which stops the train will make or when. Its route and timetable are predetermined by the railroad, and the passengers' only navigational responsibilities are to make sure that they are ready to board and leave the train at the appointed time and

JAMES R. AKERMAN

place. During the trip, then, the single most important navigational tool for the rail passenger is the timetable, which of course is nothing more than a modern itinerary showing the station stops in order of appearance on the route, along with the times when the train will arrive at each. In the heyday of passenger rail service in the United States, railroad maps were well suited to the passengers' limited navigational requirements. The most elaborate of these maps often simplified the geography of the railroad system. Individual rail routes were drawn schematically as nearly straight lines, eliminating all but the most important bends in the route and often distorting the geographic scale. Yet railroad map designers usually took great care to name virtually all of the stations along the route, because it was essential for travelers planning a trip to know which towns were served by a given railroad and in which order passengers would encounter those towns. Mapmakers placed the names of towns at right angles to the route, maximizing the number of station names that could be included. The resulting map was typically not very good for measuring distances or determining the location of places off the route of the railroad, but it was a superb counterpart to the timetable.

As a motorist traveling to Yellowstone, my wayfinding needs are entirely different. I am the pilot of my automobile. I control the speed and direction of the vehicle as well as the specific location, time, and duration of any stops that I want or need to make along the way (for food, fuel, sleep, and diversion). The design and content of the Wyoming road map I would use for the final leg of the journey anticipates this (fig. 6). The map offers not one or two but dozens of ways of reaching the park from the eastern boundary of the state. The route via Interstate 80 to Cheyenne and then north on Interstate 25 to Casper, and northeast on U.S. Route 20 is probably the fastest route from Chicago. But the northern route via Interstate 90 through Gillette and Cody has its attractions, not the least of which is the opportunity it would present my family to visit the Black Hills region of South Dakota en route. If I have unlimited time and the inclination to avoid the interstates altogether, the choices of routes entirely via two-lane U.S. and state highways seem almost endless, even in a state with fairly few trunk roads. Anticipating my wanderlust, red squares on the map mark points of interest that might distract me from my path, such as Devil's Tower National Monument in the far northeastern corner of the state and South Pass State Historic Site in the middle. The map indicates differences in the sizes of cities, so I might be able to guess where we might find the most restaurant or motel choices, or where we might be able to make an emergency shopping excursion for a new bathing suit or hiking boots.

Sorting out our first two questions ("Where am I?" and "Where am I going?") is no less contingent on contextual considerations. If I decide to create a custom digital map of my route from Chicago to Yellowstone from an online service, I begin by typing the verbal answer to the first question ("Where am I?"), my home address. Next, I type in the answer to the second question

("Where am I going?"), Yellowstone National Park. The software wants me to be more specific about my destination, so I'll type in Old Faithful Lodge as my preferred final destination. The online software then proceeds to provide the answer to the third question ("How do I get there?") with detailed turn-by-turn verbal instructions—again a variant of the ancient itinerary format—and a general map of my route, which I can customize to my needs by zooming in to and printing specific critical sections—a detailed map of Sioux Falls, perhaps, where I might spend a night on the way.

Now, as a wayfinding scenario, more is assumed here than it seems. As I write these words I am at home typing on my computer, and if I were to leave for a trip to Yellowstone National Park at this instant, this would be my starting point. But one of my kids is sick, my wife isn't home from work yet, we're not packed, and it is the middle of the school year, so this isn't a good time to go to Yellowstone Park. Let's suppose that in anticipation of a trip to Yellowstone this summer, I actually created my Web maps and directions on my office computer. The proper starting point for the maps and directions might still be my home, because I intend to leave on my trip from there. But now "Where am I?" becomes more theoretical than concrete: "Where am I?" is where I expect to start this journey, not actually where I am. Things become even more theoretical when I confess that we are going to Disney World this summer instead of Yellowstone. I *plan* to go to Yellowstone *someday*, but I have not made any concrete plans to do so. "Where am I?" is now as much a theoretical future state of being as it is a geographic location; it is a presumption of good physical and financial health and other favorable circumstances that would make the trip possible, including the presumption that my family will indulge me by participating in yet another road trip.

Where I want to go, at the opposite end of my theoretical journey, is Yellowstone Park. This is a real, tangible place, but a place with which I have no direct experience, since I have never been there. I feel that I know a great deal about it, perhaps a great deal more than the average tourist, since I have read a few histories of the park and its development along with guidebook descriptions, and I have studied many maps, historical drawings, and photographs of the region. But in many respects, Yellowstone remains a fictitious place to me, a place dwelling only in my imagination. Like most tourists, I am curious to see how the "real" Yellowstone will stack up against my imagination of it. In the meantime, my reading of the Wyoming road map supports a purely imaginary wayfinding.

MAPPING JOURNEYS OF THE HEART AND SOUL

About 750 years ago, Matthew Paris drew a map that, it appears, involved much the same kind of imaginary journey. Paris was a thirteenth-century Benedic-

JAMES R. AKERMAN

tine monk and chronicler who spent most of his adult life in the Abbey of St. Albans, north of London. His chronicles and biographies, particularly *Chronica majora* (Major Chronicles) and *Historia anglorum* (History of the English), are among the most celebrated and valuable of surviving medieval English historical texts. Paris was also an unusually prolific cartographer for his time, having produced a map of the world and of England. He may be best known, however, for his extraordinary road map drawn about 1250, showing the routes pilgrims might take from London to Rome and on to the Holy Land, which survives in four manuscript copies. One copy preserved in the British Library (fig. 9) was attached to Paris's *Chronica majora*, a general history of the world up to the mid-thirteenth century, and may have been made for England's King Henry III. This remarkable map is without parallel in the Middle Ages. Nothing like it would be seen again in European culture until more than four centuries later, when John Ogilby published *Britannia*, his compendium of road maps of Great Britain (fig. 15), in 1675; perhaps only the elaborate Chinese river scrolls (fig. 12) dedicated especially to the Yangtze River are comparable among early wayfinding maps in their detail and execution.

The first five pages of Paris's map form a road map from England to Italy. Each of these pages represents a journey of several days, giving the distances between major stages and illustrating the more prominent or religiously important cities one would encounter along the way. The first of these pages, for example (fig. 9), shows the primary route from London (at lower left), the presumed starting point of the reader, to Dover, then across the English Channel to three possible French ports, and on to Beauvais, north of Paris. The final two pages comprise a map of the Holy Land, the final destination of the journey. But what kind of a journey? That several copies of this unique map survive suggest that it enjoyed at least a limited distribution among the monks of St. Albans, who probably had use for a practical guide for travel to the holiest sites in Christendom. Recent studies of the map agree, however, that despite its inclusion of apparently practical wayfinding advice, it was more likely used to support vicarious contemplative and spiritual pilgrimages. Few English monks at the time could have afforded to make the journey—Paris himself apparently never traveled to Rome or to the Holy Land. Rather, he hoped that by laying out the route in visual and book form, providing images of the cities a pilgrim might encounter along the way before finally reaching the spiritual capital of the Christian faith, monastic readers could embark on "an imagined pilgrimage in which the monastic soul could 'fly across all the kingdoms of this world, and . . . penetrate the depths of the eastern sky'" (Connolly 1999; see also S. Lewis 1987). The British Library copy may have been intended specifically for the king, to help him fulfill his obligation to go on a Crusade—in spirit at least, if not in actuality.

Pilgrimage, of course, is not unique to the Christian tradition; as noted earlier, guides intended to provide both practical and spiritual guidance to shrines and other sacred sites were produced by Islamic and Asian religious tra-

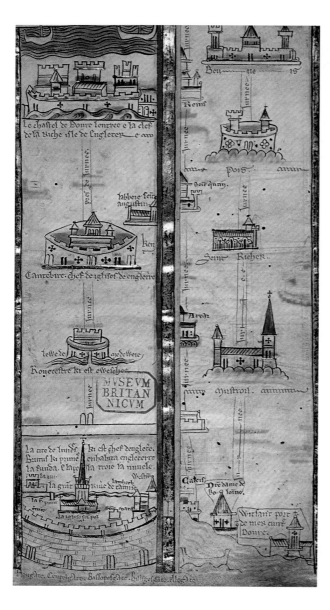

FIGURE 9.

Matthew Paris, itinerary from London to Beauvais (ca. 1250).

FIGURE 10.

Rubbing of a stone map of Hua Shan (Taihua Shan Tu) (1585).

ditions as well. Mountains have been held in particular esteem by many Eastern faiths, including Buddhism, Taoism, Hinduism, Shintoism, and shamanism from ancient times to the present day. They are powerful and easily visualized religious symbols, and since they present particular physical challenges to the pilgrim, it is perhaps not surprising to find that maps have been produced for this special class of traveler. The Field Museum holds some of the oldest of these, several early modern Chinese pictorial maps for visitors making the journey to the summit of one of the five mountains sacred to Taoism. The particular importance of these mountains in Chinese culture—emperors were expected to ascend them to gain the blessing of deities—may account for the

JAMES R. AKERMAN

fact that the maps were engraved on stone in order to preserve them. Rubbings, producing a "negative" image of the map, could then be distributed to travelers. A map dating from 1585 of Hua Shan (Mount Hua) in Shaanxi Province (fig. 10) includes a prayer at the top written by a Ming dynasty emperor who made the ascent. Undoubtedly, it was meant, like the Paris map, to be an object of contemplation. However, it also describes itself as a practical guide to the mountain: portions of the path, including stairways and bridges and wayfinding landmarks and structures. are clearly delineated on it (Yee 1994a, 63–66).

Though these wayfinding maps for pilgrims depict the geography of the "real world," their other-worldliness, their depiction of spiritual pathways, align them with traditions of allegorical cartographic representation concerned with wayfinding through the "interior landscapes" of mystical and religious experience and human desire (Certeau 1982, cited in Peters 2004, 86). These allegorical maps draw on established cartographic conventions and look strikingly like maps of "real" or exterior landscapes, but their destinations are entirely spiritual rather than geographical. Since Dante first "mapped" Hell (figs. 150–52), Christian thinkers have been drawn in particular to the notion that maps might be a persuasive way of dramatizing the manner of achieving salvation and avoiding the road to perdition. Though it was not initially illustrated with maps when it was published in the late seventeenth century, John Bunyan's popular evangelical fairy tale, *Pilgrim's Progress*, inspired a number of lavish cartographic illustrations and even separately published maps in the eighteenth and nineteenth centuries, both in Britain and in the United States. The London commercial publisher John Wallis even produced a jigsaw puzzle, or "dissected map," on this theme (J. Wallis 1790). Temperance activists in the United States published other variations on this cartographic theme. One clever example of these is the "Gospel Temperance Railroad Map" published in 1908 by G. E. Bula, which looks very much like the typical American railroad map of its day (fig. 11). It presents the traveler with three main lines diverging from Decisionville in the State of Accountability at the left-hand side of the map. The routes of the lower two lines, the Way That Seemeth Right Division and the Great Destruction Way Route, pass at first through towns representing relatively minor vices and self-deceptions of alcohol use, but lead inevitably to more serious "states" of Depravity, Intemperance, and Bondage. A River of Salvation offers hope for some, but those who stubbornly remain on the path of drink and debauchery end, without escape, in the City of Destruction. The upper line from Decisionville, the Great Celestial Route, is not without its trials, represented by such station stops as Bearingcross, Abandonment, and Long Suffering; but the final destination, The Celestial City, is clearly more desirable than its counterpart (Post 1979, 33–35).

Another class of allegorical wayfinding map charts the tortuous path of courtship and love. The earliest of these that survives was prepared for the French romantic novel *Clélie* by Madeleine de Scudéry in 1654 (fig. 149). The

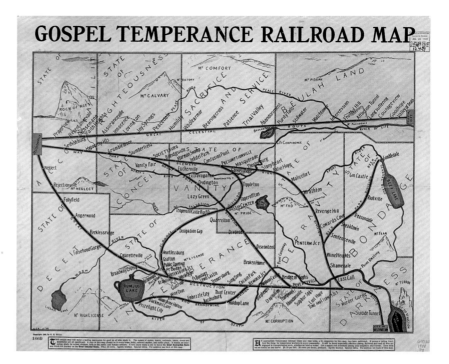

FIGURE 11.

G. E. Bula, "Gospel Temperance Railroad Map" (1908).

starting point for the presumed male suitor traveling the landscape depicted by Scudéry's "Carte du Tendre" (Map of Tenderness) is the town of Nouvelle-amitié (New Friendship) at the bottom center. According to Clélie, the heroine of the novel, to reach *tendre* (tender friendship), represented by three towns so named at the top of the map, the suitor must follow the river at the center or the two lines of towns on either of its banks. Places on these pathways are named for admirable qualities and gestures, such as Jolis Vers (pretty verse), Sincerité, Petits Soins (attentiveness), and Sensibilité. The unsuccessful suitor, in contrast, follows side routes passing through such towns as Indiscretion and Oubli (forgetfulness), and consequently finds himself in the landlocked Lake of Indifference or the troubled waters of the Sea of Emnity (Peters 2004, 83–186).

Maps depicting a more generalized island or region of Matrimony appear to have been especially popular in eighteenth- and nineteenth-century Britain. Most of these have a maritime theme and represent the (apparently) male lover as the navigator of a ship who must steer clear of hazards such as the Shoals of Fickleness or avoid landfalls in the Gulf of Despair or the Isles of Jealousy before finding safe harbor in the Gulf of Sincerity or the Gulf of Matrimony (Barron 2006; Hill 1978, 53–58; Post 1979, 35; Yale Map Collection 2006).

Despite differences in their design and cultural setting, the popularity of pilgrimage and allegorical maps suggests that wayfinding can encompass more than merely practical navigation. In a broader sense it is concerned with the choices we make, which pathways we find most expedient or desirable. In

JAMES R. AKERMAN

the spiritual and allegorized worlds described by the maps we have just considered, these choices represent broad life-changing decisions. For more mundane journeys such as automobile trips, these involve choosing routes that are most suited to the particular needs of the trip and one's temperament—how fast one wishes to go, what one wishes to see, and so on. In the next section we will return to these generally more mundane concerns to consider how the two main classes of wayfinding map approach the question of route choice, and the practical and historical implications that proceed from this question.

ITINERARY AND NETWORK MAPS

Virtually all wayfinding can be roughly classified either as *itinerary* maps, which are primarily concerned with the representation of a single route or corridor of movement, or as *network* maps, which describe an entire system of routes or pathways within a place, region, or country. Specific wayfinding maps may not always fit neatly into one or the other category; think of these broad types rather as responses to two often complementary but occasionally contradictory demands we make of wayfinding maps: (1) the need for detailed information about specific routes that helps keep us on our appointed paths; and (2) the need to understand how various routes stand in relation to one another and other geographic features, and to possible starting points and destinations, so that appropriate paths can be planned or chosen.

Itinerary maps are the direct cartographic descendants of ancient verbal itineraries. They are often very detailed, and since the routes they describe are linear by nature, they are potentially awkward in shape. The ancient solution to this problem was to put the maps on scrolls that could be rolled upon themselves for easy and portable storage. This was the apparent format of the ancient Egyptian map of the routes to gold-mining regions (fig. 3). In traditional Chinese and Japanese cartography it was a preferred format for beautiful maps of rivers such as the Yangtze (fig. 12) and celebrated routes such as the Tokkaido.

Maps were also mounted on scrolls for the use of mid-nineteenth-century pilots and travelers on the Mississippi River. The Newberry Library possesses one such map, published in 1866 (fig. 13). Ten feet (3 m) long but only about 2 inches (5 cm) wide, it was originally mounted on a spool housed in a small metal case from which the map was extracted as needed. Many airplane pilots from the era when cockpits were open to the elements also found it most expedient to mount their charts on rollers (sometimes encased in waterproof boxes) that allowed them to scroll through a map as the flight progressed with a minimal amount of effort (fig. 14). Early pilots often had to cut and prepare these scrolls from maps originally designed for other purposes, but in time military and civilian publishers issued entire series of standard aeronautical charts showing key "airways" in elongated formats (Ehrenberg 2006b).

Ribbon Map
OF THE
FATHER OF WATERS

ST. LOUIS, MO. JULY 1866

MYRON COLONEY
SIDNEY B. FAIRCHILD
Inventors & Patentees

COLONEY & FAIRCHILD'S
PATENT
RIBBON MAPS

PATENT APPLIED FOR

St. Louis, Mo.

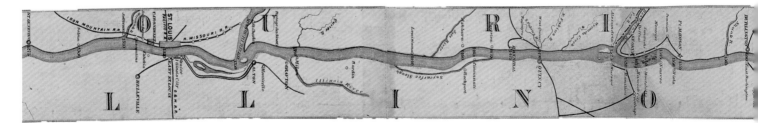

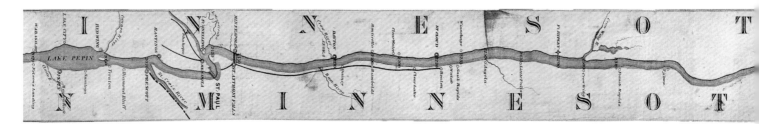

FIGURE 12.

(facing, top) *Changjian Wanli Yu* (Ten-Thousand-*Li* Map of the Yangtze River), section 9 (13th century).

FIGURE 13.

(bottom, spread) Coloney and Fairchild, "Ribbon Map of the Father of the Waters" (1866).

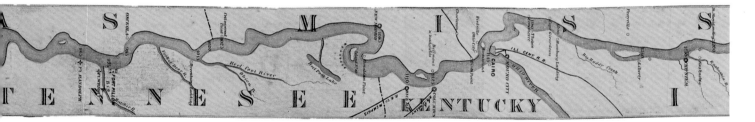

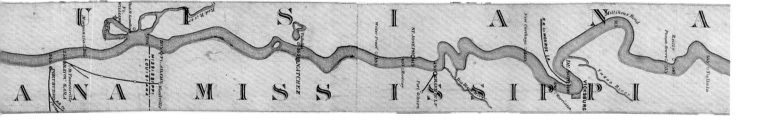

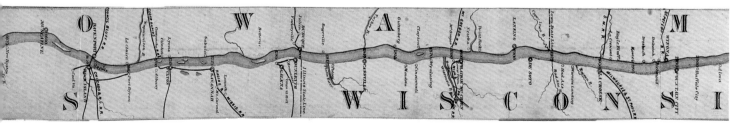

FIGURE 14.
Portion of an aviation chart of the Kaiser Wilhelm
(now Kiel) Canal (1915).

More commonly in modern times, a map of a single route was divided into sections, which could then be mounted side by side on one or more sheets of paper suitable for binding in a standard-sized book. In his road atlas of England and Wales, *Britannia*, John Ogilby instructed travelers from London to Montgomery in North Wales to begin the final leg of their journey at a marker called the Four Shire Stone (in Gloucestershire), shown at the lower left-hand corner of his map of this route (fig. 15). Map readers then followed the course of their road from the bottom to the top of the leftmost of seven strips printed on this sheet, continuing the route at the bottom of the next strip, and so on, until Montgomery was reached at the top of the strip at far right. In some contexts, however, navigators found a scroll format much more practical. In the twentieth century it had become fairly easy for printers mechanically to cut and trim printed sheets in whatever shape most suited the purpose, making it possible for the American Automobile Association, too, to produce customized route

JAMES R. AKERMAN

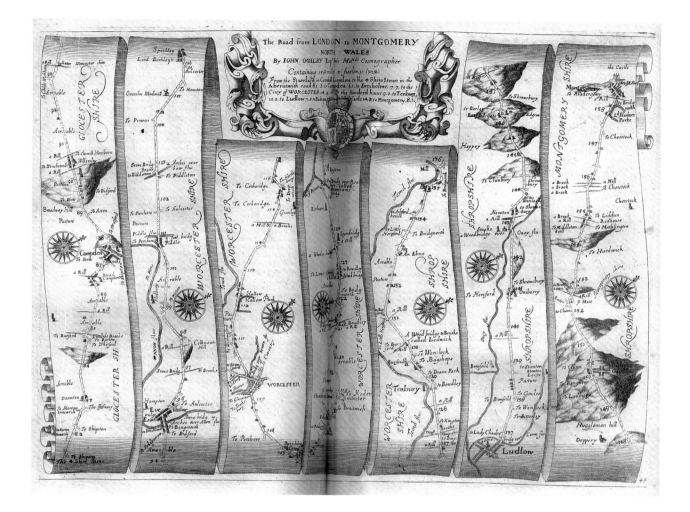

FIGURE 15.

John Ogilby, "The Road from London to
Montgomery" (1675).

maps for motorists, known as TripTiks, in an elongated format suited to the
itinerary concept.

Whatever the shape and format of the itinerary map, its visual focus remains on the route in question. Ogilby's maps provide a description only of
features of the countryside visible from the road, including local streets in
larger towns, the rivers and streams crossed by a route, the hills it must sur-
mount, and roadside landmarks and structures including mills, fences, fields,
and country estates. This emphasis on the detailed description of a single route
has practical merits that hardly need to be mentioned. The more detail a map
is able to provide about the features of a specific route, its minutest twists
and turns, confusing forks and intersections, the less likely it is that a traveler
will get lost. The more information a map is able to provide about amenities
and places of importance along the route, the more comfortable the traveler
is likely to be.

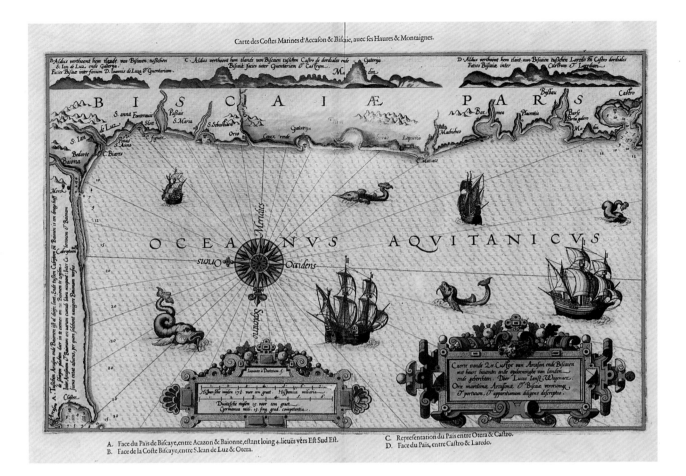

Carte des Coftes Marines d'Accafon & Bifcaie, auec fes Haures & Montaignes.

A. Face du Païs de Bifcaye, entre Acazon & Baionne, eftant loing 4.lieuës vèrs Eft Sud Eft.
B. Face de la Cofte Bifcaye, entre S.Iean de Luz & Otera.
C. Reprefentation du Païs entre Otera & Caftro.
D. Face du Païs, entre Caftro & Laredo.

FIGURE 16.

Lucas Jansz. Waghenaer, "Carte vande Zee Custen van Arcason ende Biscaien" (Chart of the Sea Coast of Gascony and Biscay) (1590).

A tradition of itinerary mapping similar to Ogilby's was established in early modern Vietnam, though the resulting maps provided a somewhat broader vision of the country appropriate to their apparent purpose. Prepared in book format, the richly illustrated maps had their origin in the desire of Lê dynasty kings (1428–1788) to reconnoiter and maintain control of key military and diplomatic routes leading from the Vietnamese capital to frontier areas or rival neighboring states. Instead of depicting a single pathway, the maps describe broad corridors of roads, rivers, and sea routes. Border fortifications are prominent features of these maps, but the cartographers also identify bridges and inns, villages, natural features, and other items of general interest (Whitmore 1994, 490–97).

The first printed nautical atlas, Lucas Waghenaer's *Spieghel der Zeevaerdt* (Mirror of Navigation), which was first published in Dutch in 1584–85 and subsequently in French and English, is a good example of itinerary mapping at sea. The charts in the atlas (fig. 16) depict the European waters most frequented by Dutch sailors: the near coasts of the British Isles, modern Netherlands and

JAMES R. AKERMAN

Belgium, and adjacent Atlantic waters stretching from Gibraltar to the Baltic Sea. The charts are large in scale and pay as much attention to the depiction of coastline features as they do to water depths, anchorages, and hazards. Towns, lighthouses, bluffs, beaches, and other coastal landmarks are pictorially represented, and on many charts Waghenaer includes profiles of the coastline as sailors would see them from the sea. His atlas is in fact an illustrated rutter, or pilot book, and its charts were meant to be read in conjunction with the accompanying verbal sailing directions printed on the backs of the charts. Closely following the tradition of verbal sailing instructions, Waghenaer organized the charts so that they could be read sequentially as one travels from The Netherlands in a particular direction. One sequence (of which figure 16 is part) traces the route south along the Atlantic seaboard of France, Spain, and Portugal to the Strait of Gibraltar; another guides the sailor east through the Kattegat of the North Sea into and around the Baltic Sea.

A counterpart to Waghenaer for wayfinding on interior waterways is Zadok Cramer's *The Navigator*, a diminutive atlas of the Ohio and Mississippi rivers and one of the most popular travelers' guides published in the United States during the first decades of the nineteenth century. It was first published in the form of an entirely verbal itinerary, but Cramer later added maps, each of which showed about 100 miles of river navigation. Cramer's maps were crude, and although they delineated the preferred course through each river, they provided little information about depths found on modern river charts. They would have been helpful to boatmen and their passengers wanting to locate themselves in relation to the rivers' many twists and turns, the mouths of tributaries, and scattered riverside settlements and landmarks. Much of the information provided in the accompanying text, however, is descriptive rather than purely navigational. Reading *The Navigator*, one learns a great deal about the economy of the area depicted, the location and nature of the plantations and settlements—potentially useful information, to be sure, but also apparently intended to satisfy the travelers' curiosity about their surroundings (Yost 1987).

Perhaps the earliest network map to enjoy widespread circulation was Erhard Etzlaub's "Das Ist Der Rom-Weg" (This Is the Way to Rome), which appeared initially in 1500 and was revised and republished under a different title the following year (fig. 17). This simple woodcut may be the first wayfinding map in any cultural tradition published explicitly for a mass travel market: the many pilgrims that were expected to travel from or through central Europe to Rome around the Christian Jubilee Year of 1500. (Rome appears at the top of the map, which is oriented to the south.) Etzlaub maps the routes leading to the capital of the Latin Church as series of small solid dots linking major cities, which are marked by hollow circles. Though we may presume that the pilgrims would still have had to ask for directions along the way, two features of the map clearly imply that Etzlaub intended it to be used on the road. The small dots each represent a fixed geographic distance, and an accompanying

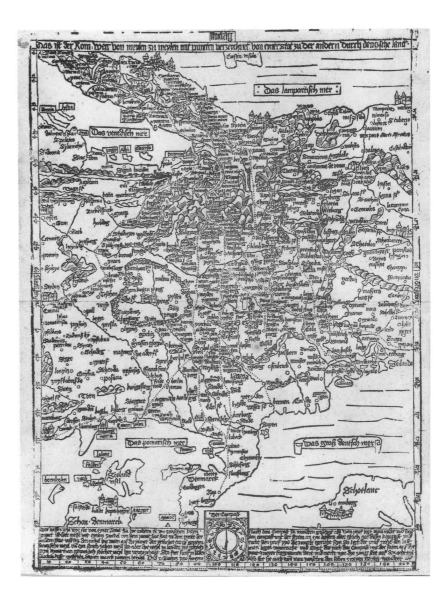

note explains how the reader may compare the number of dots on each of the
marked roads with the scale of dots across the bottom of the map to determine
distances between the major towns. A second note explains how pilgrims may
use a simple direction-finding device called the sun compass, pictured at the
bottom of the map, to orient themselves from time to time (Englisch 1996).
Most important, the several possible routes to Rome shown on the map make it
useful to the largest possible number of travelers. Unlike Matthew Paris's Lon-
don-based pilgrim, the point of origin of Etzlaub's pilgrim cannot be pinned
down geographically, though he and all the other devotees are heading for a
common destination.

JAMES R. AKERMAN

This flexibility is the chief virtue of all network maps, including our road atlas map of Wyoming (fig. 6). Their main disadvantage is that the description of any one route is relatively sketchy, because information about the entire road system of an area as large as Wyoming must be crammed into a relatively small space. This could be a navigational problem for me once I am on the road, but it usually isn't, since American highways are easily identified by route numbers that appear frequently on signs along the road itself. Other signs that give the distances to towns ahead further reassure me that I am on the right path. All I need to do is occasionally compare the route number with those marked on the map to confirm that I am heading in the right direction. The usefulness of a road map of this scale deteriorates, however, if and when my route takes me through major cities, where the density of highways and local thoroughfares increases. The bustle of traffic, the density of structures, the many important local streets, and other distractions of a city make it easy to miss route signage; even the best navigators get "turned around" in such places. This is one of the reasons why the Wyoming map includes larger-scale inset maps of Casper and Cheyenne and of the popular destinations of Yellowstone and Grand Teton national parks. These maps are acknowledgments that the network map's more generalizing, one-size-fits-all approach to wayfinding has its navigational limitations.

On the whole, though, network maps like our Rand McNallys work as practical on-the-road navigational tools in the contemporary context because modern road systems themselves are well marked and well maintained; moreover, modern modes of transportation, such as locomotives and automobiles, enable us to travel hundreds of miles in a single day in almost any direction. Technology gives us a mobility that ancient and medieval travelers never would have imagined. Yet this increased mobility means that more travelers travel more often into and through unfamiliar territory; finding our way in these circumstances requires tools that give us a more comprehensive understanding of that territory, so that if we voluntarily or involuntarily stray from our path, we can more easily find our way back to it or set out on a new course that will serve us equally well. Modern network maps give us the ability to do this, but so do underlying geographic and social conditions and technologies of movement.

MAPPING SHIFTS IN PERSPECTIVE

It is worth pondering why many of the early wayfinding maps we have considered up to this point were produced at all. It shouldn't surprise us that the scattered earliest maps of overland pathways, including roads and rivers, were produced by some of the largest and most powerful states or civilizations of the ancient world, Egypt, China, and Rome. Routes of travel having particular cultural or political significance, such as pilgrimage routes, the Yangtze River, and the Vietnamese military and diplomatic routes, also inspired mapmaking, while the countless other pathways people used remain undocumented and undescribed. Only in the later modern world were road systems developed to the extent that comprehensive mapping of entire route networks could be meaningful and practical to map users.

John Adams's huge "Angliae totius tabula cum distantiis notioribus in itinerantium usum accommodata" (A Map of the Whole of England with the More Important Distances Arranged for the Use of Travellers), circulating around the time of Ogilby's *Britannia*, is perhaps the most important network map published in Europe after Etzlaub's "Rom-Weg." It shows no actual specific roads, but rather portrays Britain somewhat dramatically as a dense landscape of road connections between cities and towns. Rectangles represent major cities and cathedral towns, while ovals represent other market towns. Double lines radiate from each town symbol to others nearby. A small circle perched on each of these lines indicates the distance between the two places in English miles. Thus, the distance from London northwest to Edgworth is 10 miles (16 km); St. Albans to Edgworth is 11 miles (17.6 km). In this way travelers could calculate distances and intermediate stops between virtually any two points in England, a great boon to persons who had cause to travel frequently to many different places. Adams's is a large map suitable only for mounting on a wall, but he published a smaller version in 1679 (fig. 18), and many similar, portable derivatives appeared before and after the turn of the eighteenth century. The map is not the first to show routes or roads with distances between towns; there are a handful of European precedents dating from the sixteenth and seventeenth centuries. This particular burst of road network mapping at the end of the seventeenth century is remarkable, however, and demands an explanation. Possibly it occurred in response to rising commercial activity and increasing traffic on British roads after the end of the recent civil war (Delano-Smith 2006, 59–61 and note 168). Certainly, as road travel continued to increase, eighteenth-century Britain (as elsewhere in Europe) would publish an unprecedented number of road maps and atlases in both network and itinerary formats.

The emergence of new modes and technologies of transportation frequently challenge our ability to grasp them and consequently seem to encourage the development of new forms of wayfinding cartography. This is especially true in the modern era, which has witnessed a rapid succession of transport innovations. Consider the way in which successive modes of transportation have competed for prominence in the cartographic representation of North America. Rivers are prominent features of early maps of the interior of eastern North America made by European cartographers, for the simple reason that rivers were the principal highways by which Indians and Europeans moved. One of the most influential printed maps of the area that became the United States and eastern Canada of the early eighteenth century, Guillaume Delisle's 1718 map of Louisiana, is typical of such early cartographic efforts (fig. 91). In contrast, maps attached

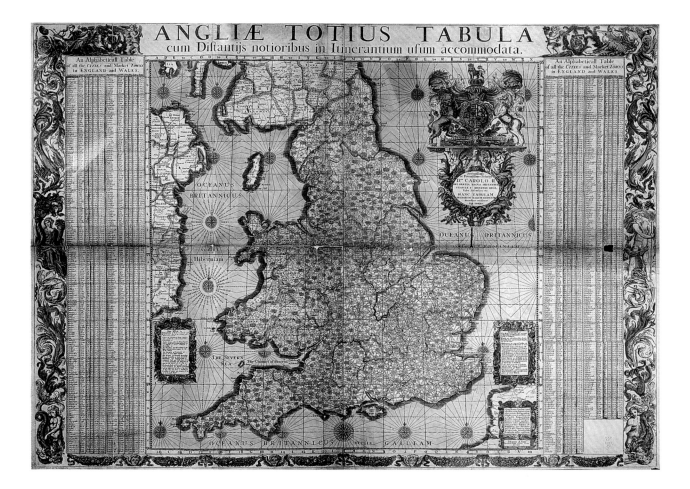

FIGURE 18.

John Adams, "Angliae totius tabula" (A Map of the Whole of England) (1679).

to popular guides to overland travel published during the 1830s and 1840s, such as J. H. Young's 1835 "Map of the United States" from Samuel Augustus Mitchell's *Compendium of the Internal Improvements of the United States*, present navigable rivers and canals as part of an integrated system of transportation routes that includes early short-line railroads and turnpikes (fig. 19). Though other geographic features are present, this map's main purpose is to create a sense that a progressive national system of "improved" internal communications is emerging—a system that was already playing a significant role in shaping the country's migratory patterns and its economic and political growth.

At the beginning of the twentieth century, Americans were already a mobile nation, and a nation in love with mobility. However, the advent of the personal automobile and the general improvement of the roads it required made it possible for Americans to imagine that they truly could travel to every corner of the country for the sheer pleasure of it, almost at a moment's notice. This almost casual attitude toward transcontinental automobile travel was not achieved immediately. Early motorists wishing to undertake long jour-

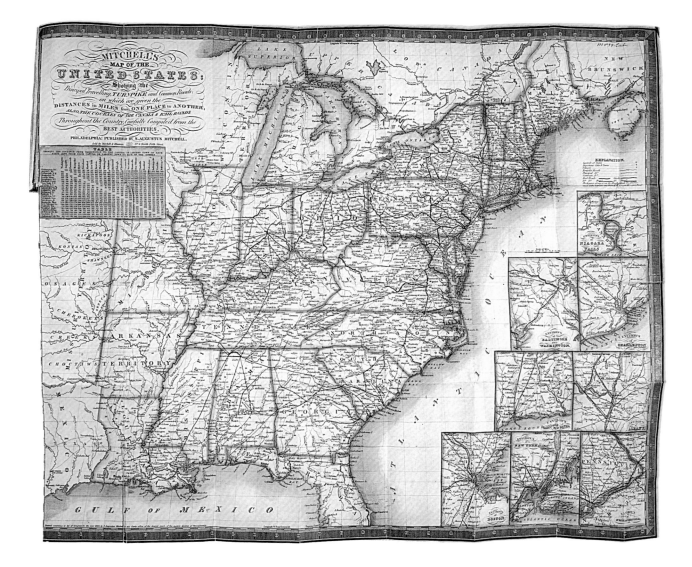

FIGURE 19.

Samuel Augustus Mitchell and J. H. Young,
"Mitchell's Map of the United States" (1835).

neys relied extensively on published written itineraries known as route books. Like modern online or in-car services, these books provided turn-by-turn instructions, kept track of the passing miles, and took note of landmarks and hazards along the way. One variant of these itineraries, Photo-Auto Guides (fig. 20), supplemented their highly detailed directions with photographs of critical intersections and forks, so that motorists might be assured that they were indeed in the right place and had not missed their turn. By the mid-1910s serious efforts to mark and improve interstate routes were under way. Federal and state funds were committed to the effort, and by the late 1920s the entire country was fully committed to the construction of paved and well-marked highways. Network road maps of the type distributed at service stations (fig. 1) became the normal type of road guide, while verbal route books largely disap-

JAMES R. AKERMAN

peared. Automobile wayfinding had become relatively simple, so road maps could afford to be simple and succinct. Freed of the need to plod along turn by turn, motorists opted for the flexibility the network map offered and the unfettered geographic mobility it represented. The rise of the network map also meant that the horizons of American motorists had been rolled back. The entire country now seemed within reach simply by choosing from an almost limitless number of automobile routes.

A similar shift in perspective at sea may be related to the appearance in the Mediterranean world of highly accurate manuscript sea charts, known as portolan charts, during the later Middle Ages. The oldest of these, the Carte Pisane, dates from the end of the thirteenth century (fig. 21). Portolan charts have long fascinated historians of cartography because they were remarkably accurate for their time in the sense we understand the term. They conform roughly to a constant scale throughout, and their depiction of coastlines and enumeration of coastal towns are remarkably detailed. Just as remarkable is the fact that so

FIGURE 20.

Two pages from H. Sargent Michaels, *Photo-Auto Guide, Chicago to Rockford* (1905).

No. 29.

TURN TO RIGHT (WEST).

This photograph was taken just around the corner from Photo No. 28.

No. 31.

TO THE RIGHT AND WEST.

White frame cottage sitting back about 100 feet on left as you turn. Next turn, Cherry Valley, 6 1-10 miles.

No. 30.

TO THE LEFT.

Brick school building on the right. Next turn 1-2 mile.

No. 32.

TO THE LEFT AND SOUTH.

Frame house with red roof on the right sitting back, just before you come to the turn. Next turn 500 feet.

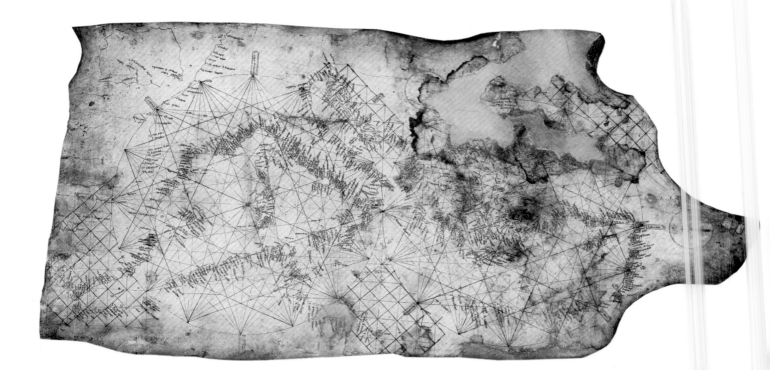

many of them survive: about 180 single charts and atlases from before 1500 alone, likely "a minute fraction of what was originally produced" (T. Campbell 1987, 373). They might be considered the first type of mapmaking, in Europe at least, to have enjoyed widespread commercial success. But who bought them, and for what purpose? The visible wear on the Carte Pisane may imply that the chart was actually used on board a ship, but more typically the surviving charts are beautifully illuminated and well preserved. Scattered evidence confirms that some portolan charts were in fact bought or commissioned by royalty and other elites who, we may presume, valued them more for the general picture of the Mediterranean region they provided than as wayfinding tools (fig. 22). But their value to practical route planning and navigation cannot be doubted.

The magnetic compass was introduced to Mediterranean navigation about the time the first portolan charts appeared. Although it was not impossible to sail across that sea (ancient and medieval sailors are known to have done so), the use of the compass certainly would have made it safer for mariners to strike out across hundreds of miles of open water beyond the sight of land in the hope of making landfall at a predetermined location. The many straight lines, known as rhumb lines, radiating from points arrayed around the margins of portolan charts clearly were meant to support compass-guided navigation. Each rhumb line corresponds to a compass bearing, consisting of the four cardinal directions and their primary divisions into as many as sixteen or thirty-two compass points, or winds (Taylor 1956). There can be little doubt that these lines are on the chart to serve as references for the bearings that sailors attempted to

JAMES R. AKERMAN

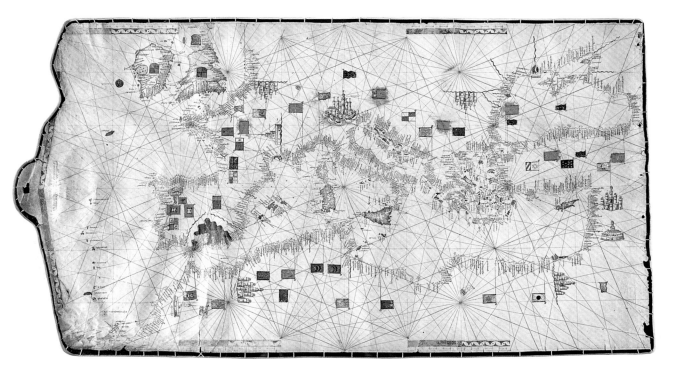

follow in navigating, say, from a port on the "boot" of Italy to a port in what is now Libya.

Though they do not represent specific routes, it is possible also to think of the rhumb lines as suggesting, like John Adams's route map of England, that there existed a network of routes leading to and from the many ports named along the coast of the Mediterranean Sea. The historical circumstances are suggestive. No one can be certain why portolan charts appeared when they did, but a general revival of trade in the Mediterranean during the later Middle Ages corresponds to the introduction of the compass and the portolan chart. This may explain why it was desirable for sailors to supplement their rutters and personal knowledge of the sea with charts that gave them a more general view. Just as John Adams's map may have responded to a sharp upturn in road traffic that required a more flexible and complete understanding of routes and distances, so, perhaps, did the portolan chart offer a much-needed flexible understanding of the geography of the Mediterranean Sea.

Exploration may be seen as wayfinding that takes the traveler off the map both literally and metaphorically, but wayfinding maps inform explorers and respond quickly to their findings. During the last decades of the fifteenth century, as European sailors and traders extended their reach into the eastern Atlantic and down the western coast of Africa, some portolan charts, such as Alberto de Canepa's from 1489 (preserved in the James Ford Bell Library at the University of Minnesota), responded to this change by widening their horizons, including more Atlantic islands and an extended African coastline.

FIGURE 22.
Petrus Roselli, portolan chart (1456).

FINDING OUR WAY

Columbus's plan to sail west to reach Japan and the Indies was very likely influenced by globes and maps such as a now lost map prepared around 1481 by the Florentine cosmographer Paolo Toscanelli, which promoted the idea (Harley 1990, 34–36). Both the Portuguese and the Spanish put great emphasis on collating geographic information gathered by their mariners in Columbus's wake. Many of the great world maps from the early sixteenth century that proclaimed and described the existence of a New World were all either copies of, or heavily influenced by, the master charts maintained by Portuguese and Spanish establishments created to manage overseas trade and the geographic information that supported it (Buisseret 2003, 71–92).

As global trade expanded for the next three and a half centuries, largely on the keels of sailing ships, the modern sea chart came into its own. Trading companies organized in The Netherlands, France, and England, such as the Dutch East India Company, were initially the main agencies that produced maps of previously uncharted waters, but commercial publishers were the primary means by which these charts were distributed. Typical of these commercially produced charts is John Thornton's "Chart of the Caribe Islands" (fig. 23) from the fourth book of *The English Pilot* (1689), a comprehensive five-part atlas of charts and sailing directions for the entire world compiled by various London publishers from the mid-seventeenth through the eighteenth century. Thornton's chart shows the approaches to the Lesser Antilles, stretching from Tobago and Granada in the south (at left) to Puerto Rico (at upper right). Larger-scale charts of some of the islands as well as harbor charts and coastal profiles may be found elsewhere in the volume along with verbal sailing directions, but smaller-scale charts like this one that provided a general context were also essential. More important, they represented a vast expansion of the navigational horizons, a fundamental shift in worldview of which the creation of portolan charts and their successors were an integral part (Blake 2004; Cook 2006; Mollat Du Jourdin and La Roncière 1984; Whitfield 1996; Zandvliet 1998).

Differences in the way human beings map and imagine movement through space are not only the result of historical shifts in the horizon. They also reflect differences in our environment and perspective. Harry Beck's design for the Underground map presents a version of London geography uniquely suited to effective subterranean navigation in this most confusing metropolis (fig. 24). Early maps of the London Underground system plotted the various subway lines in their more or less true geographic alignments and on a constant scale throughout. This fidelity to scale and direction was problematic for wayfinding underground in several respects. Several lines extended for some distance into suburban London, and if the guide maps were to show the full extent of these lines on a constant scale, the great mass of routes that converged on and snaked through central London and Westminster had to be so compressed as to make the names of the stations and the courses of the line almost indecipherable. Beck understood that cartographically, strict adherence to the geography

JAMES R. AKERMAN

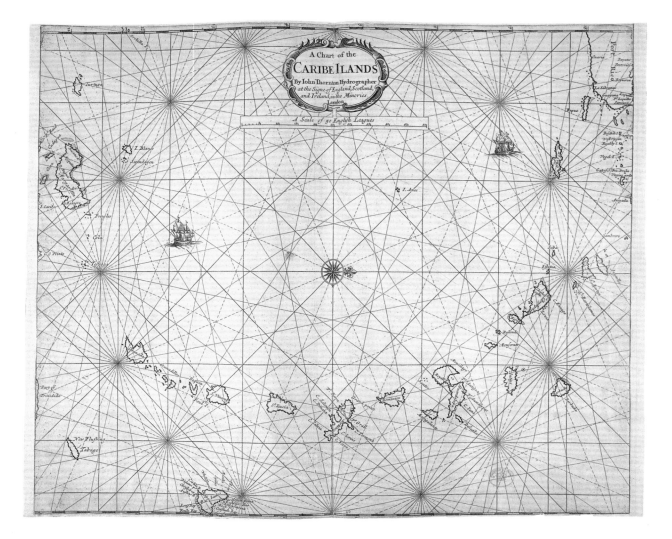

aboveground was not essential to wayfinding below. The Tube rider traveled in isolation from the surface geography, so the most essential wayfinding information for successful Underground navigation was the sequence of station stops and the location of critical interchanges where transfers between lines could be made. His solution, published for the first time in 1933, had two main elements. Beck enlarged the scale of the central area while compressing scale in the suburbs, allowing the names and relative positions of the central stations to be read more clearly. Second, and perhaps more important, he straightened the courses of the lines, so that while their general direction was shown with some fidelity, the messiest details of the geography above were eliminated, making it easier to identify stations in sequence and to understand the general relationships between the connecting lines. The design was so successful that it has become iconic, reproduced on souvenir T-shirts and coffee mugs.

FIGURE 23.
John Thornton, "A Carte of the Caribe Ilands" (1689).

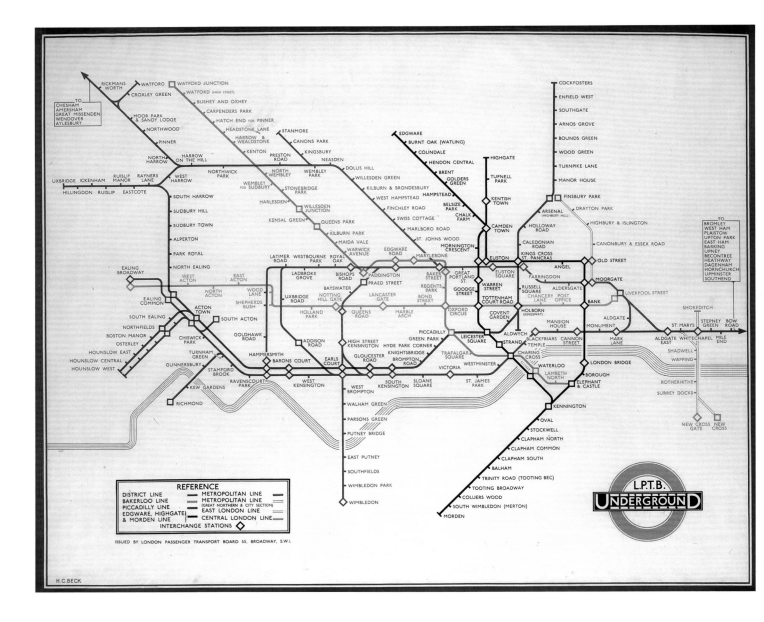

FIGURE 24.

Harry Beck, "London Underground" (1933).

Indeed, for many visitors and perhaps for many citizens of London, Beck's map represents the basic way in which they understand the city's geography, and numerous other transit systems around the world have adopted the essentials of its design for their map (Garland 1994).

Pacific Islanders, faced with the extraordinary problem of sailing over open water for perhaps hundreds of miles to make landfall on tiny islands or atolls visible to the naked eye from only a few miles distant, used traditional navigational techniques that were poorly appreciated by outsiders until the last century. One of these methods required an extraordinary ability, far more developed in the islands of the Pacific than almost anywhere else in the world, to transform the night sky into a navigational frame of reference, a wayfinding map. In the Caroline Islands, for example, the rising and/or setting positions of

JAMES R. AKERMAN

specific stars were committed to memory. Each of these positions corresponds to a direction, which in turn corresponds to the direction one must travel to a nearby island. Caroline Islanders committed to memory as many as thirty-two points, just like a modern compass; for this reason, their mental maps of the night sky are sometimes called star compasses (Finney 1998, 461–75).

Traditional sailing in the Marshall Islands relied especially on the islanders' ability to read the patterns of ocean swells as indications of the proximity and direction to an island, because the prevailing currents in their island chain are especially variable and can easily deflect vessels off course if navigating by stars alone. Swells are broad undulations on the surface of the ocean that are generated by prevailing winds many miles distant (as opposed to smaller ripples and waves created by local winds). The Marshall Islanders recognized four main swells, each coming from a different direction, and they understood that the islands and their undersea slopes bend or refract these swells and reflect them. Then the swells pass over and through one another, creating noticeable high points in the water that are detectable by practiced eyes at some distance. Sailors visualized rays to connect these elevations and pointed their craft in the direction of the islands that caused them, leading the mariners to landfall. Understanding how these principles operate required a great deal of skill and experience, and the islanders developed stick charts to help novices learn this navigational system (fig. 25). The simplest of these, *mattangs*, showed general

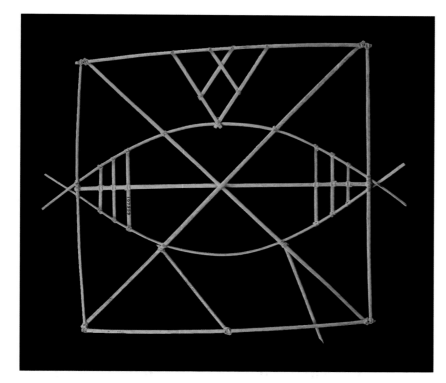

FIGURE 25.
Marshall Islands stick chart (19th century).

patterns and principles. Others, called *meddos* and *rebbelibs*, were charts of specific island groups or of the entire Marshall Island chain. Locations of specific islands were marked by shells fixed to a lattice of wood, while the lattice indicated the general patterns of prevailing swells and their refraction around specific islands. Swells and rising and setting stars exist in other parts of the world, of course, but only in this environment of scattered islands and huge oceanic skies were the technologies for mapping them so necessary and well developed (Finney 1998, 475–85).

MAPPING WHO WE ARE

European explorers of the Pacific Ocean, like Captain James Cook, who led three great voyages for this purpose during the late 1760s and 1770s, created a huge volume of cartographic material documenting their findings. The many charts and logs returned by Cook and his crews informed and guided future British naval wayfinding in the area. But the published version of his chart of the "Sandwich Islands" (fig. 26), what is now Hawaii, spoke to the British nation on broader terms. This map, showing the track Cook followed in approaching and navigating about the islands on two occasions during his third and final voyage of 1776–80, clearly was not expected to help most of its readers find their way around the islands. It was intended instead to guide readers of the published account of that voyage on an imaginary journey with Cook, his officers, and crew, to give geographic flesh to the events recounted by the accompanying narrative, which culminated tragically in the death of Cook and many Hawaiians during a brief skirmish on February 14, 1779. Although removed from its original wayfinding purpose, as a part of the volumes commemorating Cook's discoveries and achievements the map became a way of leading the entire British nation, defined at that time above all others by its naval prowess, on a memorializing and celebratory (if virtual) journey.

Many communities of what is now central Mexico, including the Culhuas-Mexicas (Aztecs), created similar itinerary-format maps that recounted historic migrations. The most famous of these "itinerary histories" is the Mapa de Sigüenza (so called because it was once owned by the Mexican polymath Carlos Sigüenza y Góngora), which dates from the late sixteenth or early seventeenth century (fig. 27). The story told by the map begins in the dominant scene at upper right. The winged god Huitzilpochtli instructs several Aztec elders to leave their homeland of Atzlan and migrate to the Valley of Mexico. The elders and a succession of leaders take the Aztecs on a journey of many years. Their path is marked in a traditional Mesoamerican manner of representing land routes, footprints walking on a double line. Along the way they stop at several cities indicated by hill glyphs, the usual sign for cities. Dots on the map indicate the number of years they remained in a particular city. Finally, the Aztecs

JAMES R. AKERMAN

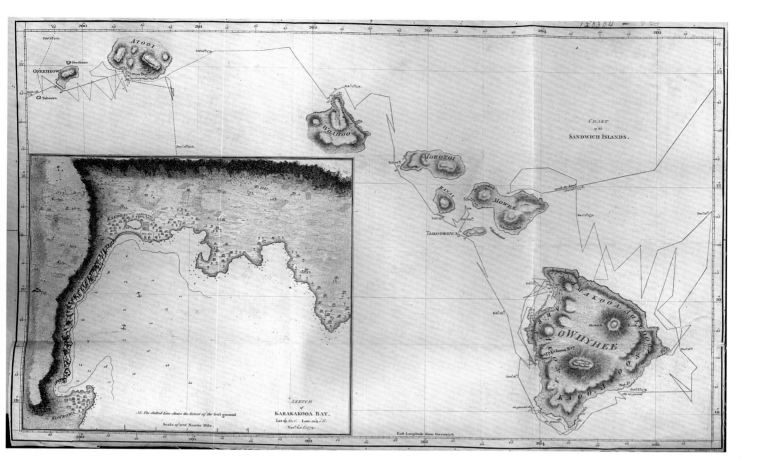

FIGURE 26.

"Chart of the Sandwich Islands." From James Cook
and James King, *A Voyage to the Pacific Ocean*
(1785).

reach the Valley of Mexico, subjugate local rivals, and settle in Tenochtitlan, represented by a cactus, in the center of the Lake Texcoco (Boone 1998, 117–31; Mundy 1998, 204–20).

Native North Americans also made itinerary maps to preserve their communal histories. The sacred medicine society, or Mide, of the Ojibwe maintained scrolls that can only be fully interpreted by initiates to the society, but a few of them have been preserved and interpreted by outsiders. These apparently represent the historic migration of the Ojibwe, or more specifically the Mide religion itself, from the Saint Lawrence River valley to the western Great Lakes (Dewdney 1975; Lewis 1998, 82–89). A map made in 1837 by an Ioway headman, Non-Chi-Ning-Ga, while attending a conference in Washington with federal officials and representatives of several Indian tribes, shows the historic migrations of his people throughout the upper Mississippi and Missouri River valleys (fig. 28). Non-Chi-Ning-Ga hoped that the cartographic retelling of his nation's movements would reinforce the historic claim of the Ioway to much of this territory, and defend them against unfavorable treaty settlements (Warhus 1997, 34–43).

In much the same way, certain modern highway maps encourage American motorists to follow specific itineraries that replicate historic journeys seen as essential to the nation's heritage and the formation of an American national

FIGURE 27.

Mapa de Sigüenza (map formerly owned by Carlos de Sigüenza y Góngora) (19th-century copy).

identity. One map published in 1932 by Esso, the forerunner of today's Exxon, laid out a track recounting the travels of George Washington on the bicentennial of his birth (Standard Oil Company of New Jersey 1932). Another map published by Rand McNally during the centennial of the Civil War encouraged Americans to retrace the course of the war's great campaigns (fig. 29). The Oregon and Santa Fe trails are frequently the subject of commemorative wayfinding maps, and even of detailed atlases for trail-obsessed motorists determined to find and follow modern traces of these famous migratory routes (for example Franzwa 1982; see Akerman 2006b, 197–205). Other, more common road maps have been deliberately marked and retained by motorists to record and preserve the memory of more private journeys. This may be why customized road maps such as AAA TripTiks and similar products have had an enduring appeal. They are, in the first instance, custom made for particular clients, albeit from a preprinted set of possible sheets, and in this sense they are very personal. But if they are preserved after the trip is over—not an unusual practice—they also help the traveler recount the route, and all the experiences

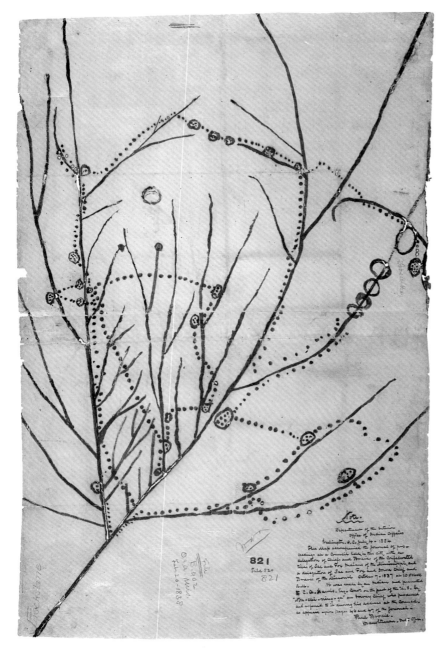

attached to it. In a very personalized way, the annotated road map, the scrapbook, and the TripTik sacralize the journey (Akerman 2000).

Wayfinding maps, then, are about much more than navigation. While they may have practical value in the way we conventionally understand them, as tools that tell us how to get from one place to another, they also reveal the habits of thought of the people and the cultures that created and used them. The

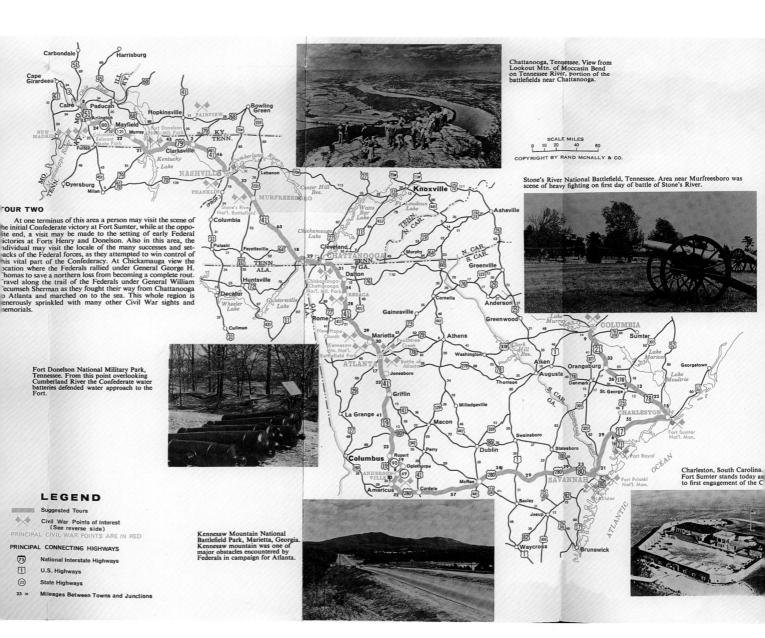

Chattanooga, Tennessee. View from Lookout Mtn. of Moccasin Bend on Tennessee River, portion of the battlefields near Chattanooga.

SCALE MILES
0 10 20   40   60
COPYRIGHT BY RAND McNALLY & CO.

Stone's River National Battlefield, Tennessee. Area near Murfreesboro was scene of heavy fighting on first day of battle of Stone's River.

TOUR TWO

At one terminus of this area a person may visit the scene of the initial Confederate victory at Fort Sumter, while at the opposite end, a visit may be made to the setting of early Federal victories at Forts Henry and Donelson. Also in this area, the individual may visit the locale of the many successes and setbacks of the Federal forces, as they attempted to win control of this vital part of the Confederacy. At Chickamauga view the location where the Federals rallied under General George H. Thomas to save a northern loss from becoming a complete rout. Travel along the trail of the Federals under General William Tecumseh Sherman as they fought their way from Chattanooga to Atlanta and marched on to the sea. This whole region is generously sprinkled with many other Civil War sights and memorials.

Fort Donelson National Military Park, Tennessee. From this point overlooking Cumberland River the Confederate water batteries defended water approach to the Fort.

Charleston, South Carolina. Fort Sumter stands today as to first engagement of the C

LEGEND

Suggested Tours

Civil War Points of Interest (See reverse side)
PRINCIPAL CIVIL WAR POINTS ARE IN RED

PRINCIPAL CONNECTING HIGHWAYS

National Interstate Highways

U.S. Highways

State Highways

Mileages Between Towns and Junctions

Kennesaw Mountain National Battlefield Park, Marietta, Georgia. Kennesaw mountain was one of major obstacles encountered by Federals in campaign for Atlanta.

FIGURE 29.
"Civil War Centennial Events 1861–1961" (1961), detail.

ascendance of a new type of wayfinding in response to technological, political, or economic changes represents not just a new way of telling us how to reach our destination. New types of these maps also reflect shifts in our perception of where and who we are and where we are going—that is, shifts in our imagination of the world, our reasons for travel, and what destinations are possible. Our usual understanding of these maps is limited by the presumption that wayfinding is essentially a pragmatic task concerned with determining route choices, identifying landmarks, properly executing a sequence of turns, or estimating distances and travel times through a visible world structured by Euclidean geometry. But dedicated travelers know that finding our way through the world is as much a journey of the spirit, a means of defining our personal and cultural identities, as it is a navigational challenge.

MAPPING THE WORLD

Denis Cosgrove

For most modern people, a world map of continents and oceans, with physical features or political divisions colored and labeled, is so familiar as to pass unremarked.[1] We may have noticed the Mercator projection's (fig. 30) enlargement of Greenland at the expense of Australia, or have been struck by the unfamiliarity of Arno Peters's equal-area world map (fig. 31) that claims to give the poorer parts of the world due prominence by hanging the continents like a line of overstretched laundry (King 1990; Monmonier 2005). But for the most part we take for granted the authority of a modern world map as a scientific document.

With a moment's reflection, however, "the world map" becomes a more complex affair. We might consider the meaning of *world*. It is different from *Earth* or *globe*. For an Abbasid imam in ninth-century Baghdad, a Mandarin administrator in Ming China, a precontact Kwakiutl chief[2] on Vancouver Island, or the Genoese navigator Christopher Columbus, the world did not encompass the globe, and *Earth* denoted an element rather than the planet. We might reflect too that the familiar global patterns of land and sea have come into certain focus only very recently. When I open my university atlas from

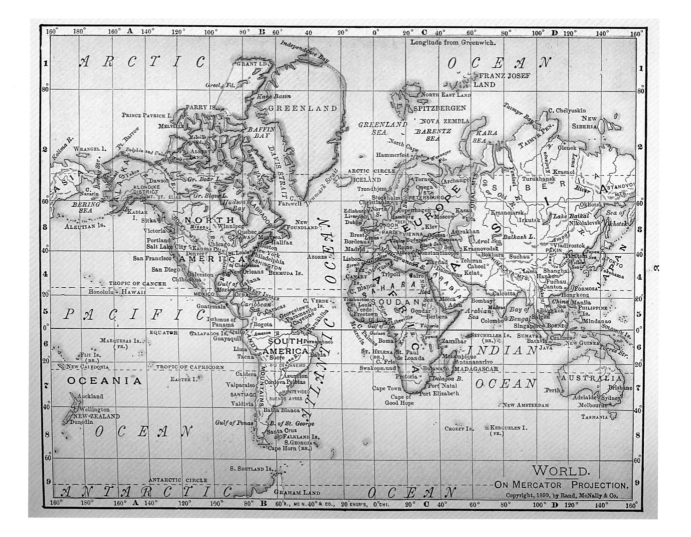

FIGURE 30.

"World on Mercator Projection." From *The Vermont Phoenix Dollar Atlas of the World* (1900).

the 1960s to the southern polar region, I find question marks rowed along the edges of the Antarctic Continent. They indicate uncertainty about the precise line of its ice-obscured coasts.

This chapter explores such aspects of the world map, discussing first world mapping in times and cultures very different from our own, and then examining the cartographic processes by which *world* became synonymous with the planetary globe, and resolved itself into the familiar patterns of today's map or satellite image. The task is of more than antiquarian interest: one of the principal arguments here is that for all their unquestioned scientific accuracy, empirical authority, and technical sophistication, contemporary world maps remain cultural artifacts, sharing—perhaps unexpectedly—many attributes of the world maps of nonmodern cultures (see Cosgrove 1999b; Edney 2005; Jacob 1996).

66

DENIS COSGROVE

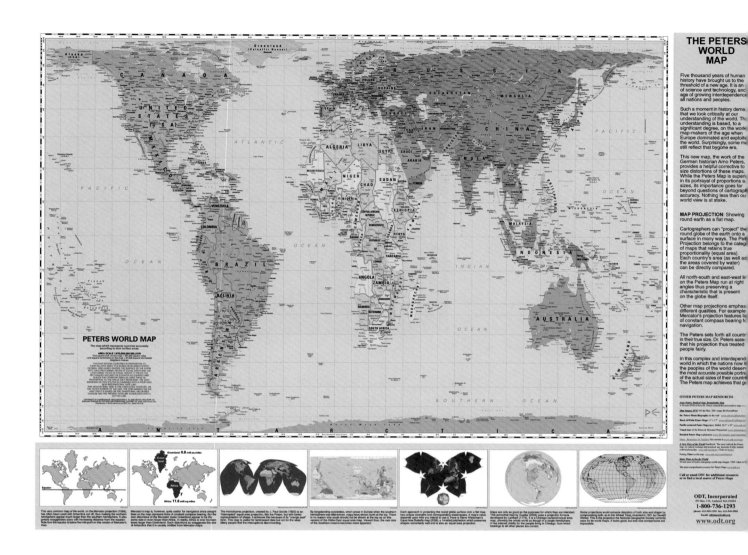

FIGURE 31.

Arno Peters, "Peters World Map" (2004). © 2005,
Akademische Verlagsanstalt, www.odtmaps.com

WORLD, EARTH, AND GLOBE

"World" is a social concept. When we claim that an autistic child "lives in her own world," we are referring to her social and communication skills as much as to her relations with her material environment. *World* is a flexible term, stretching from physical environment to the worlds of ideas, of microbes, of sin. Arguably, all these worlds can be mapped in the sense of being presented graphically according to spatial criteria, and indeed imaginative and affective relations play a significant role in world maps. But my discussion here will be restricted to human worlds that are conceived and experienced with at least some direct reference to material space. Their scale, however, may range from that of the individual body to the planetary earth. In contrast with *world*, *Earth* refers today to the planet that sustains life; its reference is elemental rather than

FIGURE 32.

View of the Earth seen by the Apollo 17 crew
traveling toward the moon (December 7, 1972).

social. The 1972 photograph taken by NASA's Apollo 17 astronauts, unique as an eye-witness photograph of humans' home planet (fig. 32), is referred to as the "Whole Earth" (although it actually shows only one hemisphere) rather than as an image of the world (Cosgrove 2001, 257–67). It is certainly not thought of as a "map," although it shares many technical aspects with world maps, and has influenced considerably the design of subsequent world maps—for example the disappearance of the graticule (grid) of latitude and longitude, the "photographic" appearance, and the use of "natural" color on many wall and atlas maps today. Nor do we think of the space it represents as "the world," in part because it does not show the whole surface of the earth, but more because it lacks visible reference to human presence. We expect a world map to present a synoptic, or all-encompassing, view with evidence of human existence, although we recognize the physical impossibility of such a unitary vision across a three-dimensional sphere.

Globe is a geometric term, another word for a sphere. Cartographically, *globe* is reserved for three-dimensional, solid models of the Earth (or other planets).[3] The relationship between the globe and the modern world map is close. Some world maps seek to maintain the visual trace of the globe by framing the world in paired hemispheres (figs. 51, 54, and 55) rather than projecting a rectangular "planisphere." Hemispheric arrangement allows, as we shall see, a wealth of other information about "the world" to be incorporated within the frame of the image, connected to, but distinct from, the mapped space proper.

The modern, scientific conception of the world extends to the whole of the globe and encompasses the whole earth, which is why these three terms

DENIS COSGROVE

are now interchangeable. Any world is a totality and has spatial boundaries, but the coincidence of the world's boundaries with the planetary globe's is a modern conception, a consequence as much as a cause of maps. Every culture creates a "world" in which it is "at home" and beyond which other spaces seem alien, exotic, often threatening. A world map pictures the totality of the lived space of a culture. There is an urge toward comprehensiveness and synopsis in all world maps, using spatial graphics to illustrate connections among the many and diverse elements that constitute worlds of experience. Locating and delimiting are key aspects of imagining the totality of a world. Marking its center and edges are structuring features of all world maps, thereby "normalizing" the center and "othering" the edges.[4] World maps are inescapably ethnocentric to some degree, and this remains true of modern maps, although the globe itself possesses neither center nor borders. A two-dimensional world map has both. From a religious perspective, the limits of terrestrial space can correspond to those of secular time, giving some world maps a spiritual dimension that reinforces their totalizing and synoptic aspects, often rendering them objects of moral reflection if not veneration.

The best way to understand these aspects of world maps is through specific examples of how different cultures have mapped their world, noting recurrent features, before turning to modern world maps and the ways they resemble and differ from nonmodern images.[5] My choices cannot hope to be comprehensive or even representative of the vast number of maps seeking to illustrate "the world," and my ordering reflects no systematic scientific, cultural, or aesthetic criteria. In attaching documents to place, every selection is a kind of mapping exercise, but I have sought to avoid privileging any one tradition of cosmographic[6] and world mapping over another.

NONMODERN MAPPED WORLDS

Mesoamerica

The Codex Fejévary-Mayer is a manuscript that dates from the fifteenth or early sixteenth century, before the arrival of Spanish *conquistadores* in Central America in 1521. Its frontispiece is a complex illustration of figures and symbols (fig. 33). Aztec records of pre-Columbian Mesoamerican belief and cultural practices allow us to make sense of this compelling image as a map of the world (Léon-Portilla 1991, 540–41). Its formal structure is that of a Maltese cross whose inner square contains a figure in profile: Xiuhtechitli, god of the hearth and fire. The radial format extending from Xiuhtechitli should be read perspectively. The image represents three-dimensional space and cosmic time: the picture is thus at once map and calendar. At the corners are the head, bones, foot, and arm of the cosmic creator, Tezcatlipoca, lord of everywhere, whose lifeblood we see flowing inward toward Xiuhtechitli. Symbols of sacrifice and worship—the

FIGURE 33.

Frontispiece, Codex Fejévàry-Mayer (1400–1521).

temple platform and altar, the disk of the sun, and a bowl of incense—surround the inner square at the junction of cosmic and domestic space. The world's surface extends toward the cardinal directions: East (red, at the top), West (blue, below), North (yellow, left), and South (green, right). The Mesoamerican world rested upon primordial waters, while the sky above and the sun's diurnal path were supported by a world tree extending to the four corners, and here represented by the T shapes at each cardinal direction. These "trees" refer to plants central to Mesoamerican life: clockwise from the South are the cacao, the maize plant, and a cactus, reflecting also "the biogeography of central Mexico: fertile croplands to the west, deserts to the north, and tropical lowlands to the southeast" (Mundy 1998, 232). East is the most sacred direction, whence the sun rises from the turquoise tree. It will set in the West (here gendered as female space by the clothing of figures there), where the crouching *tzitzitl* demon of the dusk awaits unwary souls. The dots marked along the framing lines indicate the 260-day calendar of 20 named days, counted thirteen times. The image illustrates a complex set of associations between forms, colors, and concepts in which "space—in the form of community territory—also dovetails with time, manifest as episodes of human history" (ibid.).

As a world map, the Codex Fejévàry-Mayer can only be understood in relation to the worldview or cosmology and practices of pre-Columbian Mesoamer-

DENIS COSGROVE

icans. It displays a number of structural and formal elements that recur in the world maps of other nonmodern cultures, despite marked differences in geographic location, cosmology, and graphic conventions, but bears no apparent relation to the modern world map. The world's vertical dimension—extending up toward the skies or heavens and down to a primordial underworld—is as significant as its surface and thus is mapped into the image. The center of the map is "home," the familiar space of everyday life, while the world's boundaries are the edges of known space; they do not encompass the globe. The Codex Fejéváry-Mayer map does not speculate as to what lies beyond the ends of the known space, but its makers' "home" lies at the map's center. Most nonmodern world maps thus normalize the center, banishing the strange and abnormal to the edges. Folding space into time, map into calendar, is also common in nonmodern maps—although we should remember that modern world maps too mark "universal time" in the international date line passing through the Pacific; longitude intervals to indicate diurnal hours; and the passage of the sun between the tropics to measure the year. Use of the square and/or circle to structure the map's formal design and the practice of fitting the form of the human body as a microcosm within the geometry of the macrocosm are also common.

Mapping "the world" presupposes a certain unity and coherence—even beauty—in the cosmos. Indeed, that Greek word itself implies both harmony and visible beauty (the root of the modern word *cosmetic*), marked in the heavens by the regularities of the celestial bodies and on earth in the symmetries of the ideal human body. The two scales are therefore reflected in the form, composition, structure, and decorative elements of world maps themselves, and often in their location and use. More mundanely, the Codex Fejéváry-Mayer, like many world maps, seeks to connect its abstract geometry and ethnography to the specific geographic environment (mountains, rivers, flora, fauna)—and to the customs and culture of the world it represents.

The Fejéváry-Mayer image has the quality of a sacred object that demands reverence in itself and its use. The codex was probably intended for merchants or their advisors, to determine the most auspicious days for undertaking trade or journeys. Less sophisticated and highly wrought versions of the Mesoamerican world it pictures were carved onto public monuments. The codex is now carefully preserved in a modern library, but many, perhaps most, of the world maps made by nonmodern, and certainly by nonliterate, cultures were never intended to be permanent. Nonliterate peoples across the world, including Australia and the Americas, have "performed" their spatial understanding of the forms, structures, and origins of the world in poetry, song, dance, and sacred ritual. Such performances are considered world maps insofar as they describe and illustrate a culture's singular conception of material and spiritual space. They serve similar functions to those played by the world map in modern societies: informing and educating society's members about the world in which they live. Their functions may include the moral dimension of teaching proper behavior in the world, just as the Whole Earth image is promoted today as an icon for ecologically responsible conduct.

The American Southwest

Native peoples of the American Southwest have long incorporated sand (or ground) painting into their rituals. Those produced by the Navajo illustrate a world composed of Earth (human) people, and Holy (supernatural) people, both of whom are obliged to adhere to the rules of the continued proper functioning of the universe. If disaster occurs, ritual performance by the Earth people enlists the Holy people's help in restoring cosmic order and balance. Many disasters are environmental, for example drought or flood, whose origin lies in the heavens. The performance is led by a medicine man, who chants as he sprinkles intricate patterns across a bed of sand using variously colored substances such as dried and pulverized leaves, flowers, bone, or rock. The patterns commonly include recognizably cartographic elements, especially star patterns in the realm of the Holy people, as well as representations of familiar landscape features and sacred landforms. The Navajo world picture associates each of the cardinal directions with a mountain, a time of day, a stone or shell, and a color (G. M. Lewis 1998, 110). East, for example, is represented by Blanca Peak, associated with dawn and represented by a white shell and the color white; while North is signified by Hesperus Peak, associated with night and faith and represented by jet stone and the color black. (The top of the painting normally signifies North or East.) (Lane 2002, 74–78) Sand paintings are erased after the performance, so that we cannot know today their precise nature in the years before native peoples' contact with Spanish or Anglo colonizers, or to what extent postcontact maps are hybrid creations responding to such outside influences. But early twentieth-century images transferred from the ground to woven rugs preserve some record of these precontact peoples' world maps.

Africa

Permanent artworks among nonliterate peoples frequently include petroglyphs and cave painting. Not uncommonly, these include cosmographic images. On the Bandiagara escarpment in modern-day Mali are found some of the earliest African map images, and these too share characteristics found in the Mesoamerican and Navajo examples. A sign known as *aduno kine* (life of the world) is composed of a simple cross, topped by an egg-shaped circle with a symmetrical but open curve below (fig. 34). This anthropomorphic sign has been interpreted as a skeletal map of the cosmos of the Dogon, an agricultural people of the Niger valley. In it, "the egg-shaped head signifies the 'celestial placenta,' while the legs refer to the 'terrestrial placenta.' The torso and arms

DENIS COSGROVE

form a cross representing the cardinal directions" (Bassett 1998, 26). The same form is reproduced in the ground plans of individual Dogon houses and whole settlements, another example of how world mapping often seeks to unite cosmos and hearth (see also Atkin and Rykwert 2005; Tuan 1994). In purely formal terms, this ancient African figure resembles cosmic signs in other cultures, for example the zodiacal sign for Mercury, and the Elizabethan mapmaker John Dee's *Hieroglyphic Monad*, which was intended to express graphically the totality of the created world (Dee 1947). This is not to suggest either exoteric or esoteric connections between the various images (although occult powers have not uncommonly been attributed to cosmic images; see Jung 1964); rather, anthropomorphism and the geometry of line, circle, and square appear to be graphic devices that readily occur to humans in expressing spatial form and regularity.

Buddhist Asia

World maps of literate peoples, especially those whose cosmologies are scripturally based, can be interpreted much more precisely and confidently than those of nonliterate peoples. Like the Codex Fejévary-Mayer, they are often objects intended for detailed study and interpretation. Buddhism and Hinduism present some of the most sophisticated of such world maps, again combining geometrically conceived cosmic space with pictorial or symbolic markers of location and topography, and featuring representations of the human body. Buddhism, Indian in origin but spread through China, Japan, and the Indo-Chinese peninsula, has a detailed understanding of the cosmos and the terrestrial world, which has influenced world mapping across Asia. Fundamental is the belief that the universe we know is but one among literally millions of others through which souls migrate. Each world is vertically organized atop the foundational mountain: Mount Meru (Sumeru in China, Korea, and Japan), which probably refers to peaks in the Himalayan Pamirs. Meru is encircled by mountain ranges, continents, and oceans. Vertically, the world divides into the ascending realms of desire, form, and nonform. All life experiences cycles of death and rebirth, and at each incarnation is placed on a plane determined by moral conduct in previous lives. The ultimate goal is nonform, the state of *nirvana*. The terrestrial plane (K'amadahatu), insignificant within the vast cosmos, is itself composed of four huge continents and oceans arranged symmetrically at the cardinal directions around Mount Meru, each under the authority of a specific denizen. Of these continents, Jambudvipa is southernmost and the only one inhabited.

Such a vast and complex spatial conception has yielded a considerable range of cartographic representations, many intended as aids to devotion or meditation. In Tibetan Buddhism, for example, mandalas assist the spiritual journey toward self-recognition as an incarnation of the Buddha. Like Navajo world maps, these are commonly designed as temporary forms using powdered sand, to be swept away after ceremonial use, indicating the transitory nature of all existence (Schwartzberg 1994a, 620). Complex arrangements of circles within squares within circles, colored to represent the four elements, stretch outward from earth (dark blue), through water (light blue) and fire (red) to air (yellow). Colored figures represent places, suggesting the interconnectedness of all things and a universe within which every form is subsumed into another, until only a psychic center remains.

While the mandala typically represents the whole cosmos, individual images can focus on the terrestrial plane alone, more closely resembling what a secular observer might normally think of as a world map. On the terrestrial plane, Mount Meru stands central to a flat, circular earth around which revolve the sun and moon. At its foot are seven mountains and water basins, arranged concentrically, with a brackish ocean beyond and four continents located at the cardinal direction points, each having its distinctive shape. The inhabited Jambudvipa lies to the south; its characteristic wedge (or inverted-egg) shape is fairly certainly based on concrete knowledge of the outline of the Indian Subcontinent.[7]

Buddhist maps share with the others that we have considered the formal characteristics of a vertical cosmos, geometric symmetry, and color codes. There is some reference to local geography, but the absence of anthropocentricity is distinctive: the inhabited world has no claims to significance or centrality in the spatial organization of creation.

Jambudvipa appears alone in some Buddhist maps, many produced far from India in China, Korea, and Japan. In being confined to the secular spaces of the earth's surface, such works seem closer to what we conventionally think of as a world map, although comparison is problematic, given Buddhism's unique cosmology. Nonetheless, as with the Navajo maps, the influence of external geographic knowledge is often apparent in these images. Information introduced by traders and colonizers has been incorporated without apparent upset to the conventional world picture. Thus, a 1710 map of Jambudvipa retains the continental outline and locates traditional places in Buddhist geography while incorporating European geographic knowledge, and indeed the map form of Europe itself in the upper left (fig. 35).

Buddhism's cultural influence spread through China, Korea, and Japan, although only in Japan did Buddhist cartography have lasting and documented influence. Despite scientific knowledge of the relations between solar movements that would have allowed them to do so, before 1600 Chinese scholars did not project celestial coordinates onto the earth to produce a "world" map. The influential Chinese concept of *tianxia* (*chonha* in Korean), translated as "all under heaven," and the idea of China as a "middle kingdom" are taxonomic and ethnological rather than spatial in their reference. Cordell Yee (1994c, 174) points out that the Chinese phrase "all under heaven" implies "a notion of geo-

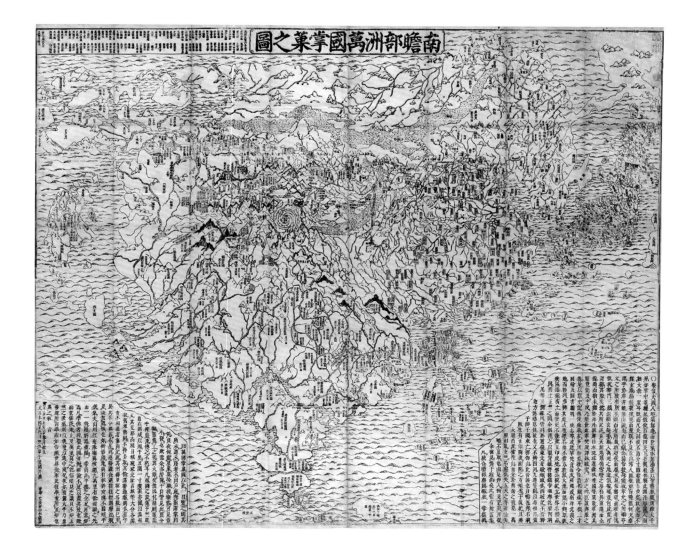

FIGURE 35.

Soshun (Hotan), "Nansenbushu bankoku shoka no zu" (Visualized Map of All the Countries in Jambudvipa) (1710).

graphical centrality, but perhaps more important, a belief that China was the center of culture, the standard for civilization"—an ethnocentrism shared with other world mappings that we have encountered. Chinese thinkers certainly saw themselves as living at the center of a world that stretched beyond their realm, whose center was placed at the great sundial of Dengfeng in Manchuria, but historically saw no purpose in mapping the world beyond the empire. The only "world maps" in China prior to the seventeenth-century Qing imperial expansion would have been Buddhist maps of Jambudvipa, which locate the Chinese empire at the continent's northern margins (see the illustrations in Yee 1994c; also Henderson 1994). The earliest Korean world map dates from the opening years of the Chosun dynasty and the introduction of Confucianism in the late fourteenth century, but as in China proper, Buddhism was persecuted and its cartographic influence consequently difficult to trace. In Japan, how-

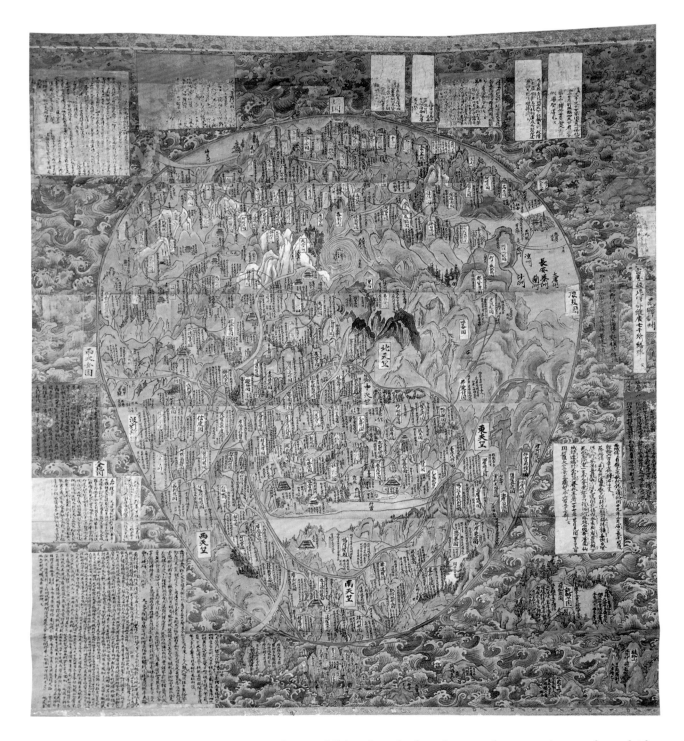

FIGURE 36.

Jukai, "Gotenjiku Zu" (Map of the Five Indies) (1364).

ever, where Buddhism dates back to the seventh century, it was tolerated. The oldest Japanese maps, "Gotenjiku Zu" (Map of the Five Indies), from the late fourteenth century and probably derived from unknown Chinese prototypes, show the five regions of Jambudvipa within an egg shape, giving China prominence while reducing the size of Korea and plotting the Japanese islands on the eastern edge of the world continent (fig. 36).

DENIS COSGROVE

Hindu and Jain India

The second great Asian tradition of world mapping relates to Hinduism and the distinct but related Jain belief. Each of the Indian faith traditions, including Buddhism, has gathered cosmological elements from the others. All posit the world as a flat disk suspended between two great bowls turned inward to form the Brahmada, or cosmic egg. Hindu world maps share with Buddhist renderings the idea of four continents arranged in a lotus pattern around Mount Meru, and often include four rivers flowing out from this *axis mundi* (world axis). Surrounding Mount Meru in Hindu and Jain world maps, the seven continents and seas are arranged concentrically, with Jambudvipa placed at the center. A mathematical ratio of increasing size is established for succeeding continents, seas, and rivers. The continents are under the tutelage of Hindu divinities, and the seas are named for foodstuffs significant in the ecology and dietary practices of the subcontinent (salt, sugarcane, curd, milk, clarified butter, wine, water).

The most intense cartographic expressions of India's syncretic world picture come from Jainism, a minority faith centered in Rajasthan and Gujurat.

FIGURE 37.
Manusyaloka (Map of the World of Man)
(19th century).

Stone-carved cosmographies here date back to the early thirteenth century, and while their overall form remains constant over many centuries, a bewildering variety of representational symbols appears in the maps. A nineteenth-century Jain work now in Washington, DC, maps Manusyaloka, or world of humans, in characteristically intense color, with the circular Jambudvipa at the center surrounded by the first (salt) sea ring and a further two continents that lie within the encircling mountain range, limiting the human world (fig. 37). At the corners are four individuals who have achieved liberation from the material world (*moksa*). Mount Meru lies at the very center; the converging elephant tusks represent other mountains, with chains of lakes, rivers, and other topographic symbols emphasized in brilliant color. Bharata (India) is probably the crescent shape in the very lowest part of the central continent, located between the rivers Ganga and Yamuna. It is crossed by a mountain chain with the Aryan land at its center. The text notes that Bharata's area is only one-ninetieth the area of Jambudvipa. But geographic fidelity is subordinate in the Jain map to a relentless symmetry that reflects its role as a contemplative icon rather than an informational document.

Medieval Western Europe

Geometric symmetry, cardinal orientation, symbolic coloration, and locally focused geographic references are also apparent in medieval European world maps, known as *mappae mundi*. Dismissed in the past as products of inadequate geographic science, to be superseded by the modern world map, these images make sense, like the Mesoamerican, Buddhist, and Jain examples, as spatial illustrations of a scriptural cosmology. Their appearance in prayer books, scholastic manuscripts, and cathedrals indicates their primary didactic and iconic functions. They represent the geographic world as the temporal part of a vertically ordered creation, whose history is revealed in scripture. The geographic content of mappae mundi is recognizable through names such as Europe, Asia, Africa, Rome, and Jerusalem. Their contents and evolution reflect medieval Christianity's characteristics as a hierarchical, proselytizing faith for which cultural accumulation and information about the material world reflected a universal mission. The legacy of classical geographic knowledge exists within the mappae mundi, but we should resist the temptation to focus on their geographic accuracy to the exclusion of the exegetical purpose of the total image. (On medieval European geographic knowledge, see Lozovsky 2000 and Wright 1925.)

Unlike Indian and Mesoamerican cosmologies, Christianity is monotheistic, with a singular creation narrative and a universalistic redemption story rooted in the idea of the creator God incarnated within historical time. Christianity anticipates a fixed end to historical time and the destruction of the earth. It is strongly anthropocentric, global, and normative in its ideal of universal salvation. The now destroyed Ebsdorf map, the Psalter map, and the Hereford Cathedral map all reflect these principles explicitly by enclosing creation within the body of God the Father. The earthly globe as the space of salvation is often shown as an orb, the T-O (*terrarum orbis*), held in the right hand of Christ, the Son of Man. Jerusalem, the *axis mundi* of the Christian narrative and the terrestrial junction with upper and lower worlds, lies at its center. Old Testament geographic features such as the Red Sea and Sinai, the terrestrial paradise of Eden, and Mount Ararat are marked, while the three continents are allocated, as the book of Genesis suggests, to the patriarch Noah's sons (Scafi 2006).

Christianity owes as much to Greek philosophy as it does to Hebraic scripture; its cosmography is based as much on Aristotle as on the Bible. The tricontinental world island of the medieval mappae mundi reflects the continued influence of an ancient Greek world picture, which locates the elemental world at the center of the cosmos. Its spheres of earth, water, air, and fire are subject to change and imperfection, while movement is characterized as linear, as objects seek their rightful place in the appropriate elemental sphere (falling to earth, for example). Beyond the zone of fire is a celestial world of seven planetary bodies, from the Moon to Saturn, whose motion is circular; beyond these are the fixed stars, the sphere of the prime mover, and the supercelestial realm of the divine. Perfect and unchanging, the celestial world is filled with the fifth element of ether.

Theoretically, earth should not protrude through the sphere of water, but the elemental world is imperfect, and eccentricity of the elemental spheres explains the world island, surrounded by ocean and penetrated by the Mediterranean, the Nile and the Don, which separate its three continents. This landmass may or may not be balanced by a southern continent. Focused on the Mediterranean, the mappa mundi is "center enhancing," portraying an increasingly abnormal and bizarre world toward its edges. When shown, the unknown southern continent particularly was the location for the monstrous, the strange, and the wondrous. The "monstrous races," inherited from Pliny and other ancient writers, appear to the right of the thirteenth-century Psalter map. On the widely copied map by Beatus of Liebana (fig. 38) illustrating his eighth-century *Commentary on the Apocalypse*, this fourth section of the world (at far right) is separated from the rest of the world by a "Red Sea." An inscription explains that the sea is occupied by "fabulous beings," which other versions of the map depict as dog-headed, one-eyed, multilimbed, or otherwise half-human creatures (Friedman 2000). The Beatus map is oval, but maintains the normally circular mappae mundi's emphasis on symmetry by, for example, distributing diagrammatic islands neatly around its edges. Christ's Second Coming, the subject of Beatus's text, is imagined as a universal event entailing the end of space and time. Correspondingly, wind heads signifying the cardinal points of the compass sometimes double on mappae mundi as trumpeting angels sounding the final day.

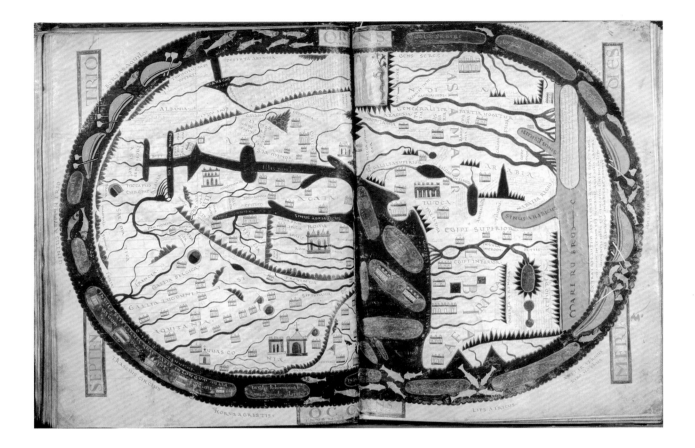

FIGURE 38.
Beatus of Liebana, world map (ca. 1050–60).

PTOLEMY, ISLAM, AND THE WORLD MAP

The mappae mundi owe much to contact and cultural interchange with Islam, initially in Spain and later in the Levant during the crusading years between 1100 CE and 1300 CE. Their geographic information increased as the reports of pilgrims and travelers were incorporated. The last great medieval world maps, produced in mid-fifteenth-century Venice, altered the balance of the sacred and secular elements (Scafi 2006, 235–40). On Fra Mauro's rendering from 1459, Jerusalem in no longer central, and the terrestrial paradise is pictured outside the map area altogether (fig. 39). The friar made explicit use of navigational charts received from King Alfonso V of Portugal, who had commissioned the map. These portolan charts, derived from coastal plotting with compass, rule, and freehand drawing, represent a distinct form of mapping that had been developed for Mediterranean navigation, but extended to the Atlantic and eventually the continental coasts across the globe. Thus the Catalan world map of 1375 represents the coasts from the Black Sea to the Atlantic coast of Africa, charted with astonishing accuracy (fig. 40). It also borrows eclectically on Islamic knowledge, mappae mundi, travel literature, and legend to create

DENIS COSGROVE

FIGURE 39.

Fra Mauro, "Mappamondo" (World Map) (1459).

an altogether unique and somewhat bizarre picture of the world. Nonetheless, it is an image that in its secularism and emphasis on earthly power (denoted through flags and enthroned rulers) anticipates an emerging purpose for the world map.

Fra Mauro's text refers to another mode of mapping, based on astronomically determined locations coordinated by a graticule (grid) of longitudinal meridians and latitudinal parallels. Geometry thus allows the sphere to be projected onto a two-dimensional surface, an idea that had been formalized at Alexandria in the second century of the common era by Claudius Ptolemy. This cartographic method was the subject of fevered interest in Fra Mauro's Italy, although the friar himself declared it unsuitable for a modern world map

MAPPING THE WORLD

because its framework is restricted to the ancient world. His own rendering, drawing on Portuguese charting of Africa and Marco Polo's reports of Asia, shows a much vaster world of which Europe, although accurately shown, is but a tiny part. Ptolemy's cartographic methods would become fundamental to the modern world map, but not as the simple outcome of their reappearance in fifteenth-century western Europe. His work hybridized in some measure with every one of Eurasia's major cartographic traditions, as it did with the portolan charts and the mappae mundi. This was principally achieved through its dominance of mapmaking in the Islamic world that stretched from its center at the crossroads of the Old World westward to the shores of the Atlantic and eastward to the Indian Ocean and the Pacific.

Claudius Ptolemy lived between 90 CE and 168 CE. His works summarize and synthesize Hellenistic astronomical, astrological, and geographical science. *The Geography* incorporates the extensive geographic knowledge of Greek and Roman empires recorded by such writers as Herodotus and Strabo, and gathered for example by Cornelius Agrippa, whose universal survey sent Joseph and Mary to Bethlehem and yielded a public world map in ancient Rome. Drawing on Greek natural science, Ptolemy could unite conceptually "world" and "globe" in a cartography that corresponded to the Roman claim of *imperium*

DENIS COSGROVE

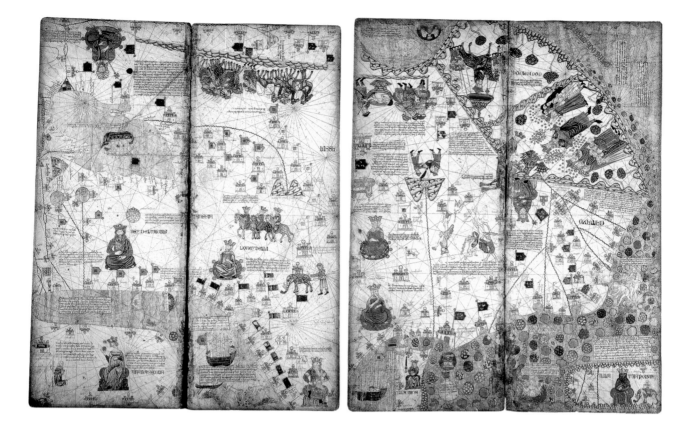

ad termini orbis terrarum (empire to the ends of the earthly sphere). He evaluated previous measurements of earth, suggested three methods for projecting the sphere onto the plane using longitude and latitude, and listed over eight thousand coordinates of known places across the inhabited earth—what the Greeks called the *oikoumene* (fig. 41). Larger-scale regional maps were systematically related to the world map. No maps drawn from Ptolemy's coordinate tables survive from antiquity, but the text's influence echoes throughout the entire history of Eurasian world mapping. (For a translation of Ptolemy's text, see Berggren and Jones 2000.)

The single most significant feature of Ptolemy's system for creating a world map is the graticule. This is the conceptual definition of 180 degrees of latitude between the poles, and 360 degrees of longitude counted from an arbitrarily determined "prime" meridian—the foundation for any mathematical projection of the sphere onto the plane. The tropic and polar circles do not correspond to whole degrees of latitude, although they were foundational to the long-enduring conception of climatic zones. Ptolemy does refer to the parallels as *klimata*, the Greeks' word for parallel climates resulting from the sun's changing angle with the earth. From these they had derived three zones of varying habitability. The middle or temperate group in the Northern Hemisphere formed the

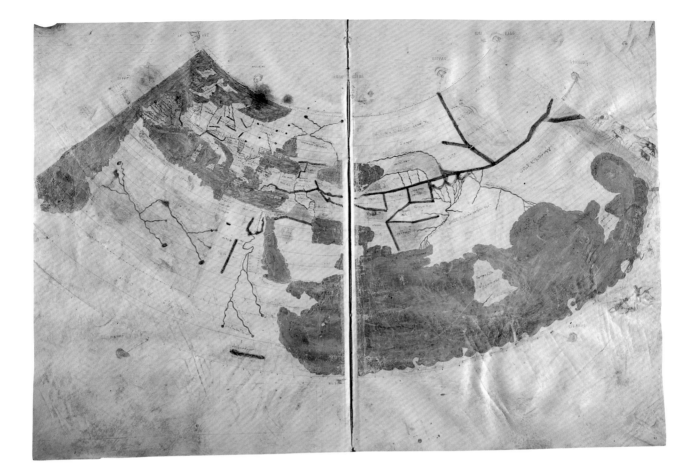

oikoumene (fig. 42); two frigid zones pole-ward of the Arctic and Antarctic circles were conventionally believed to be uninhabitable, as was the torrid zone between the tropic lines, by reason of heat. Only the two temperate zones supported human life, and only the oikoumene for certain. Its southern equivalent might or might not contain land; the ancients could not know, as the torrid-zone was thought to be impassable. Ptolemy actually challenged a number of these beliefs, mapping an oikoumene that stretches 180 degrees east from his prime meridian in the Canaries or Fortunate Isles (the Greek Hesperides and earthly paradise), and 80 degrees of latitude from the Baltic regions to about latitude 17° south. Its center was the temperate zone, and the Mediterranean its middle sea. The symmetry of a single world landmass penetrated by seas but surrounded by ocean[8] was maintained by mapping the Indian Ocean as a "greater Mediterranean," enclosed by a land bridge joining Africa to Southeast Asia.

Ptolemy's world map may itself have been influenced by cartographic conceptions imported from the margins of Hellenistic cultural space. The puzzling appearance of Taprobane as a large island occupying the location of the Indian Subcontinent may have originated in Buddhist mapping of Jambudvipa, with

DENIS COSGROVE

Lanka immediately to the south. In Islamic Indian cartography, Lanka lies at the intersection of the equator and prime meridian. Ptolemy's own world map would eventually penetrate the cartographic traditions of India, but with the fifth-century collapse of Roman administration in the West, Ptolemaic world mapping withered there.

Ptolemy's global influence was sustained by Islamic scholars who conserved, translated, and studied ancient Greek natural philosophy intensely in the three centuries following the Arab conquest. Islamic scholarship eventually spread across the breadth of the old oikoumene. Its thinkers wholly embraced Aristotelian cosmology, matching the seven celestial spheres with the seven climates that defined the inhabited earth. Their cosmographic relationship was rarely measured mathematically; coordinate tables, locations, and climates were repeatedly copied. The center of Dar al Islam (those parts of the world accepting the Prophet Muhammad) was deemed to lie in the fourth or middle northern climate. Islam's own sacred scripture, the Qur'an, along with the Hadith (recorded sayings and actions of the Prophet and his companions), the second source of Islamic law after the Qur'an, contain no systematic cosmology: Islamic acceptance of the Ptolemaic world picture was uncontroversial, and Arab maps lack the sacred status of Christian mappae mundi. This broadly Ptolemaic map was challenged only by the influence on some Islamic thinkers of Persian cosmology, in which the seven climates are patterned as six geographic circles (kishvar) arranged concentrically around the central seventh, which corresponds to Iran itself. Clockwise, the circles represented India (south and generally at the top of a kishvar world diagram), Arabia and Abyssinia (southwest), Egypt and Syria (west), Asia Minor and the Slavic lands (northwest), Turkey and the lands of Gog and Magog (east and north), and China (east) (fig. 43A–B). Both alternative schemes appear commonly as geometric diagrams illustrating Islamic books of natural philosophy and are elaborated in esoteric astrological and alchemical works exploring correspondences between the world's natural and moral orders. Such use of cosmic and world representations is similar to that practiced in many nonmodern cultures, where the harmony and regularity of the world map is believed not merely to illustrate the balance of natural forces, but also to permit its use in rituals aimed at sustaining cosmic balance and order.

Islamic cartography based on Ptolemy extends over a millennium, although the earliest surviving manuscript maps date from the eleventh century CE (fifth century in Islam). Experts today identify a group of four geographic writers beginning in the late tenth century as the Balkhi school of cartographers (Tibbets 1992a). Their texts are accompanied by or comment on sets of maps illustrating the world of Islam. These generally comprise a world map, three maps of the Mediterranean, Persian, and Caspian seas, and seventeen maps of Islam's provinces (iqlim). The Balkhi world map was generated by fitting the regional maps together to produce a speculative image of how the various

FIGURE 42.
Ambrosius Aurelius Theodosius Macrobius, world map (12th-century copy).

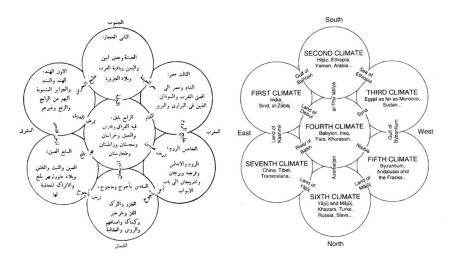

FIGURE 43A.

Abū Arrayhan Muḥammad ibn Ahmad al-Bīrūnī,
Persian *kishvar* (geographic circles) system
diagram (1021). After Togan 1941, 61.
© Archaeological Survey of India.

FIGURE 43B.

English translation of al-Bīrūnī, Persian *kishvar*
system diagram.

provinces might combine on the spherical globe as described by Ptolemy. The
result is a one-hemisphere map of a single, circular continent surrounded by
ocean and penetrated by seas. The maps are oriented with south at the top—an
Islamic convention adopted by Fra Mauro uniquely for European makers of
mappae mundi, which are generally oriented east. (Indeed, the word *orientation*, literally meaning "directed toward the east," derives from the traditional
European practice.) The Persian Sea (Indian Ocean) penetrates the continent
from the east, stretching as a great rectangle to a crescent-shaped Gulf, while
a smaller Mediterranean makes an indentation to the lower right (northwest).
Into it flow the Nile, from its sources in the African "Mountains of the Moon,"
and the Don, dividing Asia from Europe. Rectangular borders divide the Islamic provinces north of the Persian Sea, while few kingdoms are mentioned
outside the realm of Islam. The climates are marked as parallel bands on some
maps and their ethnographies noted, for example, by al-Istakhri: "The earth is
divided into two by two seas, so that we have a north or cold half and a south
or hot half. People in these two halves get blacker as you go south and whiter
as you go north" (Tibbets 1992a, 121).

Perhaps the most influential and frequently copied world map by an Arab
cartographer was Muhammad al-Idrīsī's twelfth-century work illustrating a
geographic description of the world commissioned by King Roger II, Norman
ruler of Sicily (fig. 44). Al-Idrīsī himself describes the original work, a map
engraved in pure silver, as being

of the seven climates and their lands and regions, their shorelines and hinter-
lands, gulfs and seas, watercourses and places of rivers, their inhabited and
uninhabited parts, what [distances] were between each locality there, either
along frequented roads or in determined miles or authenticated measurements
and known harbors . . . (Ahmed 1992, 159)

DENIS COSGROVE

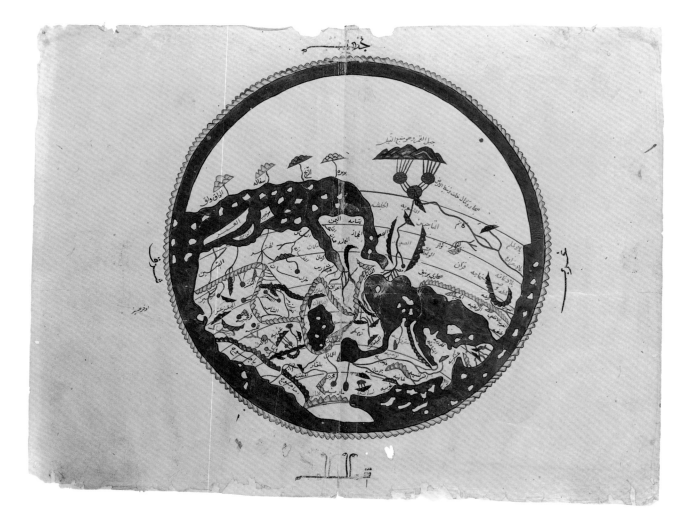

FIGURE 44.

Muhammad al-Sharīf al-Idrīsī, world map
(this copy 1553).

Al-Idrīsī offers no theoretical discussion of projection or coordinate positioning, but his textual description divides the world into regions defined by the
seven climatic bands and ten longitudinal lines, producing a genuine integration of text and map. His work has been called "one of the most exhaustive
medieval works in the field of physical, descriptive, cultural, and political
geography" (ibid., 160). Like the Balkhi maps, it illustrates a single, circular
world island surrounded by ocean and visually centered on Mecca, the center
of the Islamic world. South lies at the top of the map, and the southern coast
of the world island beyond the equator curves in a pure semicircle from the
coast of Morocco to the open waters of the Persian Sea in the east (as on the
Catalan map [fig. 40]). Islands are scattered in the three great seas and along
both coasts of the northern ocean. The Nile is a prominent feature, flowing
from the Ptolemaic Mountains of the Moon below the equator, as are mountain chains in North Africa and Asia, but there are few if any place-names. The

outline patterns of regions familiar to an Islamic mapmaker, born in Ceuta (on the Straits of Gibraltar) and working in Sicily—the Mediterranean basin, Arabian Peninsula, the Iberian Peninsula—are recognizably similar to the modern world map, although al-Idrīsī's version retains the Ptolemaic representation of an insular India and a vaguely sketched eastern Asia.

Al-Idrīsī drew upon a wide range of sources, including Latin mappa mundi, and his map is clearly a hybrid of cartographic traditions circulating with particular intensity in Sicily, a meeting point of Islamic, Greek, Norman, and Latin influences. Ptolemy's world map was thus present indirectly in Italy before manuscripts of his *Geography* were brought to Florence in the final years of the fourteenth century, and his influence would eventually reach across Asia. For example, the Marathi world map in the British Library is recognizably a rendering of the Puranic Hindu image of Jambudvipa, with Indian rivers, topographic features, and place-names marked across it, and the island Lanka at its southern apex (fig. 45A–B). But it is oriented with North at the top, while to the west the Arabian Sea with a rectangular Africa beyond, along with England, France, and a third island, of "hat wearers," appear in the upper left. The island on the opposite side of the continental world is labeled as China.

While none have survived, there is evidence also of Persianized Arab maps bringing the Ptolemaic world image to China in the early fourteenth century (Ledyard 1994, 246–47). The oldest surviving eastern Asian world map, however, was drawn in Korea in 1402, in the early years of the Chosun dynasty by ministers who are known to have spent diplomatic time in China and mention bringing maps from there (ibid., 244). Three versions of their Kangnido map, together with its prefacing text, are preserved in Japan. Such features as the Nile sources at the Mountains of the Moon, their naming, and the appearance of other Persianized Arab toponyms (place-names) make clear the map's Ptolemaic debt, however obscured by the prominence given to China as the central space of the world island, and the Korean Peninsula to its east. Korea appears the same size as Africa, Arabia, and Europe combined, all areas named and recognizable on the map. Among the islands in the encircling ocean, the largest group represents Japan, significantly smaller than Korea.

The faint influence of the Ptolemaic oikoumene may also be traceable in the popular but rather mysterious Korean *ch'onhado* world maps that remained popular into the closing years of the Chosun dynasty, but whose origins are unknown (fig. 46). *Ch'onhado* means "all under heaven," and the maps show a flat, circular earth space with a central continental island, surrounded by circular ocean and fifty-seven islands, and an outer land ring marked with fifty-five fictional places. To the east and west are trees marking the sun's passage. The central continent bears a faint resemblance to the Ptolemaic oikoumene, but amalgamated with the Buddhist central continent, with China at the center and Korea significantly enlarged. Africa hangs as an elongated appendage west of an expanded Persian Gulf. The Mediterranean is figured as an enclosed sea.

DENIS COSGROVE

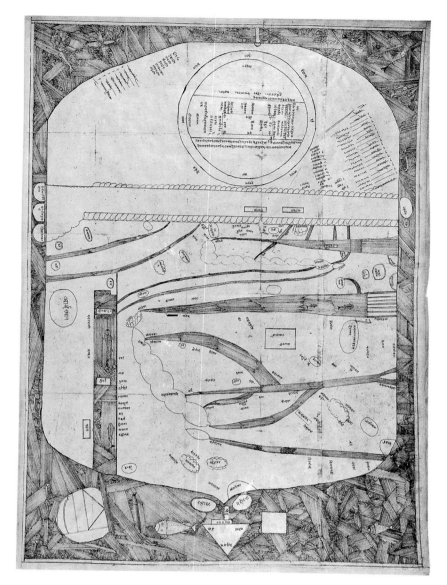

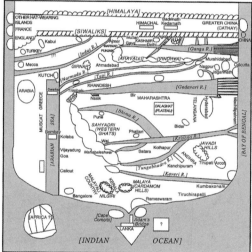

FIGURE 45A.

Marathi world map and cosmography.

FIGURE 45B.

Marathi map and cosmography; redrawing and translation from Harley and Woodward 1992, fig. 17.6.

Some researchers regard *ch'onhado* maps as a corruption of the Kangnido map, altered to conform to traditional Korean geographic conceptions (Ledyard 1994, 256–67).

By the beginning of the modern era, largely through Islam's centrality within the Old World and its preservation of Ptolemy's work—however loose and uncritical by succeeding scientific and philological standards—his world map had influenced the cartographic traditions of every literate people in Eurasia and North Africa. The result is that for all their diversity, world maps from late-medieval India, Korea and Japan, western Europe, and the Arab world itself share certain geographic features that can be matched at least broadly to their

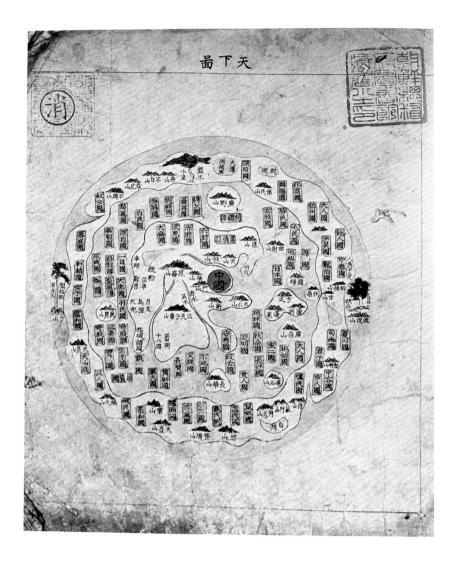

FIGURE 46.
Ch'onhado (All under Heaven) (mid-18th century).

location and form on a modern world map. This process of hybridizing intensified as the influence of Ptolemaic world mapping grew after 1400.

THE EVOLUTION OF THE MODERN WORLD MAP

Islamic mapmakers neither verified and extended Ptolemy's coordinates, nor sought to reconstruct his projections. Thirteenth-century scholars in Byzantium produced copies of the *Geography* and maps from its tables (fig. 41). It was work in Byzantium that provided the foundation for fifteenth-century German and Italian humanist studies in the very years that European navigators were pushing the limits of their geographic knowledge south and west of Ptolemy's oikoumene, and recording their discoveries on portolan

DENIS COSGROVE

charts (see chapter 1 of the present text, pp. 51–53). Relationships between the philological, mathematical, and technical work entailed by humanist studies of ancient texts, and the observational, navigational, logistical, and communication processes involved in exploration were neither simple nor direct. But within a little more than a century they yielded a geographic outline of the earth's lands and seas in which "the world" came to correspond, conceptually at first and increasingly on actual maps, to the globe—maps that are instantly recognizable to a modern viewer. The variety of mathematically accurate projections developed by the late sixteenth century to map the world has not been radically extended since that time (Thrower 1999, 69–80).[9] Yet, however close to a modern atlas map an image such as Abraham Ortelius's "Typus orbis terrarum" (Figure of the Earthly Sphere) may appear (fig. 58), we should not assume that the cosmological, iconic, ethnographic, and speculative aspects that so characterize nonmodern world maps smoothly gave way to a secular, scientific image. In some respects those aspects intensified over a complex historical process that involved both inclusions and exclusions within an evolving world map.

Technical aspects

Numerous technical demands faced early Renaissance scholars wishing to reconstruct maps from Ptolemy's *tabulae*, the tables of geographic coordinates that occupy the largest part of the *Geography*. Principal among these were understanding its projections, correctly identifying the measures used for latitude and longitude lines, and correctly identifying the place-names and coordinates of locations whose names had changed over the fifteen hundred years since Ptolemy had written. In this endeavor, Latin humanists drew upon previous studies made by their Greek predecessors, while bringing to the *Geography* their philological skills to reconstruct the original text accurately. Geometrical and optical sciences were also a matter of widespread intellectual curiosity among the Latins (Woodward 1990). And the world map soon moved beyond the studios of humanist scholars: fifteenth-century Europe saw a

blending of cartographic traditions that had remained distinct in the Middle Ages. The *mappaemundi*—previously structured on a strictly religious and symbolic space—started to incorporate the information and methods of Ptolemy. . . . Likewise, the portolan charts—previously structured solely on a network of lines of constant compass bearing for the aid of the navigator—also began in the early sixteenth century to bear latitude scales derived from the idea of Ptolemaic coordinates. (Woodward 1991)

To this rational mode of representing global space, the invention of printing at midcentury was a further stimulus, allowing maps and scientific diagrams

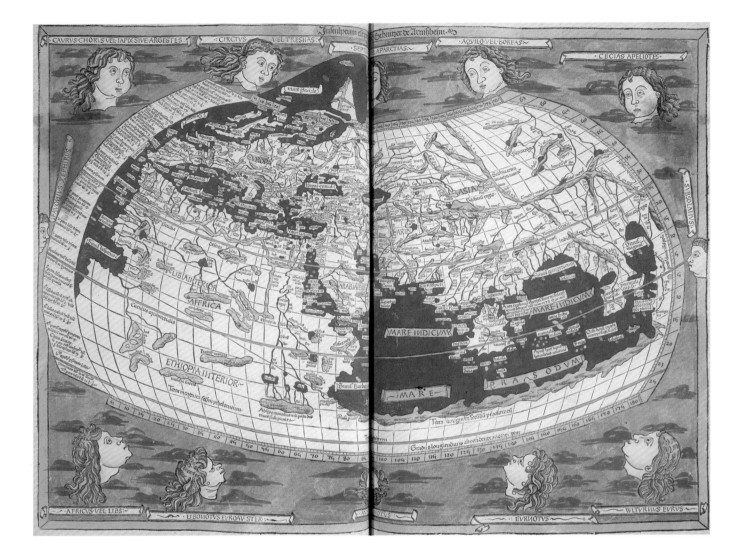

FIGURE 47.

Claudius Ptolemy, the *oikoumene* (inhabited earth) (1482).

to be cheaply and consistently reproduced and circulated—and criticized and rapidly corrected. By 1500, world maps based on Ptolemy's tables had been reproduced in all three of his projections. They took the form of beautifully colored manuscript copies made for courts in Europe and Ottoman Turkey, and copperplate or woodcut renditions circulated in printed editions of *The Geography* from presses in Italy and Germany (fig. 47). The popularity of the Ptolemaic world map by the end of the century may be judged from the appearance of a simplified version in Hartmann Schedel's commercially successful and widely translated *Nuremberg Chronicle* of 1493, some pages before a brief reference to the previous year's discovery of "islands in the Ocean Sea" (fig. 48).

But as Schedel's reference anticipates and Fra Mauro had recognized, for all Ptolemy's classical authority, his second-century world map accorded ill with the eyewitness evidence of navigators sailing beyond the limits of Ptolemy's geographic knowledge. By the first decade of the sixteenth century, maps based on his techniques were incorporating a vastly increased idea of the hab-

DENIS COSGROVE

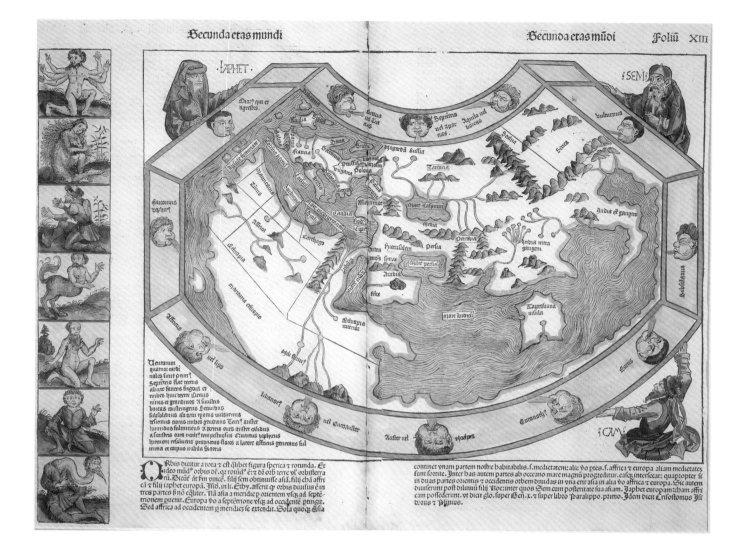

· IAPHET ·   ↯SEM↯

↯CAM↯

itable earth. Martin Waldseemüller's 1507 map—based on the second projection, but extending it through 130 degrees of latitude and a full 360 degrees of longitude and showing a sea-girt Africa, an open Indian Ocean, and a western continent which he named "America"—was designed as an explicit revision of Ptolemy by modern autopsy: eyewitness knowledge (fig. 89). The global extent of the new world map is emphasized by the insertion of two hemispheric global maps, while a portrait of Amerigo Vespucci (for whom Waldseemüller named America and whose report he printed in the accompanying *Cosmographiae introductio*) shares pride of place with Ptolemy himself as authoring spirit of the map. About a year later, the Florentine Francesco Rosselli produced an even more radical image, arguably the first to truly represent the world as global space in two dimensions (fig. 49). It comprises an oval projection, simple in conception but technically demanding to draw, marking 36 degrees of longitude at 10-degree intervals and eighteen parallels covering all 180 degrees between the poles. Rosselli's map is increasingly distorted with distance from

FIGURE 48.

Ptolemaic world map. From Hartmann Schedel, *Das Buch der Chroniken und Geschichten* [The Book of Chronicles and Histories [Nuremberg Chronicle]] [1493].

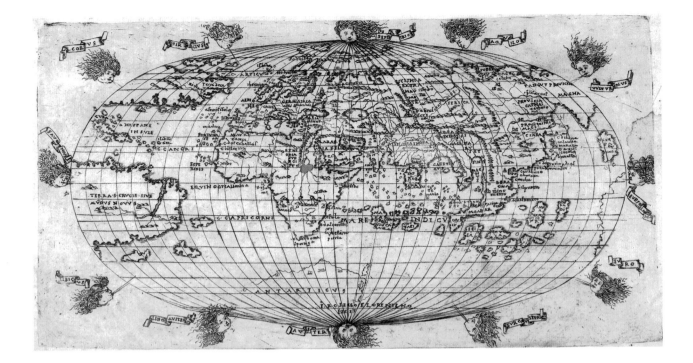

its central axes, which overly flatten the poles. But "as an image of the spherical earth it had a naturalness which made it the most intellectually satisfying model yet devised. It undoubtedly confined the Ptolemaic model to history, and pushed the plane chart of the seafarers . . . into its own magnificent cul-de-sac" (Whitfield 1994, 50). While the coasts of Asia on these maps remain largely unchanged from the Ptolemaic image, today's world map is clearly foreshadowed by these Renaissance cartographers.

Of equal significance in aligning world map and whole earth was the production of terrestrial globes. The first securely identified (and still preserved) globe dates from 1492. Martin Behaim's highly colored *erdapfel* (earth-apple) offered Nuremberg merchants updated geographic information gained from Portuguese navigation. But the production of terrestrial globes began in earnest a quarter century later, stimulated especially by the 1519–22 Magellan circumnavigation. That feat reemphasized the demand for truly global world maps that could no longer evade (as at the edges of a flat map) the question of where east and west met. Globes played a significant role in resolving the first dispute of globalization: how the pope's 1494 demarcation line of Spanish and Portuguese spheres in the Western Hemisphere should be extended into the Eastern Hemisphere (Brotton 1997). In terms of geographic contents, the history of globes parallels that of flat planispheres; technically it is distinct, principally concerned with the difficulties of constructing spheres of sufficient size to show appropriate detail, and of designing and accurately matching gores.[10]

DENIS COSGROVE

Practical problems of scale, portability, and use have meant that globes have largely been made for display, entertainment, and education in mathematical geography rather than as informational cartographic instruments.

Representing a global earth on a flat world map by means of projection inevitably distorts space: in shape, area, or direction. This has always been recognized by cartographers, and remains a matter of contention today. Altering the graphic image of the world affects the way we apprehend, understand, and think about geographic relationships. This can be manipulated for various purposes and gives rise to fierce ideological debates over the world map. Globe-spanning twentieth-century empires and cold war ideologies exploited map projections to pursue geopolitical goals, and the "Peters projection" (fig. 31) argument in favor of an equal-area world map continues the debate. Peters's explicit target was the Mercator projection, criticized as an imperial image of Western global dominance and long used (despite criticism) on U.S. world maps (Schulten 2001; see also Monmonier 2005). In fact the Flemish cartographer's 1569 projection was devised to resolve specific problems arising from the distinct Ptolemaic and *portolano* methods of mapping and charting, and the inadequacy of globes for shipboard use. In 1516 Waldseemüller had published a "Carta marina" (Mariner's Map) of the world based on the "plain charts" of Iberian navigators. Across the map's surface, compass roses and rhumblines showing bearings replaced the graticule. Unlike Ptolemaic maps, sea charts were rarely printed, remaining for the most part practical images of an oceanic world that ignored its spherical properties and the consequent gathering of longitudinal lines toward the poles.

The genius of Mercator's projection is that by proportionally increasing the spacing of parallels toward the pole on a cylindrical projection, thereby giving a constant width to any given interval of longitude, he effectively amalgamated map and chart in a new image of the world (fig. 50). Straight lines on the map indicate lines of constant bearing (loxodromes); both these and the graticule appear on Mercator's original map. For centuries, Mercator's projection had little immediate impact on either printed world maps or sea charts and marine atlases; its value lay, as Mercator recognized, in much more local charts. But it did find favor in nineteenth-century thematic maps to illustrate world distributions of natural phenomena, peoples, and states. The significance of its distortions on these maps justifies criticism of ideological bias.

From the late sixteenth to the late eighteenth centuries, the double-hemisphere construction was the favored style for European world maps. It was first used by the French cartographer Jean Rotz in 1542 and independently developed by many late sixteenth-century mapmakers. Parallels curve away from the equator, while meridians converge toward the poles. Double-hemisphere maps reduce the distortions of the single-sheet planisphere, give the visual impression of a global earth, and have the graphic appeal of presenting "old" and "new" worlds separately.

FIGURE 50.

(over) Gerard Mercator, "Nova et aucta orbis terrae descriptio ad usum navigantium emendate accomodata" (New and Accurate Description of the Terrestrial Globe, Amended to Suit the Uses of Navigation) (1569).

It is tempting to think of the modern world map as a European achievement, subsequently exported to other cultures and achieving dominance through the superior accuracy of Western mapping. But such a triumphal story is misleading at best. There is no question that cartographers in western Europe first undertook the technical work of recreating Ptolemy's projections and that European navigators were the first to reconnect the Americas, Oceania, and Antarctica to the "Old World" in a systematic and recorded fashion. But from the earliest years of its creating, the modern world map was circulated, appropriated, and developed in the context of existing cartographic cultures from the Ottoman Empire through India to China, Korea, and Japan.

The example of the Jesuit Matteo Ricci's world maps with Chinese texts, printed in large quantities and widely copied in Qing-era China and beyond, is well known. Intended by Ricci to aid the Jesuit missionary project, they appeared at a critical moment when the new Qing dynasty was strengthening and expanding its empire, undertaking its own large-scale geographic surveys and thus hungry for geographic knowledge. By 1600 Chinese renderings of both single- and double-hemisphere world maps were being made, and henceforth, Chinese cartographers were connected to the international circuits of knowledge in which the modern world map evolved (fig. 51). A similar history exists in Japan, where sixteenth-century Nanban maps often recentered European originals onto the Pacific to better represent the Japanese perspective. During the "closed" Tokugawa centuries, trade with the Dutch kept Japanese scholars in contact with the evolving world map, as the many Japanese renderings of Dutch maps indicate (fig. 52). European models were not simply copied but often subtly altered to reflect the knowledge and customs of their makers and users.

While Renaissance cartographers solved many of the technical problems of mapping a global earth, critical scientific questions continued to impact on the world map. Of these, the most significant were heliocentricity, terrestrial magnetism, and the practical matter of determining longitude at sea. For the world map, the seventeenth-century collapse of the Aristotelian world picture and quasi-universal acceptance of a heliocentric (sun-centered) universe had little technical significance. And as *world* increasingly denoted the planet Earth rather than the cosmos, the numinous aspects of the world map were yielding to a secular scientific image. Empirical studies of global magnetism, including Edmund Halley's 1702 isoline[11] world map of magnetic variation, demonstrated that the Earth is not a perfect sphere (fig. 53). Isaac Newton theorized that it is oblate: flattened toward the poles, as demonstrated by the measurement of meridian arcs in Lapland and Peru by the French in the 1730s and 1740s. Geodesy, the measurement of the shape and size of the earth, became a vital technical aspect of eighteenth-century mapping as the European states developed topographic surveys for military and bureaucratic purposes and sought to coordinate these with smaller-scale geographic and world maps. Determining

DENIS COSGROVE

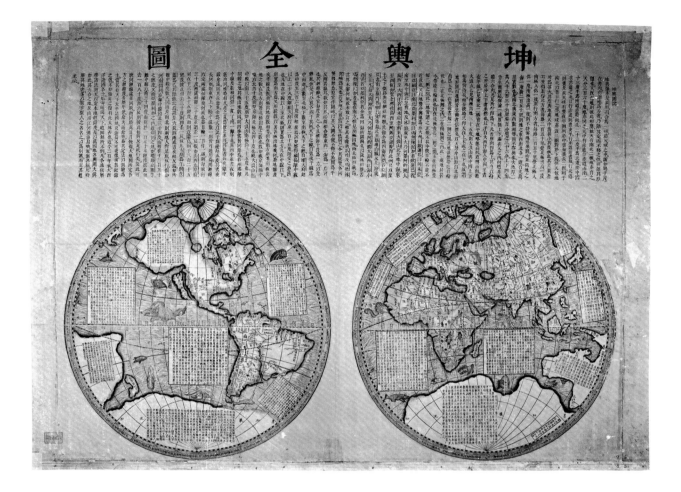

accurate longitude at sea presented technical problems quickly recognized by Renaissance navigators but not resolved until the late eighteenth-century invention of the chronometer. Growing sophistication of mapping instrumentation is reflected in the graphic design of the Enlightenment world map, which quite swiftly shed the decorative, pictorial qualities of its baroque predecessors in favor of the "plain style" that proclaims the modern world map a serious instrument of geographic science.

Since 1800, the principal technical developments in mapping the world have come from remote sensing rather than direct examination. Easy and regular oceanic navigation, especially since the invention of the steamship, has allowed data on a wide variety of natural phenomena to be mapped at the global scale: ocean currents and temperatures, atmosphere and climate, marine life and undersea geology. International agreement that the Greenwich meridian should be the 0° line for world maps was an important step in their standardization, although it has in recent years been criticized as Eurocentric. Laying telegraph and other cables across the ocean floor assisted its inclusion

FIGURE 51.
Ferdinand Verbiest, "Kunyu quantu" (Complete Map of the Earth) (1674).

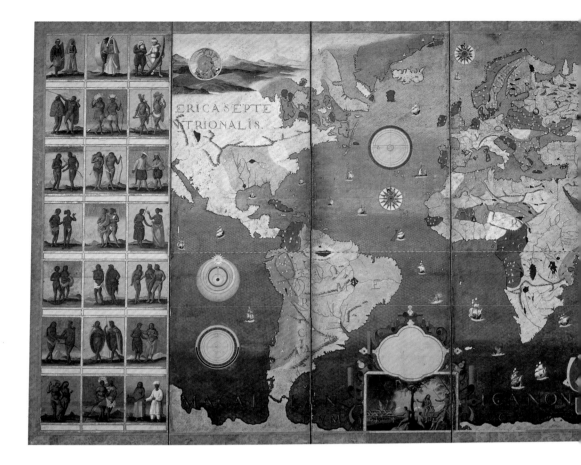

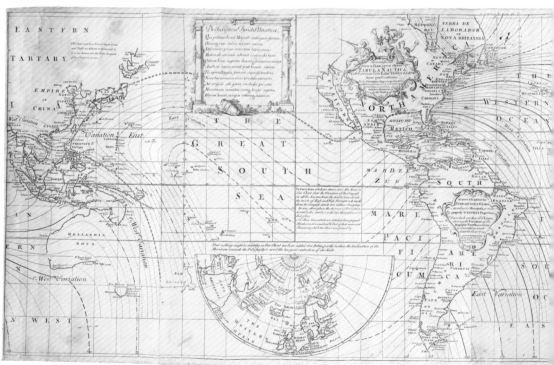

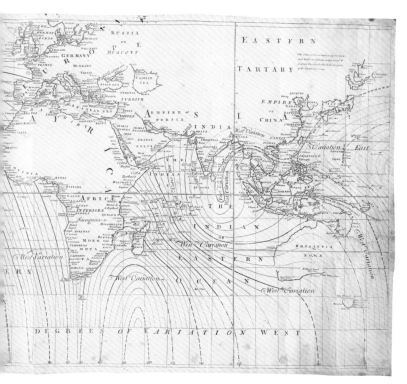

FIGURE 52.

(top) "Bankoku Ezu" (Pictorial Map of All Countries), eight-fold screen, color on paper, 178.6 × 486.3 cm, Momoyama to Edo period, 17th century. The Museum of the Imperial Collections, Sannomaru Shozokan, Tokyo.

FIGURE 53.

(bottom) Edmund Halley, "A Correct Chart of the Terraqueous Globe / Accuratissima totius terrarum orbis tabula nautica" (first published 1702; revised and updated edition, 1756).

into the world map by the 1970s (Doel, Levin, and Marker 2006), while aerial photography hastened the accurate mapping of inaccessible polar and interior continental regions. Satellite images of earth since the 1960s have contributed to an increasingly flexible world map, whose fundamental pattern is no longer defined solely by the junction of land and water across a terraqueous globe.

Contents

These technical and instrumental advances have permitted virtually any phenomenon to be plotted onto the world map with a high degree of empirical accuracy. In this the map is heir to the urge for plenitude apparent on the Christian mappa mundi. When first translated into Latin, Ptolemy's *Geography* was titled *Cosmography*, an ambiguous term that referred to both the mathematical discourse of the spheres and to narrative or pictorial description of the earth's surface. Thus, Fra Mauro's map with its beautifully wrought miniatures of cities and landscapes, its detailed toponymy, and its extended texts (many of them based on Marco Polo's *Milione* [literally, "the million"; in fact, "The Travels of Marco Polo"]) is a cosmography, as are Waldseemüller's largely mathematical text and Sebastian Münster's illustrated description of global geography.

The modern world map long maintained the cosmographic role of providing a graphic description of all that is in the world. This is an impossible task, of course, but it offers an alternative understanding of the map's evolution to a narrow focus on increased scientific accuracy. Accuracy certainly mattered to cartographers, and over the centuries we can observe the slow replacement of a speculative global geography with an empirical one: for example in the replacement of *terra australis incognita* (unknown southern land) as an extension of Tierra del Fuego by the continental islands of Australia and Antarctica; the detachment and reattachment of California to the mainland of North America; the disappearance and reappearance of the Straits of Anian separating America from Asia; and the disappearance of Mercator's Arctic Continent with its four outflowing streams and open polar sea. Cartographers drew upon the reports of navigators and travelers, assessed their reliability, and compiled the information in graphic form.

Mercator's work reflects the cosmographic mode: on his 1569 map, blank spaces of oceans and continental interiors are filled with detailed narrative descriptions of nature, ethnography, history, and other subjects (fig. 50). Graphic inserts explain the projection or justify the appearance of the Arctic Continent. The double-hemisphere map was well suited to cosmography, offering space around the hemispheres as well as on the map itself for displaying a vast range of textual, graphic, or pictorial information about the world (fig. 54). Astronomical diagrams of the competing world systems: eclipses or the surface of the sun and moon; iconic representations of the seasons, the elements, or the continents; and illustrations of world history, cities, flora, fauna, or the

DENIS COSGROVE

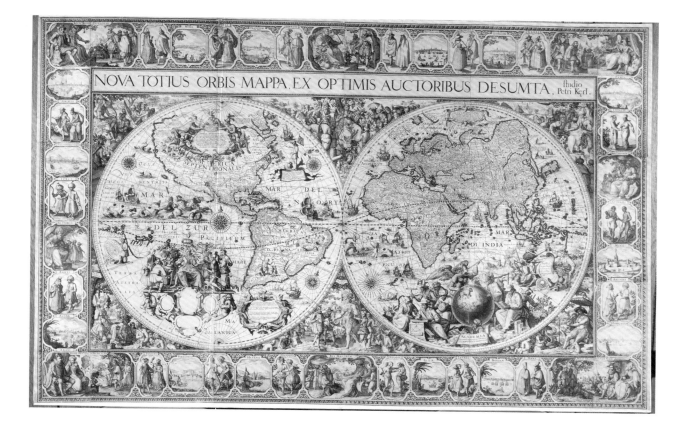

NOVA TOTIUS ORBIS MAPPA, EX OPTIMIS AUCTORIBUS DESUMTA, studio Petri Kerl.

costumes of different peoples all appear on the world map. These pictorial elements should be seen not as decorative additions to a disinterested spatial description but as integral parts of a totalizing world picture, conveying to the eye a marvelously diverse world, exotic but ultimately ordered and intelligible, and perhaps therefore controllable. But as the scale and the informational content of world maps, atlases, and globes increased to giddy proportions the cosmographic world map became increasingly elaborate and unwieldy. Vincenzo Coronelli's huge globes made for Louis XIV, Willem Blaeu's *Atlas Major*, the atlas of the Great Elector in the State Library in Berlin, or the King Charles II atlas in the British Library reflect both the desire for global synopsis and its impossibility (Cosgrove 1999c, 33–66; Pelletier 1982; H. Wallis 1969).

By the late 1700s this cornucopian aspect of the modern world map had given way to a severely disciplined graphic style. On Aaron Arrowsmith's double-hemisphere "Map of the World on a Globular Projection" of 1794, the frame is largely blank, apart from a delicately rendered apotheosis of Captain James Cook (whose oceanic discoveries the map records), and a dedicatory portrait of the Royal hydrographer Alexander Dalrymple (fig. 55). The map itself traces a fine coastal outline, leaving unexplored stretches blank; marks major mountain chains with restrained hachure lines; and names only verified locations.

FIGURE 54.

Pieter van den Keere, "Nova totius orbis mappa ex optimus auctoribus desumta" (New Map of the Whole Earth Drawn from the Best Authorities) (1611).

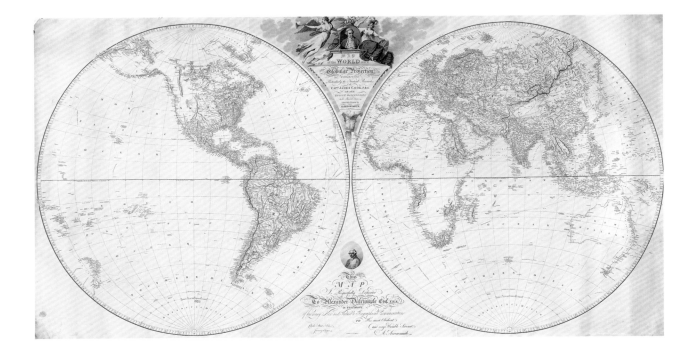

FIGURE 55.

Aaron Arrowsmith, "Map of the World on a Globular
Projection" (1794).

The paucity of pictorial information and the removal of speculative elements
proclaim the map's empirical, scientific credentials.

The Enlightenment ideal of cartography as a scientific practice is apparent also in thematic world maps. An iconic figure is Alexander von Humboldt,
whose isoline map of average annual temperatures in the Northern Hemisphere
(1817) graphically disposed of the classical climatic zones. Yet the claim to scientific objectivity is a treacherous one. The thematic maps of von Humboldt's
cartographer, Heinrich Berghaus, support "scientific" race theory (fig. 61). By
the late nineteenth century, school geography texts such as Arnold Guyot's
Physical Geography (1873) used Mercator-projection world maps to teach that
geographic distributions of natural phenomena (climate, for example) corre-
lated to ethnographic characteristics ("race," language, religion, literacy, and
so on). Although much more cautious, educational maps coloring the world
in simple, bounded patterns of difference remain common. Their persuasive
graphic authority and their tendency to generalize the complexity, flexibility,
and mobility of human society and culture in fixed, territorial patterns, make
these world maps controversial.

The dangerous seductions of the modern world map may be illustrated
using the one produced by the British geographer and politician Sir Halford
Mackinder to illustrate his influential 1904 paper, "The Geographical Pivot
of History" (fig. 56). With the world mapped to the poles, he foresaw "closed
space," wherein he believed enduring, spatially determined social relationships

would determine international politics. His world map indicated a landlocked "pivot area" occupying the central parts of the Eurasian continent. Centrality and ease of movement allowed its occupants to exert constant pressure at any point along the enclosing crescent of peninsulas and islands, from Scandinavia to Kamchatka. Insufficiently united ever to dominate the pivot, these regions' historical mission was to resist its ceaseless push toward world empire. On the map and in the historical narrative, Africa, the Americas, and Oceania play a totally subordinate role. Cartographically, the argument is persuasive. But it reflects principally British fears of an expanding Russian empire, for the graphic argument collapses completely if we alter the cartographic frame or projection or use a globe to examine its claims. Yet geopolitical maps claiming to illustrate global strategic necessities were produced throughout the twentieth century. In contrast with their geographic fixity, the scientist Alfred Wegener's 1915 map of Pangaea illustrates his theory of "continental drift." Today, plate tectonics is a widely accepted scientific theory that completely undermines the picture of a world map of seas and landmasses that are stable over geologic time.

Since the 1960s we have been able to create world maps from images directly imprinted onto a photographic film or digitizing light sensor. Moreover, combining remote-sensed satellite images with the power of twenty-first-century computing and the widespread availability of home workstations allows for an almost infinite variety of cartographic images to be streamed directly into our everyday world. The cosmographic dream, in fact, has been realized through the Internet. The twenty-first-century world map is perhaps best represented by Google Earth, a Web site that allows the user immediate access at

THE NATURAL SEATS OF POWER.
Pivot area—wholly continental.   Outer crescent—wholly oceanic.   Inner crescent—partly continental, partly oceanic.

any selected scale to satellite and cartographic images of locations anywhere on the globe. Zooming in and out, "surfing" from one location to another, the modern "world map" of preselected scale and projection seems redundant. Indeed, the ready availability of satellite imagery finally put to rest a twentieth-century project to produce a 1:1,000,000-scale map of the whole world. First proposed by the geographer Albrecht Penck to the International Geographical Union in 1891, the map never achieved more than minimal coverage, largely because of the failure of the international community to agree on a common design and key. Instead, the Enlightenment goal of coordinating the image of the world at scales ranging from the topographic to the global has been achieved by a commercial search engine.

Cosmographic and iconic aspects

The secular nature of the modern world map is apparent in some of its earliest incarnations: Waldseemüller's and Rosselli's work, for example. But the iconic aspects of the world's image, so apparent in nonmodern cartography, did not disappear. The globe, with its signature graticule or outline pattern of continents and oceans, frequently appears on the sixteenth-century moralizing emblem (fig. 57), combined with epigram and motto, to signify "the world" with all its vagaries and temptations (Cosgrove 2001, 139–75; Hofmann 2001–2). The world map itself could be mobilized for similar purposes. In the widely reproduced but anonymous "Fool's Cap" image of the 1590s, it replaces the face of a jester holding a sphere with the motto "Vanity, vanity, all is vanity." The image refers to the pre-Socratic philosophers Heraclitus and Democritus, laughing and weeping at the folly of the world. As the masks of Comedy and Tragedy, they represent the theatrical theme commonly associated with mapping the world. Typical is Abraham Ortelius's pioneering atlas project, *Theatrum orbis terrarum* (Theater of the World), first published in 1570 in Latin and subsequently translated into other languages. Its opening world map, among the most widely reproduced early modern maps, features a subscript quotation from the Roman jurist Cicero reminding the viewer of the transience of the material world: "Quid ei potest videri magnum in rebus humanis aeternitas omnis, totiusque mundi nota sit magnitude?" (For what can seem of moment in human affairs for him who keeps all eternity before his eyes and knows the scale of the universal world?) (fig. 58)

A notable example of a world map combining mathematical sophistication with iconic intent is Oronce Fine's cordiform projection of 1519, published in 1536 with a dedication to François I of France (fig. 59). This equal-area projection is named for the heart-shaped form into which it casts the globe. Today the human heart is viewed scientifically as a pumping mechanism, and associated symbolically with romantic love. For the medically trained Fine and his contemporaries, however, the heart was a complex and mysterious organ:

DENIS COSGROVE

ORBIS. HODIE SIC VERTITUR ORBIS

Quid mirum, cancri mundum consistere tergo,
Sic hodie quoniam vertitur orbis iter.

the primary receptor of sensation, the seat of memory, the portal to wisdom, and above all the governor of the microcosmic human body. Cosmologically, the heart was the Sun, ruler of the earth. Developments in human anatomy revealing the heart's role in the circulation of blood were further adding to its significance (Sawday 1995). The world map in the shape of a heart thus connected the image of the expanding globe to the most intimate core of the human individual, to the idea of rule, and to the majesty of monarchy and empire. The cult of the Sacred Heart of Jesus, strongly revived in the Reformation, further reinforced the connections between the Savior, the world, and the heart-shaped map. Additionally, Roman Catholic iconography made frequent use of

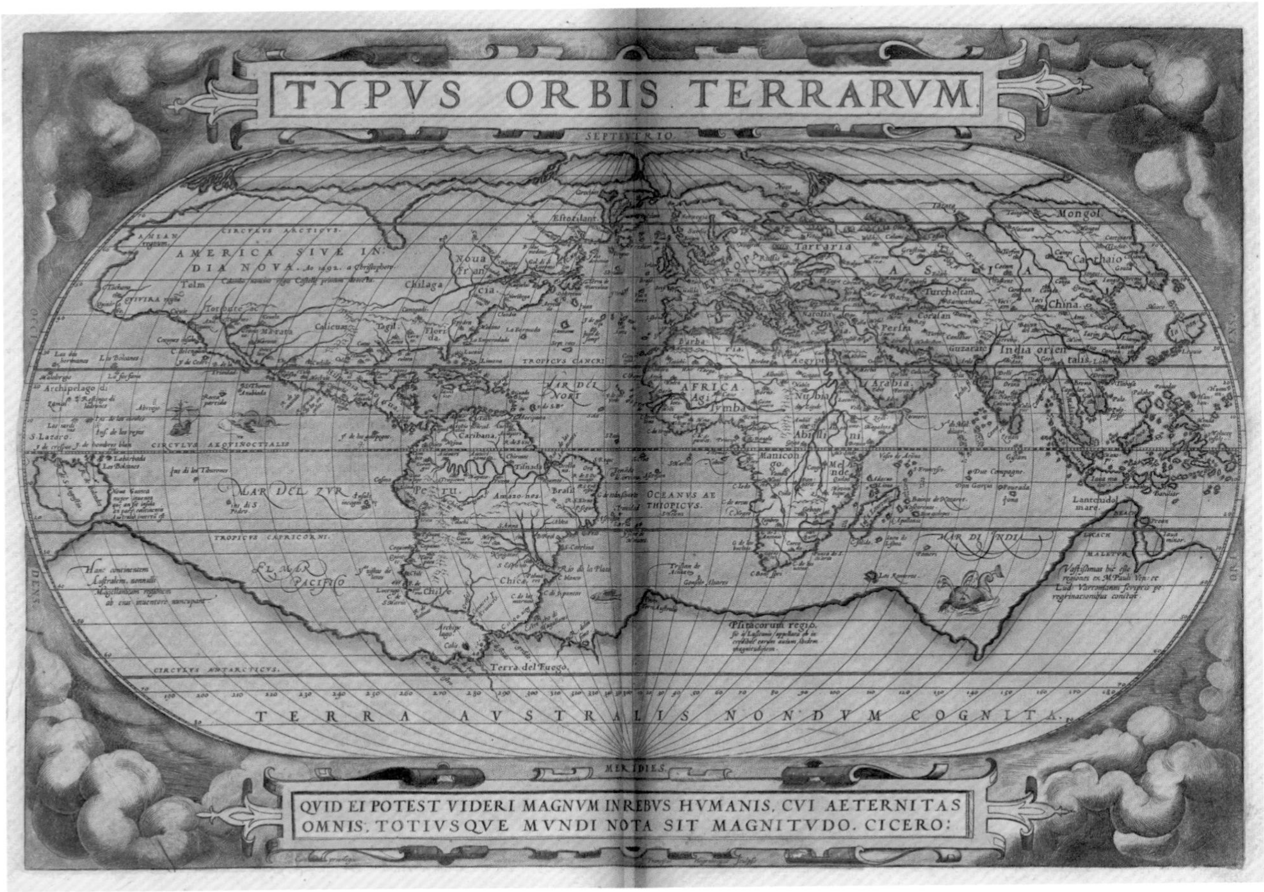

TYPVS ORBIS TERRARVM

QVID EI POTEST VIDERI MAGNVM INREBVS HVMANIS, CVI AETERNITAS
OMNIS, TOTIVSQVE MVNDI NOTA SIT MAGNITVDO. CICERO:

FIGURE 58.

Abraham Ortelius, "Typus orbis terrarum" (Figure of the Earthly Sphere) (1580).

cartographic devices such as maps and images of the continents to signify the papacy's global mission (Mangani 1998; R. Watson 2005).

World maps long retained a reserve of sacred authority even as they sought accuracy in projection and empirical content. Although printing made the world map more familiar and widespread, picturing the world long remained rare and charged with a certain wonder. This helps explain the endurance of the pictorial cosmographic map. Monarchs and princes competed for the symbolic authority of the world map, for the same reasons that priests and rulers in nonmodern cultures lay claim to the power associated with representing the world. Cosmography was the foundation of Louis XIV's royal iconography. His regime encouraged French scientific primacy in measuring and mapping the globe, and the king commissioned the Venetian cartographer and globe maker Vincenzo Coronelli to make the largest globe pair ever constructed to decorate his palace at Versailles. The celestial globe mapped the stars at the king's nativity, while its terrestrial partner emphasized the latest French geographic discoveries (Cosgrove 1999c).

DENIS COSGROVE

The emblematic qualities of world images have not wholly disappeared, despite the ubiquity of such images and the irreverence with which we treat the globe, from key fobs to beach balls. Devotees of environmentalism treat the 1972 Whole Earth photograph as a quasi-sacred image (fig. 32). Also, the world map remains a powerful symbol for any project that lays claim to universal values or concerns. Images of the globe are particularly popular among both Christian and Islamic charitable and missionary organizations. A secular equivalent is the United Nations, whose light-blue flag is marked with a polar-centered map showing the "land hemisphere" of continents gathered into close contact at the neutral North Pole. More commercially, newspapers and media

FIGURE 59.

Oronce Fine, "Recens et integra orbis descriptio" (Current and Complete Description of the World) [1536].

MAPPING THE WORLD

companies, airlines, telecommunications corporations, and other groups with claims to global reach find in the world map a simple, positive, and instantly recognizable symbol of universality.

Ethnographic and cosmopolitan aspects

The significance of mapping always extends beyond the description of physical earth, into the human implications of the "world." The ethnocentricity of nonmodern world maps, on which "other" peoples beyond the normalized culture are figured as exotic or bizarre, is easy to spot: the "monstrous races" on the antipodean continent in medieval mappae mundi or the fantasy islands of women on the margins of Japanese and Korean maps are examples. The modern world map, on which world and globe coincide, is directly challenged with the issue of ethnocentrism. We have noted how Asians very quickly reworked the modern map to locate China or Japan nearer its center, and we have remarked on the criticisms made of the Greenwich "prime" meridian. Tensions between the modern map's equalizing and universalizing effects and the focalizing qualities of any graphic construction are unavoidable. In picturing the globe as a single "world," the modern world map is implicitly cosmopolitan. It shows us the earth as a whole, from the godlike perspective of the heavens—a perspective long associated with both imperial power and a cosmopolitan vision uprooted from the parochialism of local attachments (Cosgrove 2001, 2003). These ideas underlay the Renaissance idea of the "theater of the world," the emblematic use of the world map, and the Stoic references on Ortelius's "Typus orbis terrarum" (fig. 58), which was the model for many Chinese and Japanese world maps (Hostetler 2001; Società Geografica Italiana 2002). Yet the world map is always viewed through localized lenses, and the resolution of the resulting tensions appears in the use and design of the maps themselves.

An example of this is the frontispiece of the Venetian Giacomo Franco's 1610 *Habiti d'huomini et donne venetiane* (Costumes of Venetian Men and Women; fig. 60). The map of Venice appears as a jewel within a crystal globe whose analogy with the earthly globe is intentional:

Here . . . is the design . . . of the marvelous city of Venice in spherical form, a real portrait of the world, whose resemblance is so close to nature, and made by the arts, similar to the orb of earth, those who well admire this design discover, from the height of a bird, the Arctic pole and the Antarctic and also from the East and the West with all the other parts that go around this world; similarly circled by water in a manner that seems the continent is entirely surrounded by the great ocean. (Quoted in Wilson 2005, 67)

The author goes on to explain the unusual orientation of the city, viewed from the east. It is intended to resemble the world map: the three central parts are the

DENIS COSGROVE

continents of the old world, and the Island of Giudecca to the left is "in a guise that resembles the new world." Franco deploys the universalizing qualities of the world map to suggest that the great commercial (and mapmaking) city is a microcosm of the whole world (Wilson 2005).

Habiti d'huomini et donne venetiane is a costume book, produced for an age when individual identity was more closely tied to dress than to the physical body or personality, dress codes were highly regulated, and understanding costume differences was a principal way of knowing the world. As Venice was a commercial center at the time, the costumes of many diverse peoples could be seen there; it was a cosmopolitan landscape in which costumes denoted customs (the words are the same in Italian). This directs us to a feature of seven-

1 2 **3** 4 5 6 7

MAPPING PARTS OF THE WORLD

Matthew H. Edney

By comparison with the cultural complexities of world maps, the imaginative mappings of fictional worlds, and even the challenging details of wayfinding maps, maps of particular parts of the world might appear to the modern eye to be dull and—well, mundane. Surveyors go out with their instruments to observe and measure the surface of the earth and then use their measurements to recreate that surface in miniature on paper or computer screen. Maps of the earth's surface are routinely evaluated by how accurately or erroneously they depict physical reality. Indeed, from this perspective it is possible to imagine a perfect map that completely replicates the earth's surface, point by point; all that this accomplishment would require would be sufficient effort on the part of the surveyors and sufficient money to equip and to pay them. Yet images of such perfection have long been rejected by logicians and philosophers. Over a century ago, in *Sylvie and Bruno Concluded* (1893), Lewis Carroll pointed out a fundamental problem with a map made at the same size as the earth itself:

In other words, cartographic perfection is self-defeating. Were people to fulfill the map's apparent potential, they would set aside the very real reasons why they have bothered, over many millennia, to make maps of parts of the world (see also Borges 1964, 90; Eco 1985, 1994).

What maps give us, over and above the country itself, is a means to simplify, organize, and manage the world's complexity in order to comprehend and manipulate our environment better. People have used maps for many thousands of years to create spatial conceptions that they can then use to think about, deal with, and relate to the world. Maps permit us not only to conceive of the world, but to *grasp* it as well. The preceding chapters demonstrate that people have mapped their world according to their own particular cultural worldview. It is therefore essential to understand each map in terms of the culture that created it. Even so, if we consider not so much the how but rather the why of mapping of parts of the world, we can discern some cross-cultural commonalities that give us a handle on an otherwise huge and diverse subject. This chapter accordingly presents a sampling of maps from many cultures to explore some fundamental themes about why people have variously mapped land as comprising distinct places (including property) or as constituting extensive spaces (including regions and countries).

The existence of cross-cultural commonalities does not mean that making and using maps is a universal practice in human history. All humans construct "cognitive maps," or personal senses of experienced space. But the making of a map entails a conscious and special act to create a concrete spatial image. It requires specific needs (Wood 1993). Those needs, and so mapmaking, vary by society. In "limited" societies, in which people live in small groups very close to the land, everyone shares a common experience with the world, so there is little need for them to do more than use the country itself (in Lewis Carroll's phrase). But as societies grow more "expansive"—the more involved, specialized, and articulated their economic activities; the more stratified and variegated their social structures; the more extensive their trade, political connections, and geographic horizons—then their spatial aspects become ever more complex. People in expansive societies have ever more reasons and opportunities to make and use maps in order to simplify and comprehend spatial complexity (Wood and Fels 1992, 28–47). Accordingly, most of the examples cited in this chapter originate in expansive societies.

Cartographic practices are differentiated within each society. When we realize that maps are intellectual resources akin to economic resources in that

MATTHEW H. EDNEY

they are disproportionately distributed throughout any given society, we can easily see how their production and use contribute to the creation and maintenance of social inequalities. We might observe, for instance, that cartography has until very recently been an almost exclusively male activity; at the same time, most cultures construe nature as inherently female, so that making and using maps sustains one of the most fundamental social divides. We can also easily see how different groups have used maps in negotiating or contesting their social status. After all, as conceptual surrogates for the world, maps afford their users certain powers. Maps enable people to "see" crucial aspects of the world and to conceive of the world as something that can be changed; such changes range from the comparatively simple (moving from one location to another), to the complex (building dams and highways), to the highly complex (creating new states and empires). Maps thus organize peoples' understanding of the world and their ideas for changing it; not surprisingly, people often refer generally to *mapping*, *planning*, or *plotting* their strategies (*plot* being an old term for *plan*). Furthermore, people use maps as instruments to direct and guide the actual implementation of spatial change. The making and using of maps can therefore be understood as expressions of power—of the technological, social, economic, political, military, and cultural ability to exert authority over the world and to modify it. At the very least, if ability is lacking, maps express the desire for such power.

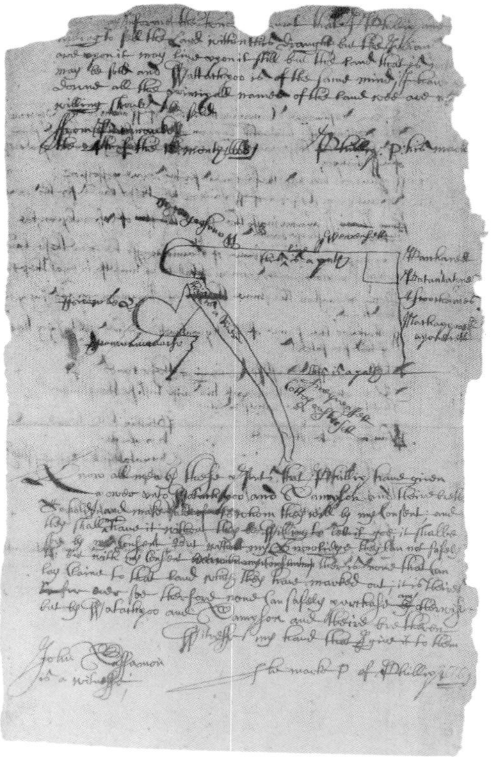

Yet neither social need nor power relations *requires* people to make maps. By and large, people have adhered to their traditional ways of depicting spatial conceptions and recording spatial information. Those techniques might be cartographic, but they don't have to be. People in past societies have used words (spoken and written), numbers, and graphic imagery for tasks which today we might expect to be handled through maps. The production and use of maps in a society depends on a certain degree of cartographic literacy, which is to say the awareness that maps can indeed be useful and effective devices to manage spatial complexity. Conversely, maps have never been used to the exclusion of other strategies for simplifying spatial complexity. Even in modern industrial societies, maps are routinely used in conjunction with other, noncartographic descriptions of space. In this respect, maps are never "stand-alone" entities; how they are made and how they are used in conjunction with other descriptions of space is a matter for cultural convention.

These three issues—social needs, power relations, and cultural conventions—underpin the production and use of all maps. As a first example, consider a map prepared as part of a 1666 property exchange in New England (fig. 62). Oriented with southwest at the top, this map shows a rectangular block of land at Sepecan, approximately 15 miles (24 km) long by 9 miles (14 km) wide, with three promontories jutting into the sea near Rochester, Massachusetts; the two diagonal lines indicate the Weweantic River. The map was made for a very specific purpose: to illuminate a land transfer between the Wampanoag

FIGURE 62.
Page from an "Indian deed" for land at Sepecan, Massachusetts (1668).

and the English of Plymouth Colony; indeed, the map is part of the "Indian deed" recorded by an English clerk in 1668 for the colony's official legal archive (Pearce 1998, 170–74 and 184–85 note 46). However economically simplified life might have been on the colonial frontier, the English were part of an expansive society that featured legal, bureaucratic, financial, and religious traditions of reading and writing. By comparison, the Wampanoag formed a limited society in which everyone, of all social grades, lived close to the land and its resources; for such societies, in which everyone is immediately familiar with the land, there is little need to record its details. The deed is thus an English legal document that records both English and Indian ways of referring to and understanding space.

It was quite exceptional for an Indian deed to include a map. For the most part, the English colonists and the Indians employed very similar cultural conventions when it came to referring to land. Both groups were part of cultures that commonly delimited property by means of landmarks, place-names, and metes-and-bounds descriptions of property boundaries (see below). The graphic depiction of property was still rare among the English colonists in this period. Accordingly, virtually every deed recorded landmarks, place-names, and metes-and-bounds descriptions in the form of words and did not use maps. The deceptively simple Sepecan map was drawn without a measured survey with the apparent purpose of clarifying an especially complex property transfer. The property itself was defined by existing features (its two long sides are both labeled "This is a path" on the map); but around the edge of the plan are several place-names that, the deed makes clear, identified specific tracts *within* the depicted area which the Wampanoag were *not* selling and on which they would continue to live.

The presence of a map in this deed has even more significance within the context of the relationship between the English colonists and the indigenous peoples. The land was sold to the English by the Wampanoag sachem King Philip (or Metacomet). The deal was probably not equitable; Indian deeds were drawn up by the colonists, mostly from positions of economic and demographic strength, and some were occasionally fraudulent. (Eventually, discontent over the voracious English appetite for new land was a major cause of King Philip's war on the colonists in 1675–76.) The Sepecan deed, with its map, was thus a special act of negotiation between unequal communities. It is very likely that the original map was drawn by John Sassamon, a Massachusett who had been raised and educated by English settlers and who in the 1660s worked as a secretary and interpreter for the Wampanoag sachems in their dealings with Plymouth Colony (Lepore 1998, 21–47). The map can be understood as a particular attempt by an Indigenous people to use a representation that the English could understand and accept in order to affirm and preserve their own land rights in the face of aggressively expanding English settlement.

From these considerations, we can conclude that the character of any map is not determined by the territory it depicts or by the resolution of its depiction. Maps are not simply the replication in graphic form of physical territories that have been observed and surveyed. Rather, each map's character is determined by the context within which the map was made and used, a context formed from an amalgam of social needs, power relations, and cultural conventions. The role of any map is to invest the world with meaning and significance. In this respect, maps are used to aid the fundamental distinction drawn by humans between places and spaces (Tuan 1977). Maps enable people to conceive of certain portions of the world as *places*, each distinct from others, and to identify with them; more particularly, maps enable people to lay claim to precise parcels of land as their *property* and to exercise their rights over them. Furthermore, maps enable people to configure and delineate specific *spaces* which they can comprehend, govern, and control.

MAPPING PLACES

A fundamental aspect of human existence is the discrimination of places. In this general sense, a place is a specific site that humans invest with certain meanings through their actions, thoughts, and writings. Conceptions of place encompass spatial extent and cultural significance, environmental and economic characteristics, and individual as well as communal experiences. They are also fluid: places are not static but are always being formed, sustained, challenged, and reformed as people variously reaffirm and reconsider their attachments to specific sites (Cresswell 2004). Mapping provides one way in which people can create, perpetuate, and reconfigure ideas of place within a wide variety of social and cultural contexts. Like poetry and art, maps fix and define particular interpretations and understandings of what such-and-such a place is, how it is distinct, and what it should mean.

The manner in which particular places are conceptualized and constructed is evident in delineations of cities, such as Jacopo de' Barbari's glorious six-sheet woodcut view of Venice, published in 1500 (fig. 63). While a few other artists had previously conjured views of that city, Jacopo's was without precedent in terms of either its size or its degree of detail (Schulz 1978; Romanelli, Biadene, and Tonini 1999). Jacopo depicted Venice as if it were seen from the southwest, looking across the city's islands toward the distant *terre firma*. This view showed the city to great effect, with Venice's great focal point, the Piazza San Marco, near the center of the image. Juergen Schulz has demonstrated that Jacopo did not derive this great bird's-eye view from the rigid geometric frame of a detailed ground plan of the city and its islands. Rather, he assembled this specific perspective from a great many particular observations of the city's

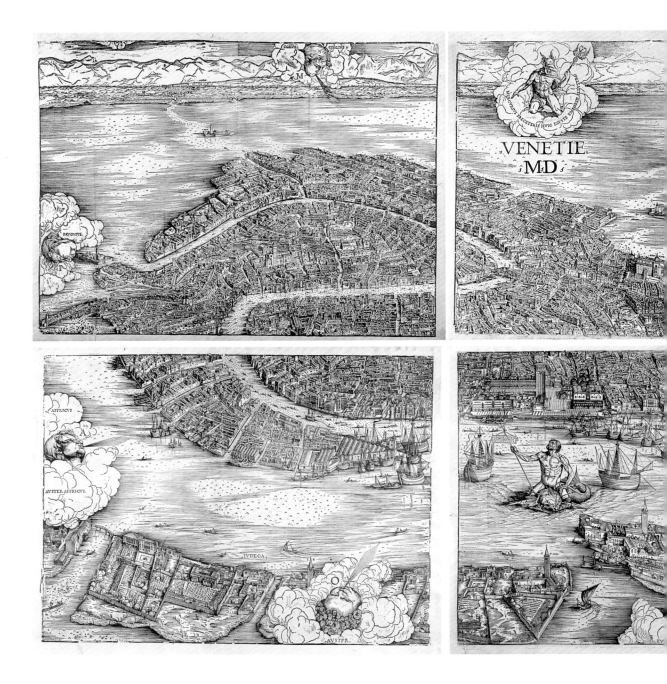

FIGURE 63.

Jacopo de' Barbari, "Veneti[a]e MD" (1500).

buildings and streets. Jacopo played with the physical spacing of the islands in order to show churches and palaces clearly; conversely, he simplified and even eliminated buildings of low status. He also exaggerated the heights of the buildings, and especially the bell towers, for greater visual emphasis. The view's title means "at Venice" rather than simply "Venice," and suggests that the gods shown at top center and in the lagoon—Mercury, god of commerce and trade (with the legend "I Mercury shine favorably on this above all other emporia") and Neptune, god of the seas ("I Neptune reside here, smoothing the waters at this port")—really do reside there and give the city its essential character (Schulz 1978, 468). Overall, Jacopo wedded a truly impressive view of the physical might and glory of the urban heart of the Republic of Venice with a powerful statement of the quintessentially maritime and economic character of the city's community, in an image that "fulfilled all the needs and expectations of contemporary culture" (Nuti 1999, 101). No viewer could have misunderstood the majesty of this place. The manner in which Jacopo's point of view emphasized the city's Renaissance glory made it the preferred view of Venice ever since. When the city declined in political, military, economic, and cultural might, Venetians cast a nostalgic eye onto that past glory, while outsiders steadily converted that nostalgia to touristic desire. As a result, very few views of the entire city have been made from other vantage points: Venice has become the place envisioned by Jacopo.

Jacopo's depiction of Venice exemplifies the standard Renaissance idealization of cities as distinct places. In this ideal, cities are understood to be thoroughly artificial places—in terms of both their built environments (*urbs*) and their self-governing communities (*civitas*)—in explicit contrast with rural environments and feudal societies, which were construed as manifesting the natural human state (Kagan 1998). In their images, Jacopo and other artists constructed cities as discrete places, although in actuality urban and rural areas have rarely been sharply distinguished: cities blend into the countryside through suburbs and exurbs; rural activities of animal husbandry and food growing can reach deep into cities.

That city maps and views present particularized and idealized images of unique places can be seen by comparing different approaches taken for representing U.S. cities in the nineteenth century. City maps and views allowed the wealthier residents who bought them to share a common image of a single, unified, economically active, and civic-minded community. It was in that spirit that John Cullum prepared, engraved, and printed a map of Portland, Maine, in 1836 (fig. 64). Portland was still a "walking city" at the time, but one that had doubled in size in the fifty years since the American Revolution. It had grown all the way across the neck of land from its original site as a port town along the Fore River, and was now beginning to climb up the bounding Mount Joy (Munjoy) and Bramhall hills. Cullum bracketed the map's title with two angels, who blow trumpets and bear laurel wreaths, literally heralding Portland's

MATTHEW H. EDNEY

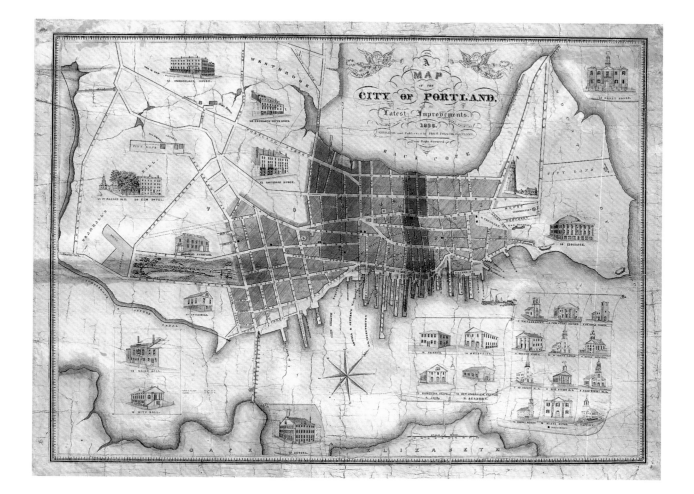

FIGURE 64.

John Cullum, "A Map of the City of Portland with the Latest Improvements" (1836).

dynamic economy and growth. He selected certain buildings to be shown in architectural profile—the customhouse, city hall, courthouse, exchange, large hotels, and more than fifteen churches and meetinghouses—to highlight the city's economic vitality, civic structure, and morality. The area of the built city was further divided into wards, indicating the presence of local democratic institutions and city government. All in all, this map presented a comprehensive and positive view of Portland to its inhabitants; it depicted a dynamic, moral, and socially coherent community.

In contrast with this "view from above" are many "views from below," which were made for the same wealthier audiences as the city maps and views, but which depicted particular aspects of urban life. These partial views focused on the underside of the modern industrial city, specifically its less desirable structures (slums, taverns, gin mills, and brothels) and inhabitants (laborers, immigrants, criminals, moral deviants, and the diseased). They took the form of illustrated urban ethnographies, dictionaries of cant, exposés of poor sanita-

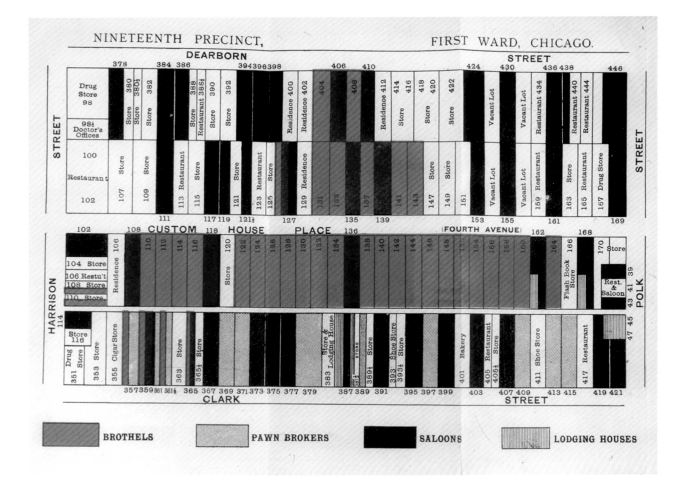

FIGURE 65.

"Nineteenth Precinct, First Ward, Chicago." From William T. Stead, *If Christ Came to Chicago!* (1894).

tion and disease, and, occasionally, maps of especially noisome neighborhoods. One such map depicts a particular site of vice and iniquity in Chicago (fig. 65). It highlights specific sites of immoral or illegal activity: lodging houses, which accommodated a transient and unregulated population that was a source of great anxiety for the city's more sedate and decorous populace; pawnbrokers, who were both a source of quick cash and a ready home for stolen goods; saloons, which were intrinsically nefarious at a time when the temperance movement was strong among the sober middle classes; and brothels, which naturally were sites of moral depravity and turpitude. The map labels but does not emphasize the more morally neutral economic activities of the legitimate stores and restaurants. Thus, while the views from above gave their consumers—members of the city's upper and middle classes—a reassuring image of the economic vitality, morality, and communal wholeness of their hometowns, the views from below worked to tame the anxieties about social dysfunction held by the same people, by isolating each dysfunction, comprehending it, and

MATTHEW H. EDNEY

so enabling it to be corrected and eliminated (Kasson 1990, 70–80; also P. Gilbert 2004).

The mapping of cities is a specific kind of place construction. More generally, places are defined through maps that combine specific features of the physical environment with social and cultural attributes in unique spatial depictions. Each place is thus constructed as a discrete entity, entire unto itself. This practice is a common feature of cultures that accord a sacred dimension to landscapes and places. In this respect, maps of places are closely related to the cosmographic maps discussed in the preceding chapter, in that they seek to organize and explain the relationships between humans (past, present, and future), a particular physical landscape or site, and the spiritual realm (Woodward and Lewis 1998).

Aboriginal peoples of Australia have traditionally used maps to conceptualize their own communal places in an environment thoroughly imbued with religious, historical, political, and economic meanings (Sutton 1998). One such map, or *dhulaŋ*, from the Yolngu of eastern Arnhemland in Australia's Northern Territory, manifests a Dreaming (fig. 66). The background pattern symbolizes the Yolngu's "fire Dreaming," which intertwines the present land with the Yolngu's ancestral beings, the material with the spiritual. The saltwater crocodile is an ancestral being that defines the Gumatj clan's specific territory; the equivalence is in part spatially realistic in that the crocodile's rear legs describe the line of the coast and its tail Caledon Bay, although the general relationship of the rest of the crocodile's parts to particular parts of the clan's lands is much more metaphorical. To read the *dhulaŋ* fully requires the reader to be deeply intertwined within Yolngu culture. More generally, *dhulaŋ* are part of the community's internal economy of knowledge, to be preserved and traded among the Yolngu but not normally to be shared with outsiders. This particular *dhulaŋ*—of a style used for teaching Yolngu children—is one of three made in 1985 explicitly to be given to outsiders. The published explanation of this *dhulaŋ*'s cultural significance is limited to unexceptional knowledge; the image bears further meanings that the Yolngu were not willing to divulge. As such, it must be recognized as a unique product of interaction and negotiation between two cultures—to be more precise, of the resistance of one culture to the demands of another—and as such it is only emblematic of the kinds of *dhulaŋ* that the Yolngu make and communicate among themselves (H. Watson and Yolngu community 1993, 31–33).

Even without overtly magical or religious overtones, places are mapped as particular conjunctions of physical landscape and sociocultural values. In this sense, maps of places are closely intertwined with landscape views, according to each culture's traditions. French landscape studies in the later 1700s often combined view and plan on the same page (fig. 67). Together, these images of the main harbor on the Aegean island of Milos neatly display the common basis of modern landscape imagery in the spatial geometries of perspective. The im-

FIGURE 66.

Thelma Yananymul, "Crocodile and Sacred Fire Dreaming" (1985).

VUE DU PORT DE MILO,
prise du Cap noir.
A.P.D.R.

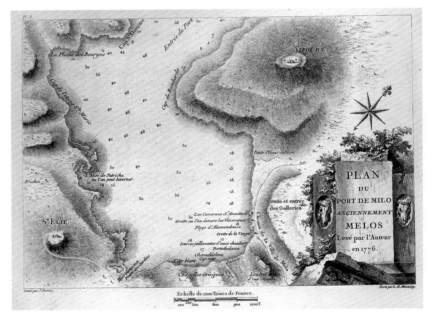

FIGURE 67.

Marie-Gabriel-Florent-Auguste comte de
Choiseul-Gouffier, "Vue du port de Milo, prise du
Cap Noir / Plan du port de Milo anciennement
Melos" [View of the Port of Milo, Taken from Cap
Noir / Plan of the Port of Milo, Known as Melos
to the Ancients] (1782).

ages differ only in their representational conventions. The view conformed to
the established strictures of the picturesque landscape imagery then popular
(three fields, with animals or human figures in the foreground, human structures in the middle ground, and sunset over hills in the background), whereas
the plan conformed to the established strictures of the landscape map (no single point of view; naming of particular sites and features; formalized sketching

MATTHEW H. EDNEY

of slopes; symbolization of other features, such as vegetation; and the exclusion of temporary or mobile features, such as people and ships). Such images promote a sense of control over landscapes, permitting their users to understand and manage those locations. Places and their landscapes become objects to be physically as well as intellectually manipulated and reconfigured.

The military concern with fortifications has produced detailed mappings in support of building, maintaining, managing, defending, attacking, and destroying those structures. An eighteenth-century Korean plan (fig. 68), for example, depicts the fortress of Ch'ŏrong, situated about 60 miles (96 km) north of P'yŏngyang, in what is now North Korea. The "Iron Jar Fortress" itself is at the extreme left of the image, together with the district seat and major military base of Yŏngbyŏn, all surrounded by a 13-foot-high (3.9-m) fortified wall atop cliffs. The image reads outward from the middle, and was perhaps intended as an aid to governing the overall military complex (Ledyard 1994, 280 and 342–43). Such conceptions of power also play an important role in the design and construction of parklands and gardens according to the latest fashions in landscape architecture, and in civil engineering works generally.

FIGURE 68.

"Ch'ŏrongsŏng Chŏndo" (Complete Map of Iron Jar Fortress) (18th century).

Finally, and less commonly, there has been the mapping of places by antiquarians, who have sought to define the uniqueness of particular sites through the situation of historical remains and relics in the landscape (fig. 67). While antiquarian mappings might be thought purely intellectual, they can in the right context be expressive of power, as one society gives meaning to and so appropriates the landscapes of another. As the landscape painters Thomas and William Daniell observed in 1810, after reflecting on the British appropriation of the wealth and cultural artifacts of India and on the British use of South Asia as a site of scientific investigation, "it remains for the artist to claim his part in these guiltless spoliations, and to transport to Europe the picturesque beauties of these favored regions" (quoted in Dirks 1994, 211). By extension, it remained for British mapmakers to transport India's landscapes to Europe as well (Edney 2003).

MAPPING PROPERTY

Maps of property form a distinct subset of maps of places in that they too are concerned with discrete and definable portions of land. But if the mapping of places is multivalent, encompassing many social and cultural values, then the particular mode of property mapping is monovalent. Every society has developed complex systems of property rights, in which different groups often exercise shifting privileges over certain resources within the same areas. Over time, as some societies have grown in economic activity and complexity or have been subordinated to other, more powerful societies, such inclusive systems of property rights prove unwieldy. Or rather, they prove unwieldy for those who seek to exert their authority and control over the land's resources. In these circumstances, the powerful in a society have sought to rationalize and regularize property rights by breaking up communal lands and allocating to certain individuals or institutions *exclusive* rights to the resources of each parcel of property. Maps have played important roles in establishing exclusive property rights. The fundamental purpose of the property plan, as with any map, is to bring a distant site to hand and to permit its spatial structure to be comprehended and modified. More specifically in this case, property plans have been used to simplify both spatial organization and property rights. This is the sense in which such maps are monotonous: they accord a single sociocultural value—ownership—to a precisely defined parcel of land.

In limited, traditional societies, communities have physically marked boundaries in the landscape when they have felt the need to demarcate property in the face of competition for resources from other communities. Existing landscape features might be used, or special markers placed into the landscape. Either way, the preservation of these boundary markers was an important task

MATTHEW H. EDNEY

for the community. Ancient Jewish lore was perhaps most succinct in this respect: "Cursed be he that removeth his neighbor's landmark" (Deuteronomy 27:17). In societies that have grown sufficiently complex that they have needed to manage property rights—to register or grant ownership, or as part of legal disputes—the common strategy has been simply to record the circuit of boundaries around each parcel of property. For example, in about 100 CE, a Roman surveyor reflected on the registration of tribal lands:

We often find in public documents [that] territories [are] distinctively designated as follows: "From the small hill called such and such, to such and such a river, and along that river to such and such a stream or such and such a road, . . . and from there to the cross-roads of such and such a place, and from there past the tomb of such and such to" the place from which the description began. (Quoted in B. Campbell 2000, 79; also 468–71)

Such metes-and-bounds descriptions are still the legal standard for describing properties in much of the modern industrial world.

These verbal descriptions directly replicate the linear process of tracing property boundaries between the all-important boundary markers; in case of dispute, they readily permit a boundary to be retraced and its markers relocated. But they also require direct reference to the landscape. By themselves, metes-and-bounds descriptions do not enable the easy comprehension of a property's shape and structure or its connections to neighboring properties. Such comprehension is, however, permitted by graphic plans. Property plans have proved especially appealing to desk-bound bureaucrats, benched judges, and absentee landlords.

Government officials have mapped properties to aid in the assessment and collection of land taxes, the economic foundation of every centralizing state in history. All such states have compiled cadastres: written lists of properties, their owners (already manifesting a high degree of exclusivity), and the taxes assessed. Great states have been built atop such lists. South Asian states all featured complex systems for the assessment and collection of land taxes, but without the use of maps, from at least the time of the Mauryan Empire, in 400–300 BCE, until modern times (Schwartzberg 1992a, 317). In some states, bureaucrats have made cadastral plans (maps showing property), which have allowed them to keep track of and to manage the tax system more effectively. These plans have a great antiquity: Mesopotamian clay tablets with simple field plans survive from 3000 BCE to 2000 BCE, with more complex plans surviving from 2000 BCE to 1000 BCE. The clay tablet in figure 69 dates from about 1500 BCE and depicts a main watercourse (the U-shaped parallel lines) with smaller irrigation ditches branching off; each field between the watercourses is named and identified as belonging either to the king or to a particular temple

FIGURE 69.

Plan of fields and watercourses from Nippur,
Mesopotamia (ca. 1500 BCE).

(Finkelstein 1962). Archival records also indicate that cadastral plans were well understood in China in 200–100 BCE and probably earlier, although none have survived (Yee 1994b, 75).

Cadastral plans have also been integral to the creation of new property through the regulated reclamation of land—as in medieval Japan (Unno 1994, 356–62) or the early modern Netherlands (Kain and Baigent 1992, 11–24)—or through centrally organized colonization. A recurring theme in the settlement of new lands, evident for example in ancient Rome's *coloniae* (Dilke 1987a; B. Campbell 2000), pre-Columbian Peru (Gartner 1998, 264), Spanish America, and the modern United States, has been the use of cadastral plans to guide and record the regular division and distribution of new property. In the United States, federal lands were divided into (ideally) square townships 6 miles (9.6 km) on each side, with each township comprising thirty-six sections, and each section being further quartered and quartered again, with little regard for the contours of the landscape (fig. 70). Each quarter-quarter-section of 40 acres (16

MATTHEW H. EDNEY

FIGURE 70.
Official plat of T15N R11E (Marquette),
Wisconsin (1858).

ha) was the basic unit by which the federal government sold off lands acquired from the Native Americans (White 1983).

After 1500, cadastral plans were increasingly used by Europe's centralizing states as they variously sought to reform their tax systems in order to enhance and regularize their revenue. One way to increase the flow of taxes into government coffers was to eliminate the sticky fingers of the local tax collectors, but this required central government officials to act in place of those authorities, and that required the central government to collect detailed cadastral information—and especially maps—about distant properties. This factor was very much in the minds of British bureaucrats, for example, when they tried after 1800 to reform the land taxation systems in their South Asian empire (Edney 1997, 110–13 and 334–35). A further factor in Europe was the desire to head off popular unrest by making tax assessments equitable; this required abolishing the privileges enjoyed by the nobility and reassessing land taxes according to the actual quality and quantity of land. These reforms, together with their cadastral surveys, were first undertaken in some parts of The Netherlands during the 1530s, but they did not become widespread until the 1700s.

Perhaps the most important early cadastral survey in this category was that undertaken by the Austrian Hapsburgs of the small but rich duchy of Milan. An inequitable tax assessment in 1568 had left the greatest share of the tax burden on the shoulders of the artisans and peasantry, who by 1700 were emigrating in significant numbers. To stabilize the duchy's economy and to increase its revenues, the Hapsburgs undertook a new assessment in the 1720s (the *censimento*), complete with a detailed cadastral survey. An important innovation of the *censimento* was that it encompassed *all* of the land and not just the productive fields—a policy that explicitly extended the state's control over the agricultural lands which might in future be reclaimed from wastelands (Kain and Baigent 1992, 181–90).

The second general class of property plan are those undertaken for legal purposes. This sort of plan became commonplace in Europe with the progressive centralization of the courts after the later medieval period (Dainville 1970). As judges increasingly stayed behind the bench rather than touring their district, they began to rely on maps to understand the spatial dimensions of each case. For example, a dispute between St. Mary's Abbey in York and the earls of Lincoln over some very profitable peat beds in the marshes of Inclesmoor—covering some 240 square miles (624 km²) on the border of Yorkshire and Lincolnshire in England—led to the preparation of two property plans. After a century of conflict, the case was finally referred to Henry IV in 1402–3. The king directed his clerks to make copies of all the relevant deeds and documents. Some royal clerks also went to Inclesmoor itself and drafted a large and ornate plan of the area, highlighting the points in contention; a smaller and plainer copy of that plan (fig. 71) was then made and bound prominently in the book made from all the relevant records, where it served as a graphic summary of the case (Beresford 1986; also Harley and Woodward 1987, plate 37). To this day, legal cases have led to the production of a great many property plans to inform distant judges.

The third general class of property plan are those commissioned by landowners of their own estates for use as a tool of management and agricultural improvement and also as a symbol of ownership and control. Such privately

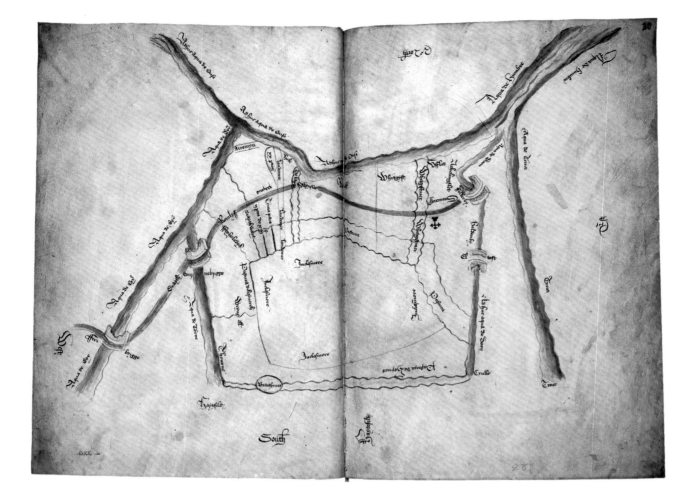

FIGURE 71.

Anonymous plan of Inclesmoor, West Yorkshire, United Kingdom (ca. 1407).

commissioned property plans are strictly a phenomenon of European culture since the sixteenth century (for example, Bloch 1966, 131–33). They range from relatively simple plans based largely on approximation and sketching to minutely surveyed ones, which often bear symbols of ownership and possession. Thus, when the officers of Christ Church College, Oxford, were uncertain in the mid-eighteenth century about the lands they owned within the parish of Woodnorton in Norfolk, they undertook a noncartographic survey and review of land tenures in the parish; the ambiguities exposed by this review led in turn to the detailed measurement of the college's lands and the preparation of a complex plan (fig. 72). However, the surveyor seems not to have been thorough in his work, and he was soon criticized for failing to include some 120 acres (48 ha) of college land. His final map nonetheless gives the appearance of care, complete with a mock picture frame and the college's coat of arms top and center. Not that all landowners and their stewards embraced the property plan in early modern Europe. A study of the estate stewards employed by the

MAPPING PARTS OF THE WORLD

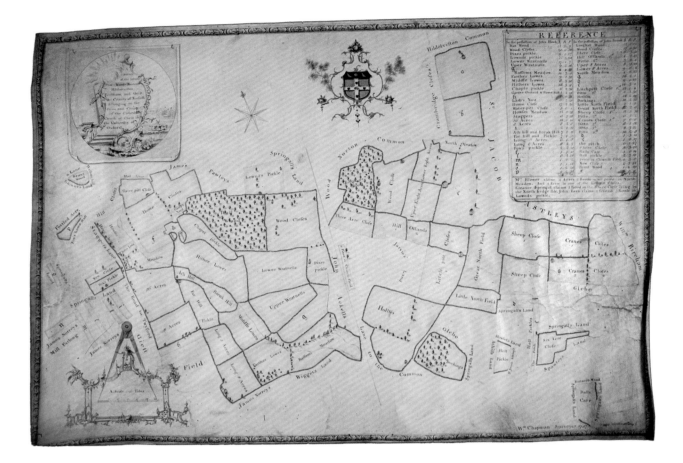

FIGURE 72.
William Chapman, plan of Woodnorton,
Norfolk, United Kingdom (1757).

same Oxford college reveals that they were quite conservative and preferred to use the older, written methods of terriers (rent rolls) and metes-and-bounds descriptions rather than any newfangled graphic maps; they would not uniformly accept the conceptual benefits afforded by the graphic property plan until well into the nineteenth century (Fletcher 1995, 14–15 and 25–29).

Published manuals of land surveying played a significant role in teaching early modern landowners to appreciate graphic plans of their property. These manuals explained the principles of surveying in an appropriately educated and mathematical manner; they described instruments that were the height of technological innovation; they were long on theory, but short on practice. Their authors, land surveyors all, used the manuals to demonstrate their intellectual credentials and so secure patronage, advance their career, and eventually improve their social standing. These points are evident in the decorative title page to Aaron Rathborne's 1616 text, *The Surveyor* (fig. 73). Rathborne used the highly literate language of allegory to claim an educated standing for himself and his craft. The statues represent the mathematical principles of land surveying, arithmetic, and geometry; their columns support celestial and ter-

MATTHEW H. EDNEY

FIGURE 73.

Title page to Aaron Rathborne, *The Surveyor* (1616).

restrial globes, symbols of learning and erudition. In the upper vignette, the surveyor stands atop two figures—a fool, distinguished by his coxcomb, and a faun, with pointed ears—that represent precisely those things that the surveyor seeks to supplant through his erudition and craft: unreason/ignorance and nature/dullness, together suggesting the peasantry itself (Bartolovich 1994, 260). The surveyor's art, flagged by the upper vignette's label of *artifex* (from *ars* [art, craft, skill] and *facere* [to make]), is set against nature. Finally, the Latin phrase *inertia strenua* (restless idleness), set below the lower vignette, is a tag from Horace and declares Rathborne's claim to a classical education.

Surveying manuals also promoted the graphic plan of property as a crucial and empowering tool for the landowner or steward. John Norden explained the potential of such a plan in his *Surveyors Dialogue* (1607). His imaginary tenant asks, "Is not the field itselfe a goodly map for the lord to looke upon, better than a painted paper?" Norden's surveyor responds with a ringing endorsement of the power and knowledge provided by the graphic survey plan:

A plot rightly drawne by true information, describeth so the lively image of a Mannor, and euery branch and member of the same, as the lord sitting in his chayre, may see what he hath, and where and how he lyeth, and in whose use and occupacion euery particular is upon the suddaine view. (Norden 1607, 15–16)

The property plan was thus a central element in the destruction of traditional social order. As long as the landlord lived close by his properties and could use "the field itselfe" as "a goodly map," he would remain an integral part of the social order, interacting with his tenants on a daily basis. But give the landlord a plan of his property, and he would no longer need to go into his fields; he would no longer need even to live near them.

Property plans permitted landlords to distance themselves from their tenants and so contributed to the erosion of long-established customs. And they were widely recognized as such. Elizabethan plays, in particular, were full of commentary that was strongly critical of the changing social dynamics of rural England, the imposition by landlords of new agricultural practices, and the decline of the (idealized) older practices in which rural communities possessed a significant voice (Sullivan 1998). More generally, popular lore throughout Europe has always insisted that the land surveyor is little more than the landlord's lackey and can never be trusted by tenant farmers. My favorite indication of this popular mistrust is the Polish folk tradition that dead surveyors become will-o'-the-wisps, illuminated streams of marsh gas "meandering over bogs and swamps night after night"; indeed, will-o'-the-wisps were sometimes called "measurers" or "metermen" (Kula 1986, 16–17).

All property plans, regardless of who made them and why and when they did so, provide their users with the extra capacity for clarifying and simplifying spatial relationships and so facilitate the exercise of authority over discrete parcels of property. In this respect, every property plan is "an instrument of control which both reflects and consolidates the power of those who commission it" (Kain and Baigent 1992, 344). That control functions both in an intellectual manner, permitting the tax collector, judge, or landowner to envision property and to understand it; and in a more overtly instrumental manner,

MATTHEW H. EDNEY

permitting those distant authorities to make decisions about the properties depicted and then to implement those decisions and make them real.

MAPPING SPACES

Even as people have employed maps to define and demarcate precise places and properties, they have used other maps to understand and manipulate wider geographic spaces that are far beyond the ability of any one individual to comprehend directly. Through geographic maps, people have variously demarcated provinces, regions, countries, and even entire continents. They have used maps to turn vague conceptions into concrete entities through which they have organized their lives and activities. In particular, geographic maps are used to frame regions and countries and continents as bounded spaces, each possessing a unique coherence and uniformity.

A crucial point about geographic mapping is that the reasons presumed for spatial coherence—governance by a single ruler, habitation by a dominant people, subjection to colonial desires, and so on—are proposed by the people who make and read the maps. Significantly, those people do not necessarily have any relationship to the people and institutions who actually occupy the areas mapped. Consider the region of Southeast Asia. No one in the modern West had conceptualized such a region until the Second World War. Europeans and Americans previously thought in terms either of the broad spatial concept of Asia or of several distinct spaces: independent Siam (Thailand) and the various colonial entities of British Burma and Malaya, the Dutch East Indies (Indonesia), and French Indochina (Vietnam, Laos, Cambodia). But in 1942 the Japanese army made these boundaries and distinctions irrelevant. Only then did Westerners frame this particular region—as a single theater of war—and label it *Southeast Asia*. One of the very first of the new regional maps appeared in *National Geographic* in 1944 (fig. 74). The accompanying editorial acknowledged that the map's novel spatial frame was intended to help the magazine's huge readership understand "the heart of the vast Pacific battleground" (Anonymous 1944, 449). By the end of the war, the new regional conception was firmly and indelibly fixed in military, popular, academic, and diplomatic imaginations, even though it is impossible to identify any regional coherence based on internal, formal characteristics such as economy, ethnicity, colonial legacy, political ideology or structure, language, religion, or culture (Emmerson 1984).

The manner in which geographic spaces are intellectual creations is neatly demonstrated by the allegorical representation of five continents on the title page to Abraham Ortelius's atlas, the *Theatrum orbis terrarum*, published in many editions and languages between 1570 and 1612 (fig. 75). Statues of women are arrayed around a stage (the *theater* of the atlas's title). Their attributes—

their dress and possessions—identify them as symbolizing the continents: sovereign Europe sits enthroned at the top; Asia stands at left, Africa at right; and on the floor of the stage itself, the focus of the geographic drama, reclines a naked, barbarous America next to the bust of Terra Australis, the putative southern continent (Waterschoot 1979). Ortelius thought that Tierra del Fuego, separated from the southern limit of South America by the Straits of Magellan, was the northernmost part of a promontory of a huge continent covering the South Pole, which he called *terra australis nondum cognita* (southern land not yet known; fig. 58). On the title page, the fire in the bust's plinth implies the volcanoes and name of Tierra del Fuego; the bust itself indicates that the continent is only partially known; and the plinth suggests that the remainder of the feminized landmass lies passively waiting for European geographers (all men) to come and give it shape and meaning. By recalling the origin of allegorical imagery in classical statuary, Ortelius indicates that geographers actively shape continents and other geographic conceptions just as sculptors chisel statues out of stone.

These two examples are specific to the West's post-Renaissance engagement with the wider world. Traditional societies have tended to map distant

MATTHEW H. EDNEY

lands only within the context of world maps and have rarely framed more precise spaces. A prominent exception was medieval Islam, whose wide-reaching and energetically expansive culture promoted the mapping of extensive areas. One spatial conception expressed in several Islamic cartographic traditions held that the two seas that bounded the Muslim world were completely self-contained regions (Tibbetts 1992a, 106, 125, and 1992b, 154; Johns and Savage-Smith 2003; Edson and Savage-Smith 2004, 85–98). To the west, the Mediterranean Sea is, of course, almost entirely landlocked and therefore readily

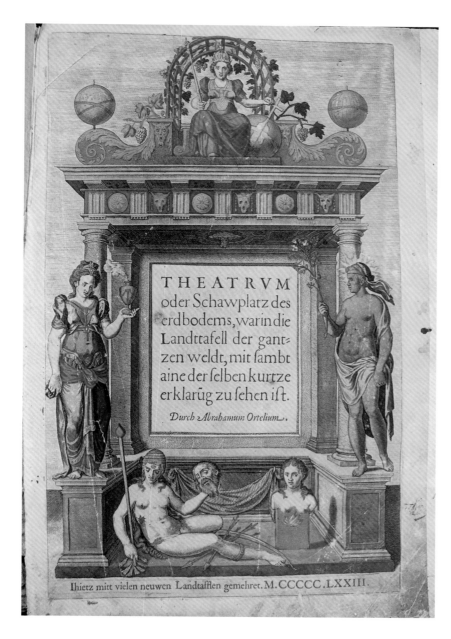

FIGURE 75.

Title page to Abraham Ortelius, *Theatrum, oder Schawplatz des Erdbodems* (Theater; or, Showcase of the World) (1573).

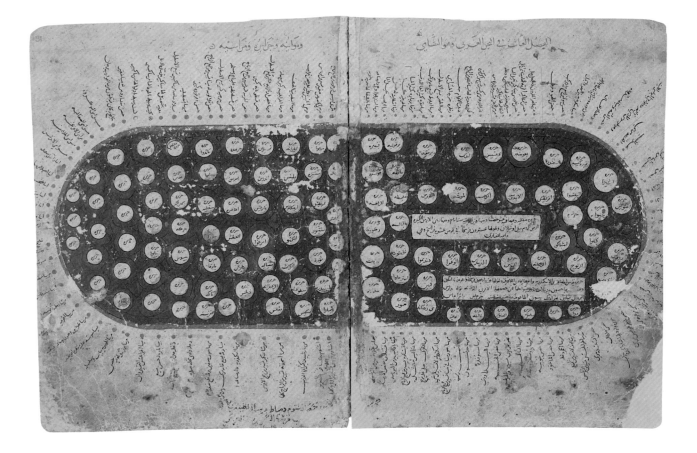

understood as a closed region; the similar understanding of the Indian Ocean, to the east, probably derived from older Hellenistic conceptions of a bounded sea (fig. 76A–B). Although they depict seas, such geographic maps cannot be mistaken for charts: they were not intended to help mariners conceptualize and navigate these waters. Rather, like all geographic maps, they were abstractions that defined the seas for a general audience as particular subdivisions of the world.

If not simply plagiarized from an existing map, each geographic map is made by arranging and compiling a variety of geographic information—other geographic maps, itineraries and travel narratives, information from Indigenous sources, sea charts and ships' logs, and even word of mouth—within a culturally approved framework. Geographic mapmakers in early modern Japan, for example, generally drew on a mixture of Dutch images and other European maps that had already been appropriated and reinterpreted by Chinese scholars (see chapter 2 and fig. 52). The map reproduced in figure 77 was created from an assemblage of a Chinese map (which itself ultimately derived from European maps) and ethnographic imagery, all further reinterpreted in accord with

MATTHEW H. EDNEY

FIGURE 76B.

The Indian Ocean. From *Kitāb Gharā'ib al-funūn wa-mulah al-ʿuyūn* (The Book of Curiosities of the Sciences and Marvels for the Eyes) (early 13th-century copy of a mid-11th-century original).

Japanese cultural traditions. We can therefore identify "parent" and derivative, "child" maps, and arrange them in genealogical trees of cartographic descent. The differences between generations of maps are often small, but they can build up over time: the maps made of English counties by Christopher Saxton (see fig. 170) in the 1570s were repeatedly copied and modified for more than two centuries, producing direct lineal descendants that appear to have little in common with their original progenitors (Delano Smith and Kain 1999, 66–81).

There has been a recurring desire on the part of scholarly, library-bound geographers to remake the world anew by carefully examining and comparing *all* the sources of information available to them, whether already published or newly received from travelers. Such work clearly requires that the mapmaker be well placed within communication networks, as were Eratosthenes and Claudius Ptolemy in Alexandria in the second centuries BCE and CE, respectively; Muhammad al-Idrīsī in twelfth-century Sicily (fig. 44); and Fra Mauro in fifteenth-century Venice (fig. 39). Alternatively, the critical mapmaker is situated at the center of expansive cultures, as were the geographers Guillaume Delisle and J. B. B. d'Anville in eighteenth-century Paris (Jacob 1999). The

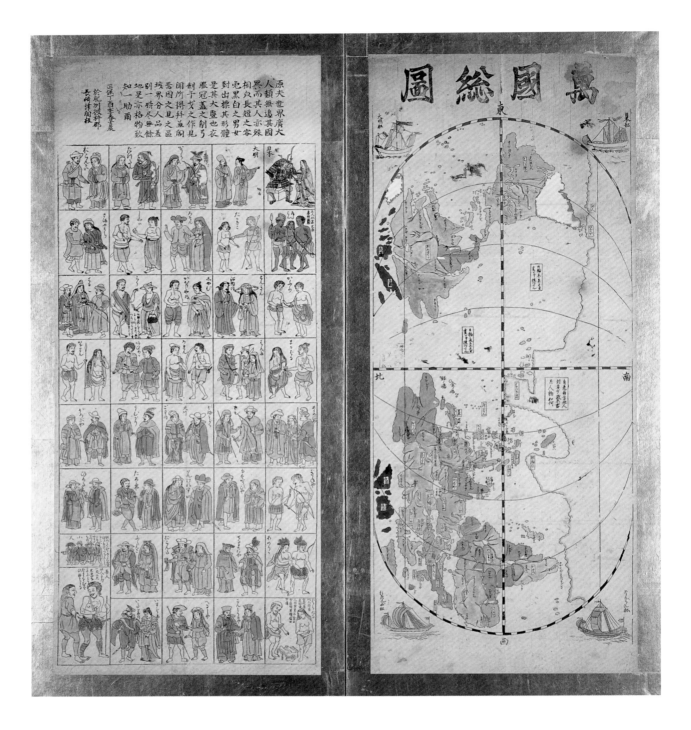

FIGURE 77.

"Bankoku Sozu" (Complete Map of All Countries)
and ethnographic depictions of peoples of the
world (1645).

FIGURE 78.

Guillaume Delisle, map of North America
(ca. 1696).

conclusions reached by these critical geographers in their studios and libraries, as they sought to reconcile conflicting or partial sources with their convictions about how the world ought to be organized, can strike us today as fantastical. But in their time, they represented careful and reputable science.

Claude and Guillaume Delisle were able to confirm the long-suspected existence in the western interior of North America of the great Mer de l'Ouest (sea of the west) by carefully collating contemporary source materials (fig. 78); Joseph-Nicolas Delisle would eventually publish maps and commentaries on the nonexistent sea in 1751 and 1753 (Lagarde 1989). In 1798, James Rennell combined his conviction that high mountains surround large watersheds with a single observation of distant hills by the deceased explorer Mungo Park to create the "mountains of Kong," which he stretched in a long, thin chain right across Africa from Senegal to Ethiopia (Bassett and Porter 1991). Once these new geographic features were established, they lived on in maps through sheer cartographic inertia: even when their nonexistence was proved in the late 1800s, the mountains of Kong persisted on maps as a region of West Africa, a meaningless region which can still be found within Côte d'Ivoire in the latest edition of *Goode's World Atlas* (Veregin 2005, 230).

Geographic maps are a staple commodity for educated and literate groups. They have been published for a wide and nonspecialist readership among merchants, professionals, and the gentry. The allure of these maps, according to many commentators, is that they allow their readers to comprehend a wide extent of space without having to experience the physical hardships, dangers, and expense of real travel. As Georg Braun wrote in the preface to the *Civitates orbis terrarum* (Cities of the World [1572]),

What could be more pleasant than, in one's own home, far from all danger, to gaze in these books at the universal form of the earth . . . and, by looking at the pictures and reading the texts accompanying them, to acquire knowledge which could scarcely be had but by long and difficult journeys? (Quoted in Skelton 1966, vii)

At the same time, a good grounding in geography was a desideratum for anyone with claims to a decent education. As an Englishman wrote in 1635 (and he was just one of many who expressed such sentiments), no work of poetry or history "can be well read with profit," or effectively and properly interpreted and explicated, "withouth the helpe and knowledge of this most Noble Science" (Saltonstall 1635, xi). It was in the same spirit of geographic knowledge elucidating and illuminating texts that Protestant reformers added maps to their Bibles (Ingram 1993). Moreover, this desire for general geographic knowledge supported the extensive printing and distribution of maps, atlases, and geographic texts in Europe and east Asia. It fed the development of the great Dutch map-publishing houses of the later sixteenth and seventeenth centuries that supplied Europe with atlases (figs. 75 and 79) and large wall maps for display in the public spaces of houses, as well as a host of smaller and cheaper products. Printed maps were also widely consumed in China, Korea, and Japan in a variety of forms, ranging from the expensive to the comparatively cheap.

This consciously cultivated geographic literacy encouraged the formation of new social identities. In part, these identities were centered on the formation of what might be called a protonationalism: as people carefully studied maps of their own countries, they saw (or imagined they saw) their land and began to comprehend the existence of other people who shared a common attachment to the same country. Conceptions of several European states, of the character and extent of each, were debated with the aid of regional maps (for example, Helgerson 1992; Conley 1996; Padrón 2004). Maps of distant regions, especially *cartes-à-figures* which bore ethnographic images of people in local dress (figs. 52, 77, and 79), signaled the foreignness and even exoticness of those regions and peoples in comparison with the maps' readers, thereby reinforc-

MATTHEW H. EDNEY

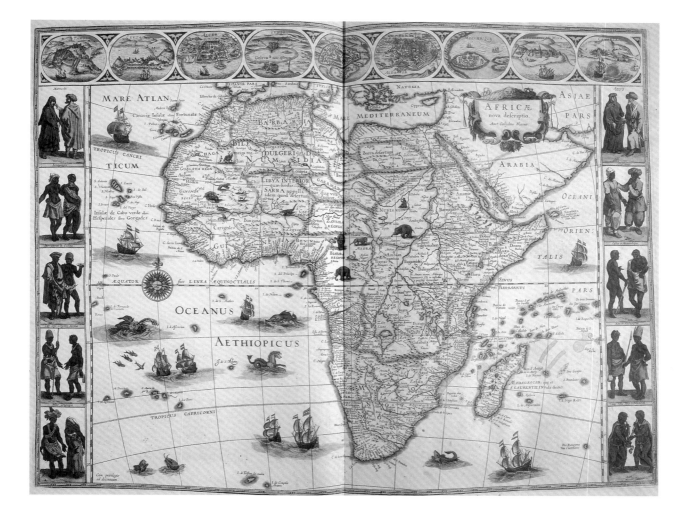

FIGURE 79.

Willem Jansz. Blaeu, "Africae nova descriptio"
(New Description of Africa) (1640–43).

ing the burgeoning imagined sense of community. This development characterized cultures in both Europe (Schilder 1986–, 4:85–86 and 301–4) and East
Asia (Hostetler 2001; Yonemoto 2003, 69–100). Eventually, such geographically
minded conceptions of community—conceptions of an us/self that contrast
with a them/other—would displace more traditional identities rooted in family
and clan ties and in feudal allegiances. Yet these protonations were not extensive. Map readership did widen significantly in the sixteenth and seventeenth
centuries, but even the cheapest printed maps remained beyond the reach of
most people. The cartographically imagined affinities therefore remained socially limited. Moreover, the sense of common identity remained weak in the
face of the coercive power and cultural authority of central governments. This
was especially the case in Japan, where the Tokugawa shogunate and the aris-
tocracy it supported remained unchallenged until the mid-nineteenth century
(Yonemoto 2003, 173–78).

But in Europe, the religious wars of the seventeenth century significantly loosened the authority of the royal courts and permitted private individuals to engage with, critique, and even regulate cultural and political affairs. For the first time, the lesser gentry and the middling sort claimed a say in the affairs of the mighty. The loose sense of community fostered by regional maps before 1650 began thereafter to form what has been called the public sphere, an association and sociability built upon the reading and discussion of the same published works. Public debates gave new meaning to the utility of regional maps both as persuasive devices and as essential knowledge for any educated person. In the early 1750s, for instance, London and Parisian map sellers published many maps of North America as contributions to public debate over Anglo-French imperial tensions on the continent (fig. 93) (Pedley 1998). After 1800, Europe's public spheres promoted new democratic movements, which in turn steadily embraced the newly industrialized working classes. The mass literacy engendered by modern states transformed the socially restricted "public" into the inclusive "nation" and gave rise to truly nationalistic cartographies that pervaded all levels of society. Such cartographies have often reduced national territories, or the territories claimed by nations, to iconic outlines. They have been crucial in postcolonial states—first in the early 1800s in independent America (Brückner 2006; Craib 2004), and later in twentieth-century Asia (B. Anderson 1991, 163–85; Ramaswamy 2004; Winichakul 1994)—when colonial-era icons were reconfigured as symbols of new nations. This reuse of older icons is evident, for example, in a 1955 Tamil-language schoolbook from southern India, in which a poem lauding "our nation" is presented with a picture of a student gazing intently at an iconic map of pre-Independence British India (Ramaswamy 2001).

MAPPING TERRITORIES: TREATING SPACES AS PLACES

Expansive, complex societies have expended a great deal of effort in mapping and delineating their own and their neighbors' territories. Basic geographic information—the locations of towns, the routes followed by roads and rivers, the ranges of mountains—has long been recognized as crucial to the administration of territory and for guiding the deployment of military forces. Certainly, the military applications of geographic maps were understood in China as early as the third century BCE, and specifically military maps of Hunan Province survive from the second century BCE (Yee 1994a, 40–46 and 1994b, 73).

Most geographic mapping for military and administrative purposes has been undertaken within early modern and modern states, in both Europe and Asia. As a member of the English parliament opined in 1571, just as Saxton's government-sponsored survey of the counties of England and Wales got under way,

MATTHEW H. EDNEY

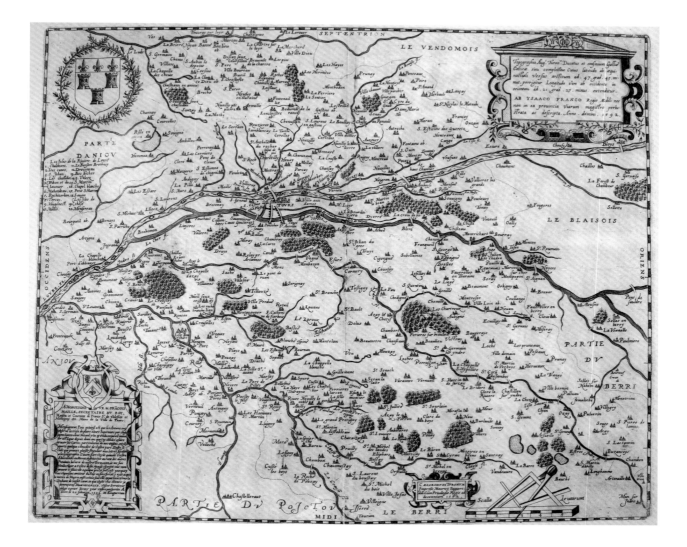

FIGURE 80.

Ysaac François, "Topographia Augt. Turonis.
Ducatus" (Map of the Duchy of Touraine) (1592;
edition of 1619).

How may her Majesty [i.e., Elizabeth I], or how may this Court [i.e., Parliament]
know the estate of her Frontiers, or who shall make Report of the Ports, or how
every Quarter, Shire or Country is in state? We who never have seen *Berwick* or
St *Michael's* Mount, can but blindly guess of them, albeit we look on the Maps,
that came from thence. (Quoted in Klein 2001, 90)

A large number of government-sponsored regional mapping projects were
undertaken from the sixteenth century onward. Perhaps the most famous in
Europe are Saxton's surveys of the English counties (see fig. 170), but we can
also point to a large number of particular and occasional maps made by government officials, such as the map of Touraine prepared in 1592 by Ysaac François,
the duchy's royal overseer and magistrate (fig. 80), and many other similar
projects (Buisseret 1992; see Woodward 2007).

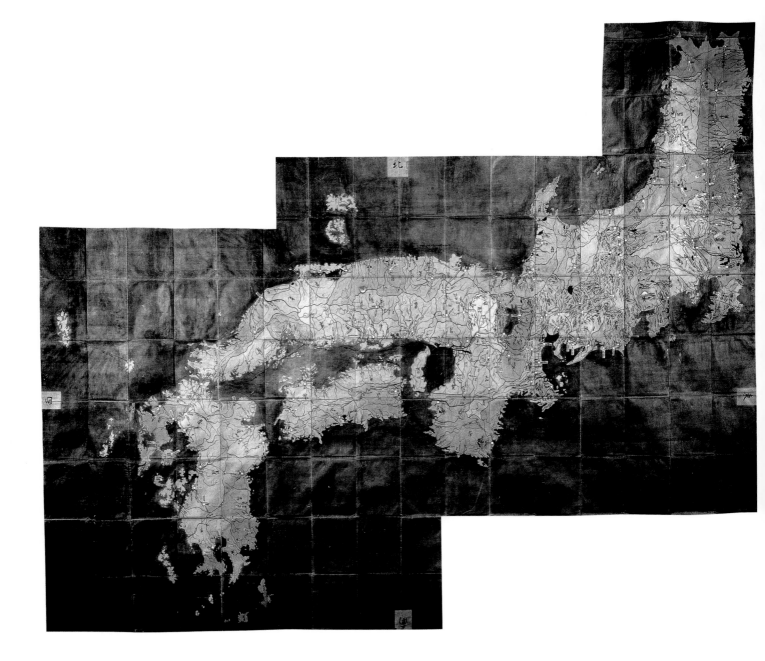

FIGURE 81.
Keichō map of Japan (ca. 1653).

In Japan, the comparatively high degree of centralization and bureaucracy led to projects to map each province as early as the late eighth century CE. Several provincial surveys were undertaken in the later sixteenth century. Then, in the seventeenth, the new Tokugawa shogunate ordered no less than four surveys of the entire country. All four decrees required the chief clans in each province to construct a detailed provincial map; these could then be combined into very large maps of all Japan (fig. 81). The repeated directives for detailed maps and other information, such as population figures and rice yields, effectively constituted one more form of indirect control by the shoguns, because

MATTHEW H. EDNEY

each provincial daimyo had to expend significant time and resources to provide valuable information which the shogunate could then use to control and regulate still further each daimyo's activities (Unno 1994, 394–404 and 472; Yonemoto 2003, 10).

Military interest in spaces significantly intensified in Europe in the seventeenth century, when military strategists slowly abandoned their focus on the defense and attack of static fortresses and increasingly stressed the mobility of armies and open-field battles. This shift in strategy featured the expansion of the generals' traditional emphasis on the landscape and character of particular places (fortresses) so as to encompass the landscape and character of much more extensive regions. Far-reaching, detailed maps—often accompanied by thorough reports on the state of roads and bridges—allowed generals a much better understanding of the terrain over which their troops were to maneuver and fight. New corps of military "topographical engineers," all trained in similar ways and using increasingly standardized systems for symbolizing geographic and topographic features (see fig. 84), undertook extensive surveys of European landscapes over the course of the eighteenth century: in Scotland after the failed Jacobite Uprising of 1745; in the many provinces of the sprawling Austrian Monarchy after 1763, once the Seven Years' War had demonstrated the utility and necessity of extensive terrain maps; and throughout the German states. The results were hybrid mappings of both place and space: large areas of space were mapped in great detail. Inner Austria alone required no less than 250 sheets (fig. 82). To make such detailed surveys, the military engineers worked quickly, perhaps simply sketching from horseback, with only a minimal geometrical framework to control and contain their errors (Nischer-Falkenhof 1937).

The eighteenth-century public was quick to grasp the utility of extensive surveys. After 1750 the English counties were mapped at relatively high resolution by commercial entrepreneurs who sought to fulfill local demands for the spatial information required to plan new canals and turnpikes, to organize the reclamation of marshes, and to implement other improvements (Delano Smith and Kain 1999, 81–101). In France between 1744 and 1789, a commercial enterprise headed by the Cassini family of astronomers and largely funded by the king himself, surveyed the entire country in conjunction with provincial authorities to produce the 182-sheet *Carte de France* (fig. 83). This deservedly famous map sought to depict all of France in a uniform manner, permitting the French public, bureaucrats, and generals to envision any part of their state. The significance of the project was recognized during the Revolution by the new government, which appropriated all the materials of the survey and map in 1793; cash strapped, it went ahead and published the last plates, still unfinished, such as that shown here (Konvitz 1987, 1–31; Pelletier 1990).

To map all of France as a single territory, the Cassinis needed to use a rigorous foundation of triangulation. Done with sufficient care and skill, the deter-

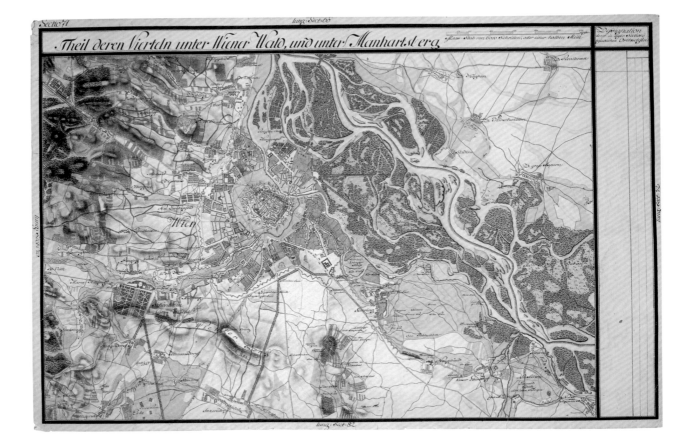

Sectio 71

Theil deren Vierteln unter Wiener Wald, und unter Manhartsberg.

FIGURE 82.

Survey of Inner Austria, sheet 71, Vienna and environs (1784–87).

mination of the precise geometry of the triangles made by sightlines between high places (see fig. 84) can be used to measure with great accuracy the length of a meridian of longitude or parallel of latitude across the earth's curved surface; that information is in turn used to calculate the size and precise shape of the earth. Such high-quality surveys were the period's equivalent of atomic science. They were massively expensive, laborious, and necessarily underwritten by the state, but they contributed to basic science and particularly to the proof of Isaac Newton's theory of gravity (Greenberg 1994). In France, members of the Académie royale des sciences had intermittently pursued a system of triangulation since the 1660s to measure the lengths of the meridian and parallel running through Paris and so measure the earth itself (Levallois 1988). The academicians also, and for the first time, extended their triangles over very large areas, to determine the positions of many locations across almost all of France. It was this secondary triangulation that formed the basis of the *Carte de France*. When used for territorial mapping, a triangulation constrains the errors endemic to detailed surveys of the earth's surface, preventing them from propagating, thereby enabling the surveys to be fitted together cleanly into a

MATTHEW H. EDNEY

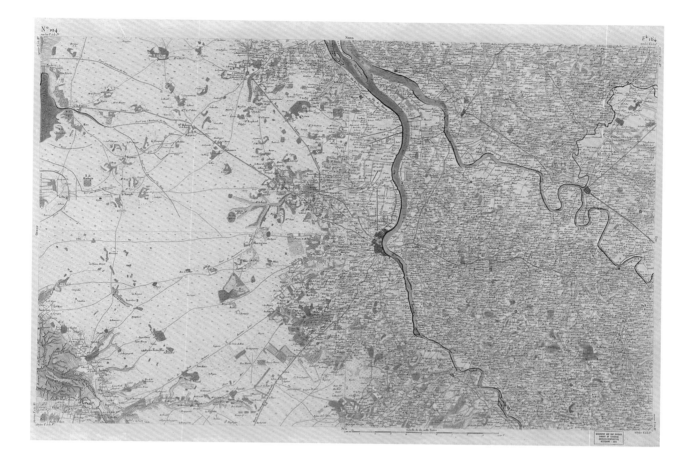

FIGURE 83.

César-François Cassini de Thury, Bordeaux. From "Carte de France" (Map of France) (ca. 1790).

single image of an entire country (Edney 1997, 16–22, 104–10, and 179–84). The success of the *Carte de France* fueled a general desire among modern European states after 1790 to undertake similar surveys in order to produce uniform, high-resolution, and extensive maps of their territories.

In practice, however, few states in the nineteenth century had the economic, intellectual, or technical wherewithal to undertake such ambitious surveys; many were begun but soon ended. After the Revolution began in 1789, a reformist French government sought to implement a single, statewide cadastral survey to underpin its land reforms, but the necessary effort and political coordination was simply too great, and the state soon had to resort to having each municipality undertake its own local project (Konvitz 1987, 32–62). Other state surveys limped along, chronically underfunded and understaffed, unable to implement a systematic project and resorting instead to the appearance of system. This was the case for the French in Egypt in 1798–1802 (Godlewska 1988), for Portugal (Branco 2005), and for British India (Edney 1997, 261–89), as well as for the post–Civil War surveys of the American West (Bartlett 1962). George Wheeler's attempted military survey of the western United States, in 1871–79,

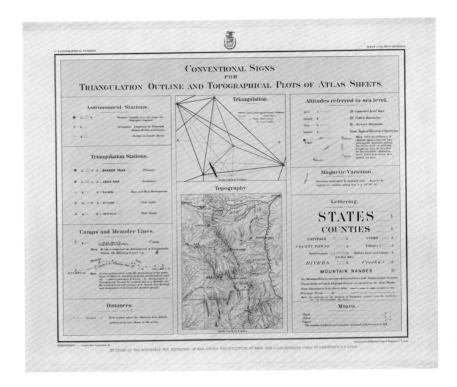

FIGURE 84.

(above) "Conventional Signs for Triangulation
Outline and Topographical Plots of Atlas Sheets."
From George M. Wheeler, *Topographical Atlas
Projected to Illustrate United States Geographical
Surveys West of the 100th Meridian* (1873).

FIGURE 85.

(facing) Ordnance Survey of Ireland, detail of sheet
18 of *Dublin* volume (1844).

was intended to be based entirely on a triangulation, but as Wheeler's published explanation of symbols indicates, he and his teams also relied upon the older, less precise mix of astronomical observations for latitude and longitude and measured itineraries (called "meander lines" in figure 84).

The successful comprehensive surveys were those whose superintendents were careful to secure sufficient resources, not overreach themselves, demonstrate continually the worth of their work to their political and financial masters, and delimit precisely what they could and could not do. The British successfully completed the Ordnance Survey of Ireland between 1824 and 1846 because the survey's superintendent, Thomas Colby, was able to prevent the project from having to do everything at once. Colby undertook a basic cadastral survey for the purposes of tax reform, at a relatively low resolution sufficient for recording rural boundaries and with a careful division of labor in which most of the detailed features were surveyed by a large pool of unskilled, low-cost labor (fig. 85). These plans were then reduced in scale to serve as the basis of separate and less detailed surveys of the terrain for military purposes. The British also undertook still more detailed cadastral surveys of urban areas. By breaking up the project's different requirements, Colby was able—just—to keep to a budget that his political superiors could sustain (Andrews 1975).

Yet over time, the bureaucratic desire within industrializing and imperial states for ever more precise and comprehensive knowledge of territories,

MATTHEW H. EDNEY

peoples, and economies was complemented by steadily developing technical and financial capacities. New needs prompted new surveys. By the 1840s, sanitary reform in the sprawling industrial cities required the construction of water delivery and sewage systems, leading to a whole new round of massively detailed urban surveys in Europe, North America, and Asia. The development of long-range artillery in the First World War required the reconfiguration of military maps onto common projections and the imposition of reference grids to permit artillery officers to calculate easily distances and bearings from gun to target. In the United States, from the 1880s to the 1930s systematic surveys were undertaken as collaborative ventures between the federal and state governments, ostensibly to aid in geologic investigations, and at a relatively low resolution; during the cold war, the demands of organizing and regulating the industries necessary for sustaining a nuclear power led to a further, comprehensive round of much more detailed topographic mapping enabled by the techniques of aerial photography, which had been perfected during the Second World War (fig. 86A–B). Furthermore, the cold war engendered satellite surveillance systems, which began as a means to keep a close eye on the opposing side, but which have proliferated into a wide variety of platforms that monitor and

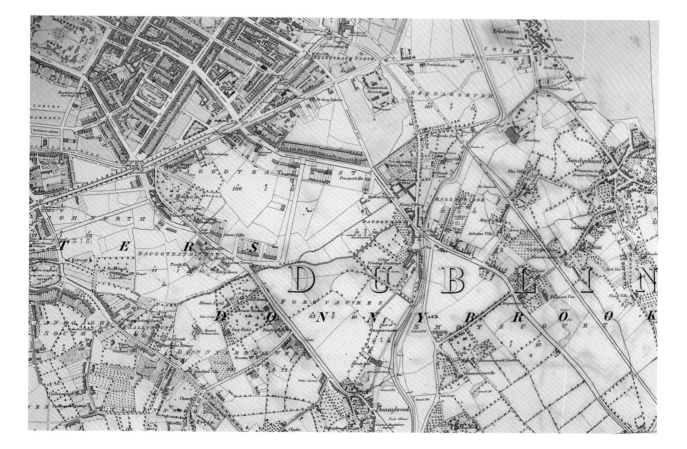

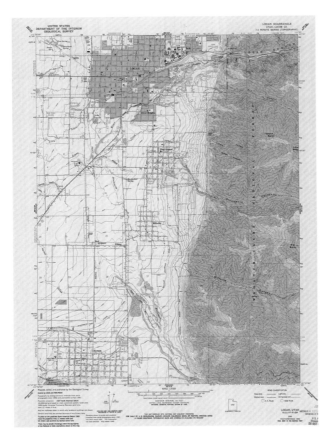

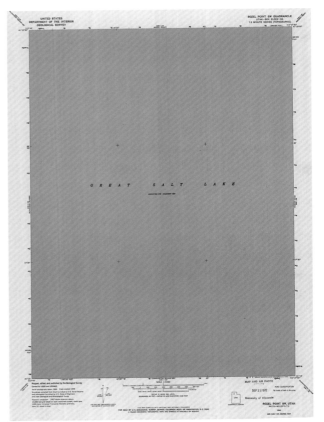

FIGURE 86A. U.S. Geological Survey, Logan, Utah (1961).

FIGURE 86B. U.S. Geological Survey, Rozel Point SW, Utah (1969).

map all aspects of our environment; satellite data are today increasingly accessible to a wide public through the Internet, whether raw or in a more digested form through Web applications such as Google Earth.

The net effect of all these expansive mappings of state territories has been the apparent collapse of the traditional distinction between mapping places and mapping spaces. Since the eighteenth century, spaces have increasingly been mapped as places. The space of each state, its territory and community, has been progressively broken down into ever smaller pieces, each piece being mapped and regulated with ever greater care and detail (fig. 86A). But this process, which is so important to the basic logic of modern industrial states, can be taken to the point of absurdity. Once set in place and anchored with bureaucratic controls, territory can be reduced to pieces so small that they effectively lose significance (fig. 86B).

Yet the manner in which these huge maps collapse space and place together is chimerical. We can conceive in principle of assembling all the individual sheets of a territorial map into one giant image, but we can neither comprehend nor grasp the result. If one were to assemble all the 1:24,000 series of U.S. Geological Survey topographic quadrangles (like those shown in fig. 86A–B)

MATTHEW H. EDNEY

for the United States' forty-eight contiguous states, the complete map would measure some 783 feet (234.9 m) in length by 383 feet (114.9 m) in breadth. Such an endeavor would be like mapping interstellar dimensions: we can represent them, but can we truly appreciate a map of that size? (This perhaps explains the twentieth-century fancy of calling maps "god's-eye views," because only a god can truly comprehend such huge images.) Humans can only comprehend and grasp these vast maps in portions. The larger spaces of the world still need to be brought down to a human scale, whether on paper or a computer screen, to permit each user to have a personal relationship to the whole. In practice, individual map readers can only take in, at most, a few sheets at a time.

And so we return to the issue with which this chapter began: the delusional nature of the promise of cartographic perfection engendered by the detailed mapping of extensive territories and, most recently, by digital computers and satellite imagery. We can now see that this perfection entails a significant re-direction of the cartographic impulse to control the world. Specifically, it configures maps to be records of the space of the physical world, records which can be made complete and thorough. But the purpose of maps, even maps of territories, has always been to break down the human world—the world we live in—into portions that can be intellectually digested; all maps are devices through which humans have invested each portion with meaning as a discrete place or coherent space. As a recent novel set in an only slightly parallel universe reminds us, maps are devices through which humans organize their activities, act in the world, and make the world conform to humanity's needs:

Owing to a mistake in Wellington's maps of Spain, the city of Pamplona was not exactly where the British had supposed it to be. Wellington was deeply disappointed when, after the Army had marched twenty miles in one day, they did *not* reach Pamplona, which was discovered to be ten miles further north. After swift discussion of the problem it was found to be more convenient to have Mr Strange move the city, rather than change all the maps. (Clarke 2004, 361 note 7)

In other words, maps are not records of what each part of the world actually is; regardless of historical and cultural context, maps are careful imaginings of what people have wanted the world to be.

▲

I am much indebted to Jim Akerman and Bob Karrow for all their contributions to this project, and to the Field Museum staff for all their help with editing and illustrations. I must thank Catherine Delano Smith, Amy Derogatis, William Gartner, Mary Pedley, Paul Rogers, and Lydia Savage for making me think in new ways about this chapter after reading a first attempt, and also Kathryn Edney, Jeff Hylan, and the anonymous reviewers for their comments on the complete draft.

MAPPING PARTS OF THE WORLD

1 2 3 **4** 5 6 7

MAPPING AMERICAN HISTORY

Susan Schulten

Near the west stairway of the chamber of the U.S. Senate, located in the Capitol Building, in Washington, DC, hangs a portrait of Abraham Lincoln with his cabinet (fig. 87). Painted by Francis Bicknell Carpenter during the Civil War, it solemnly commemorates the first reading of the president's Emancipation Proclamation. The painting is familiar to many, yet within it lies a detail that has gone virtually unnoticed in scholarly discussions. In the lower right corner, a map leans against a table, bright white against the somber hues that surround it. The map documented the population density of slaves in the Southern states, and as it was one of Lincoln's favorites, Carpenter decided to include it in the painting. Yet while its presence in the picture is undeniable, its placement is slightly obscure: at the edge, nearly off the canvas. Similarly, while this map structured Lincoln's ability to *see* the South, it is difficult to pinpoint the influence it had over his understanding of the region, or the war. On a more general level, Carpenter's painting captures the complex relationship between maps and American history. It is not hard to see that maps document contemporary politics or reflect prevailing ideas, but ascertaining their influence over these

FIGURE 87.

Francis Bicknell Carpenter, *First Reading of the Emancipation Proclamation of President Lincoln* (1864).

ideas and politics is tricky. While historians know much about the creation of maps, we have much less evidence about their meaning to users. For instance, Carpenter's memoir tells us that Lincoln relied heavily on the slave map, but beyond that we know fairly little. In this respect the relationship between maps and history is reciprocal, but also fundamentally elusive.

For our purposes, it is useful to think of maps as brokers of information that both shaped and were shaped by history. They show us what people knew, what they thought they knew, what they hoped for, and sometimes what they feared. In each case, maps had the unique ability to translate a three-dimensional reality onto a two-dimensional plane. Before the modern era, mapmakers were limited in their access to information, but they also chose to stress certain features over others, and thus were telling stories about the territory they were depicting. Without any claim to being exhaustive, this chapter discusses a few maps that reflected and mediated historical change in North America. This does not limit us to maps of discovery and exploration, or those which established imperial or national boundaries. It also includes maps that both expressed and shaped people's imagination of the landscape, the region, and eventually the nation. The maps discussed below are an eclectic mix; some are deservedly well known, while others may help readers see American history in a new light. I have included a range of genres, such as maps of diplomacy and

SUSAN SCHULTEN

wartime strategy, maps of exploration, maps of the nation and its past, and maps that disrupt conventional perspectives.

A chapter covering "maps and American history" involves a serious intellectual problem. From our vantage point it is tempting to treat the British colonies as simply a prelude to an independent United States. In other words, we tend to assume the nation was inevitable. Yet this was just one of many possible outcomes, so we must reinvest in the contingency of history, a quality which—ironically—maps generally downplay through their graphic fixity and stability. Related to this is a problem of geographic definition: how do we define *America*? Does this term extend to all of North America, or simply that which became the United States? Again, there is no satisfactory answer, and certainly not one that scholars can agree upon. In fact, both of these problems are endemic to historical inquiry. The best we can do is to stress that America—whether it denoted a territory or a concept—was fluid.

The chapter proceeds chronologically, and readers will notice recurring themes over time. First, especially in the earliest years of North American settlement, the most consequential maps were those of discovery and exploration that uncovered the outline and interior of the continent. This effort to *know* the land extended from the first appearance of the territory on Juan de la Cosa's map of 1500 through the nineteenth century. But if exploration was scientific work, it was also a deeply political enterprise, and this points to our second theme: the degree to which maps were demonstrations of influence, ownership, and control. When Guillaume Delisle issued his masterful map of "La Louisiane," which claimed the wide axis of the Mississippi River for the French, English rivals immediately countered with maps signaling their own robust presence on the eastern seaboard. After eighteenth-century wars settled some of these competing claims, maps of the newly established United States became statements of the nation itself. Here the vision of cartographer John Melish in 1816 strikingly anticipates the continental expansion that came later. By 1850, the nation had nearly assumed its modern territorial form, but how the physical country translated into a more abstract nation depended on the power of cartography to *show* Americans their shared identity. This project was disrupted by sectionalism, secession, and war, which made cartography all the more critical for its ability to document the persistence of the Union.

The restoration of the Union settled the question of national identity, in fact if not necessarily in sentiment. In subsequent decades maps recreated, integrated, and consolidated the nation, which exemplifies our first two themes but also points to a third: the relationship of maps to land use. North American experiments with land organization and measurement included ranges and townships, parcels, long lots, and ranchos. These methods were shaped and extended by virtue of their appearance on maps. With far more ambition, the Corps of Topographical Engineers set out to map the interior and Far West, while Francis Amasa Walker created thematic maps based on data provided by the Office of the Census. Here cartography was not just an aid to governance; rather, the nation became impossible to imagine, understand, or govern *without* it. Maps gave disparate regions and provincial states the appearance of national coherence. After the Civil War, Rand McNally and other companies flooded the market with detailed, mass-produced maps of the railroads, and in the process created a cartographic style and view of the landscape that proved remarkably enduring.

Twentieth-century maps continued these trends of increasing cartographic precision, the identification of the continental outline of the United States with the nation, and the use of cartography to plan and shape local, regional, and national development. Yet at the same time, advancements in transportation and communication prompted some cartographers to depict distance in less absolute and more "imprecise" terms. For instance, as mapmakers assimilated the realities of aviation, they began to picture a world of heightened immediacy and mobility. This was particularly important in navigating the geopolitical dynamics brought by the nation's rise to international power. Many maps in the 1940s explained global warfare through nontraditional perspectives and orientations, and in a curious way they echo the relational ideas of space that appeared in Native American cartography encountered by explorers in the sixteenth century.

There are innumerable other ways that maps influenced and documented American history, yet these themes indicate the range and power of cartography to shape and reflect the most important trends in the nation's past. When examined together, these maps suggest uncertainty, which is precisely the quality that individual maps tend to obscure. The history that repeatedly emerges is not one of clarity but rather complexity and contingency.

CARTOGRAPHY IN THE ERA OF CONTACT,
DISCOVERY, AND EXPLORATION

Our story begins with the first European map to identify land in the Western Hemisphere, Juan de la Cosa's portolan chart of 1500 (fig. 88). Portolan charts were navigational aids that featured winds, shoals, and coastal detail but gave little detail of the interior of a land. In this age of aggressive overseas exploration, they were the vanguard of geographic knowledge, and thus it is unsurprising that the New World would first appear on a navigational chart rather than a traditional map. Cosa was a Basque pilot and cosmographer who traveled repeatedly to the New World from 1492 to 1504, perhaps most notably as the owner of the *Santa Maria*, Christopher Columbus's flagship. At the far left of the chart, Cosa used the figure of Saint Christopher to represent Columbus as the bearer of Christianity to the New World. Cosa's repeated voyages west prepared him to map the islands and surrounding lands that take up the left-

SUSAN SCHULTEN

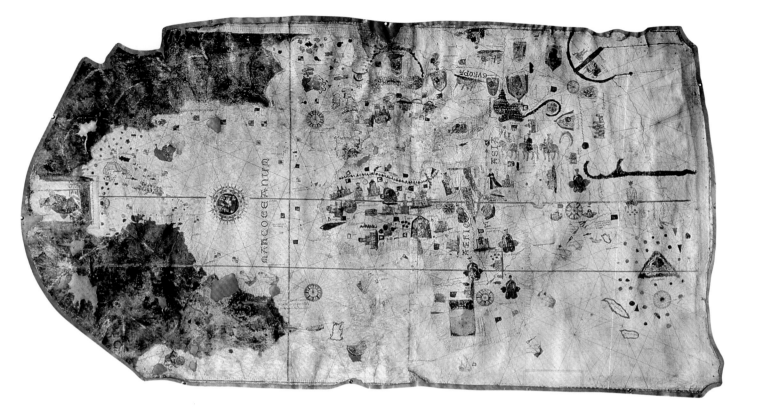

FIGURE 88.

Juan de la Cosa, world chart (ca. 1500).

hand portion of this chart. He dotted the Caribbean with flags from Aragon and
Castile, while to the north he placed English flags on the New England coast
noted the "Mar descubierta por los Ingleses" (sea discovered by the English), a
crucial notation that would later give evidence of John Cabot's disputed voyage of 1497.

Perhaps the most jarring aspect of the map is the size of the continents,
caused by Cosa's use of two separate scales for the Eastern and Western Hemispheres. Yet this should not diminish the achievement of this chart. It is unknown whether Cosa knew he was mapping the eastern portion of Asia or else
a separate continent altogether, which means that the first picture we have of
the New World was perhaps inadvertent. To add to the mystery of the map, for
many years its authenticity was disputed, in part because it was lost from the
time of its creation until the nineteenth century, which also meant it had little
to no influence on subsequent maps (Mollat Du Jourdin and La Roncière 1984,
212–13; Schwartz and Ehrenberg 1980, 18–20; Short 2001, 28, 32).

Perhaps unwittingly, Juan de la Cosa gave us the first cartographic glimpse
of a New World, but geographic knowledge about the Western Hemisphere
quickly advanced. Just a few years after Cosa's chart, some believed the new
lands to the west were not part of Asia, and one of the first to identify this
separation on a map was Martin Waldseemüller (fig. 89). His 1507 "Universalis

FIGURE 89.

Martin Waldseemüller, "Universalis cosmographia" (World Map) (1507).

cosmographia" (World Map) also used latitude and longitude in order to integrate the world into a single picture, which gives his map a substantially different appearance from that of Cosa. Ironically, Waldseemüller is most widely known not for his use of projection or cartographic skills, but for placing the name "America" on the Western Hemisphere in an effort to honor the voyages of Amerigo Vespucci. After a change of heart, he removed the name from his 1516 map, but by then it had begun to stick in the minds of cartographers, and the territory would never be widely known by another name. Thus, the continent bears the name of an explorer who was neither the first nor the most important European to sail to the New World. Waldseemüller also shaped the map of North America by separating the northern and southern continents; soon thereafter Gerard Mercator entered the names "N. America" and "S. America" on his 1538 world map. At the top of his map, Waldseemüller placed portraits of Ptolemy—the ancient geographer—and Vespucci, literally elevating "America" as a symbolic name for the place "discovered" on the map. The name was also used by Abraham Ortelius in his *Theatrum orbis terrarum*, which went through numerous editions and thereby codified the nomenclature of America (Buisseret 1990, 50–51; Harley 2001; Schwartz and Ehrenberg 1980, 28, 69–71).

By the seventeenth century, transatlantic exploration had been eclipsed by efforts to settle the North American continent, and these changing imperatives can be seen in contemporary cartography. One of the most fascinating

SUSAN SCHULTEN

examples of this new phase of mapping is John Smith's 1612 rendering of Virginia, printed to accompany his description of the failed Jamestown colony and widely reproduced and disseminated as one of the earliest and most influential maps of the Chesapeake Bay (fig. 90) (Harley 2001; Blansett 2005). At first glance, the map is disorienting to modern eyes, for north is at the right and west at the top, showing the perspective of Europeans coming from east to west, from the ocean to the interior. Furthermore, it is filled with imagery and information, though it retains an appealing artistic quality. Smith had dual goals for this map. On the one hand, he designed it to represent the terrain and its features, both human and natural. But he also used it to encourage and instruct future settlements, giving it a persuasive character. Ironically, the map projects power, certainty, and success—notice the monarchical seal—even as it depicts a colony that failed to survive. It also minimizes the cultural distance and differences between the colonies and England: Smith marks the landscape in an orderly manner, with a variety of trees, open land for farming, and a network of rivers indicating ease of transportation. In fact, the sort of landscape pictured is reminiscent of contemporary maps of English counties (fig. 170). The oversized figure of Sasquesahanoug, the hunter, strikes a classical pose familiar to the English viewer. In all, Smith's map depicts Virginia as an extension of England, neither threatening nor distant enough to discourage potential investment or settlement (Brod 1995, 91, 103–4).

Scholars have been intrigued by this map's apparent synthesis of European and Native American geographic knowledge. Smith marked the limits of his firsthand knowledge with small Maltese crosses, and noted that what lay beyond this region was known to him through Native American accounts: "to the crosses hath been discovered—what beyond is by relation." Smith also relied on Natives to map sites along the coast, and because the Powhatan did not view the English settlers as a threat—the latter were barely surviving—they shared their knowledge. Smith took this information, but frequently substituted English place-names for Indian ones, thereby shaping the landscape in the image of England (Waselkov 1998, 207, 213; Harley 2001, 172–82). We have just begun to discern the contributions of Native Americans to North American cartography, but by most recent accounts they were pivotal. Some scholars speculate that Native Americans depicted space in relational rather than absolute terms, which was certainly true among nomadic trans-Mississippi tribes that measured distance in terms of time.

Throughout the seventeenth century, the mapping of North America continued to be a deeply political as well as scientific operation. One animating force in the drive to map the continent was the rivalry between the French and English, which became an incentive to explore and map the interior rivers and valleys. The maps produced in this period were also instruments in these larger conflicts, and in this context maps became projections of power. Louis Jolliet, Jacques Marquette, and Robert LaSalle led the seventeenth-cen-

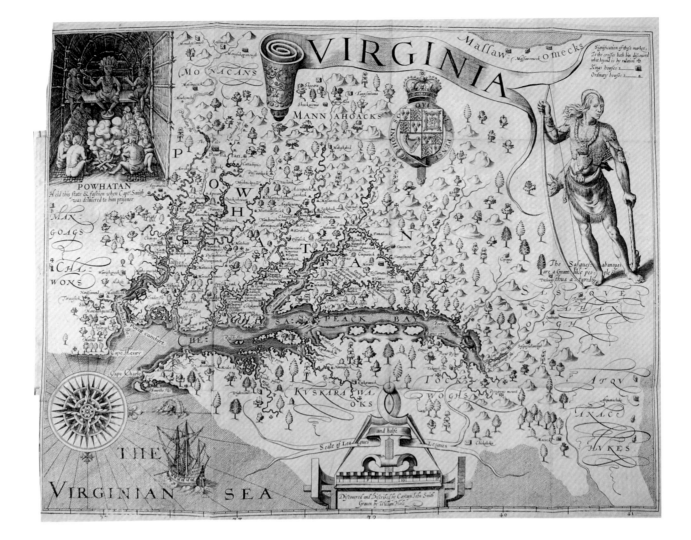

FIGURE 90.
John Smith, "Virginia" (1612).

tury French explorations through this area. In this context we can see how maps document knowledge of the terrain but also *advanced* territorial claims. For example, Samuel de Champlain (1567–1635) wore three hats—governor of New France, explorer of Nova Scotia and Quebec, and cartographer. Through each of these roles Champlain became critical to the imperial representation of North America. Indeed, it is impossible to separate these roles, for his maps, like those of John Smith, were designed to both document and market the territory (Short 2001, 54).

The most impressive maps of the first half of the eighteenth century were created by Guillaume Delisle, a cartographer for the French government. His unrivaled map of "La Louisiane" lays out an appealing view of the region that is neatly divided by the Mississippi and its tributaries (fig. 91). Delisle com-

SUSAN SCHULTEN

pleted the map in 1718, just as New Orleans was founded, and his depiction drew the eyes toward that city as the outlet for a vast drainage system. The map clarified the contours of the Mississippi River Delta and valley, traced the explorations of Hernando de Soto, and became the main source for later maps of the region. But Delisle's work was also an explicit statement of French power in the North American interior. The coloring of the map tells us all we need to know: by demarcating Spanish territory to the west and tiny British colonies to the east, Delisle claims the balance—a massive region united by its river system—for the French. This claim is reinforced by Delisle's naming of "La Louisiane" across the heart of the continent. The British colonies, which exclude everything south of Virginia, look precarious by comparison, trapped between the Atlantic Ocean and French claims to the south and west. Delisle even implied French ownership of the Carolinas, and brought part of the Spanish southwest into French Louisiana. In the aftermath of the Treaty

FIGURE 91.

Guillaume Delisle, "Carte de la Louisiane et du Cours du Mississipi" (Map of Louisiana and the Course of the Mississippi) (1718).

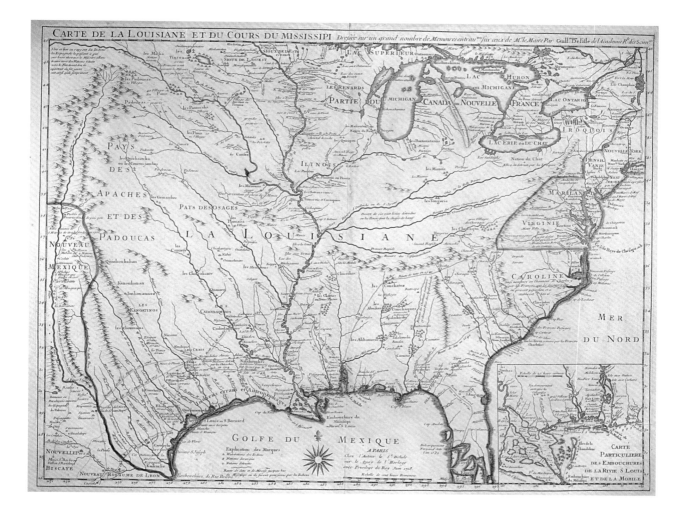

of Utrecht (1713), when imperial claims in North American borders remained uncertain, it is hard to imagine bolder use of cartography for political ends (Cumming 1998, 23–25).

Delisle's map was an opening salvo for the French, and the English responded by mapping the continent from their own perspective. First were the maps of Herman Moll, who had come to England as a young man from either Germany or Holland and who became one of the most influential mapmakers of the late seventeenth and early eighteenth centuries. Having spent most of his life in England, and as part of a literary circle that included Daniel Defoe and Jonathan Swift, Moll was disposed to seeing the world—and the emerging imperial contests—from a British perspective. In 1715 he created a map of the British dominions in North America which depicted the eastern seaboard as densely populated and powerfully situated. He relegated French claims along the Gulf of Mexico to a small inset map while the British colonies dominated the main cartographic stage. After Delisle published his map of "La Louisiane," Moll responded with a series of his own, the most important being his 1720 "This Map of North America According to ye Newest and most Exact observations" (fig. 92). Here the English dominion reaches from the Carolinas to Newfoundland, and even the ocean is named "Sea of the British Empire." By using a wider perspective, Moll reduced the French sphere to one of a series of claims, bounded by the English and the Spanish. And with a coat of arms and dedication to the English administrator of the colonies, his map projected British power onto the continent. Noticeably absent are any non-European settlements or influence, which implies the possibility—even the need—for settlement and development on the continent. Also note Moll's depiction of California as an island, an error Delisle had only recently corrected on his own maps (Cumming 1998, 23–24; Schwartz and Ehrenberg 1980, 140–42; Harley 2001, 137–39; Reinhartz 1997).

From the mid-eighteenth century to the Revolutionary War, maps of the North American continent proliferated. The intense rivalry between the French and English led the governor of New York to sponsor the first large-scale printed map of North America, drawn by Henry Popple and among the largest produced in that century (93 × 89 inches [236.2 × 226.1 cm]). But another map of North America drawn a few years later exerted even greater influence. John Mitchell's single—and singular—contribution to cartography was his 1755 map of the continent, drawn on a scale of 32 miles (51 km) to the inch (fig. 93). Not only was it the primary ("mother") map of the continent until the early nineteenth century, but it was used to arbitrate boundary disputes well into the twentieth. Reprinted twenty-one times in four languages by 1791, and reproduced many more times without credit, Mitchell's map circulated widely in both Europe and North America. Little wonder that it has been called one of the most important maps in American history (Martin 1972; Cumming 1998, 25–26).

SUSAN SCHULTEN

Mitchell had expertise in botany, zoology, climatology, and medicine, and brought this scientific sensibility to his cartography. He was intent on correcting inaccuracies prevalent on prior maps such as Popple's, and was particularly determined to identify longitude and latitude for both the coasts and the interior, which was a notable achievement. He took pains to identify topographic features, particularly in the Appalachian range, as well as the placement of creeks and settlements—all of which indicate that he had access to maps and geographic reports from the archives of the British Board of Trade (Martin 1972, 105). Yet Mitchell's goals were political as well as scientific, and included the defense of British claims against the French in the trans-Appalachian West. While Popple limited the western border of the British colonies on his earlier map, Mitchell consciously extended the reach of the Carolinas, Virginia, and Georgia across the Mississippi River beyond the edge of the map; indeed, they have no western limit. The English frontier settlements were emphasized also. Both techniques make Mitchell's map a geographical description as well as a geopolitical statement, made explicit two years later with his narrative defense of the British colonies.

The timing of Mitchell's map is crucial. Published after the onset of the French and British struggle for control over the eastern interior of the continent, it became both a reference work for the resolution of the French and

FIGURE 92.

Herman Moll, "Map of North America According to ye Newest and Most Exact Observations" (ca. 1720).

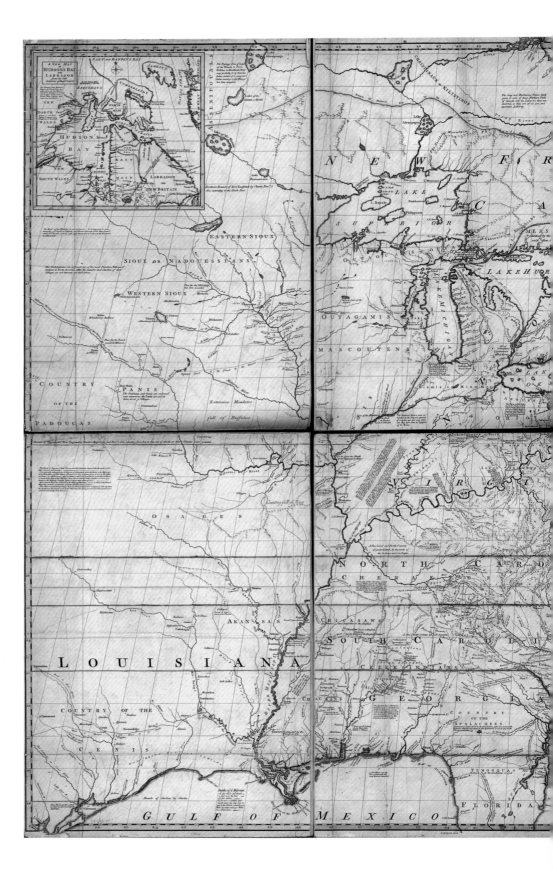

FIGURE 93.

John Mitchell, "A Map of the British and French Dominions in North America" (1755).

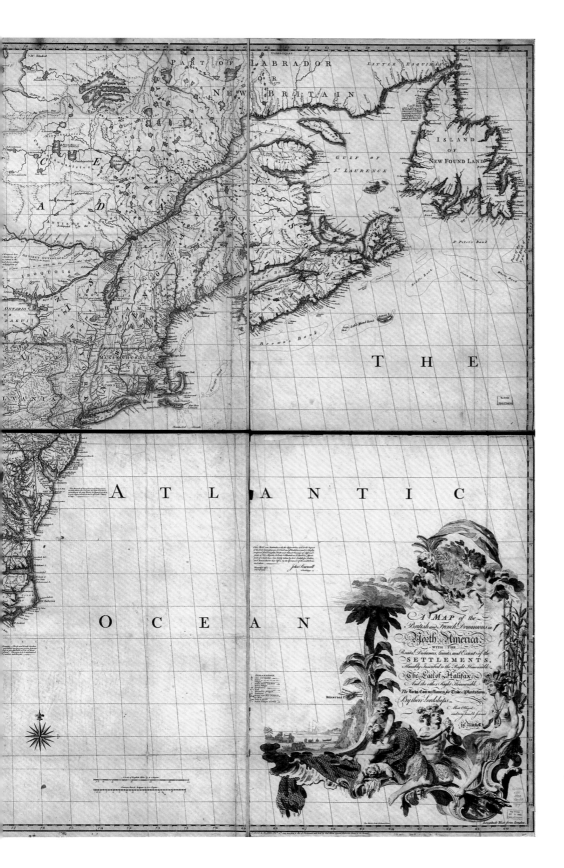

Indian War, and an argument for the superiority of British claims. The British prevailed in the war, and thus after 1763 the French ceded control over their domain west of the Mississippi River to the Spanish. Mitchell's map was also used in the Treaty of Paris in 1783 to negotiate boundaries for the newly established United States (Cumming 1998, 26). In addition, it has the unique advantage of being familiar to subsequent generations. It evokes the future of the country both in its coloring and style, but also because later generations could recognize the political form the territory would take. Mitchell's map unintentionally anticipates the shape of the United States decades prior to its independence from the British and nearly a full century before the nation's transcontinental expansion. And because it embodies the reciprocal relationship between historical forces and geographical representation, it raises the question of whether maps of the colonies catalyzed the movement for independence. Brian Harley has argued that the maps of the colonies had the power to inscribe English power. For instance, a 1755 map of the Connecticut colony recreated the geographic landscape of England and linked the colony to the mother country through place-names, historical events, and the erasure of Natives (Harley 2001, 144–45). In this case the maps legitimated and extended English power. Yet maps also may have generated unity in the colonies by making clear their physical proximity and interdependence while clarifying their distance from the Crown. Here we can see just how difficult it is to assess the influence of maps on historical change.

The American Revolution is richly documented in maps, one of the most exciting of which was recently acquired by the Library of Congress. In 2000, the library purchased a set of six maps by Michel Capitaine du Chesnoy. Capitaine came to America in 1777 as an aide to the marquis de Lafayette, who was determined to join the colonists in their fight for independence. Capitaine's training as a topographic engineer distinguished him from American mapmakers, and that expertise was quickly apparent in the maps he produced of battles in New York and New Jersey. After aiding an expedition on the Susquehanna River, Capitaine rejoined Lafayette in May of 1778, and the two traveled to Paris together the following year, in part to make a case for expanded French aid to the colonists, a case which Capitaine's maps enhanced. When Lafayette and Capitaine returned in April 1780, they brought to the colonists news of the impending arrival of troops under the comte de Rochambeau.

By early 1781, the primary theater of war had moved to the southern colonies, and in April Washington sent Lafayette with his troops to Virginia in order to halt the advance of the British through North Carolina. Washington recognized that the Americans could not match the size of the British army, so he directed Lafayette to engage in a series of skirmishes and retreats from April to October. These maneuvers between Lafayette and British general Charles Cornwallis prior to the battle of Yorktown are masterfully recorded by Capitaine on his map of Virginia (fig. 94). From April to August Lafayette

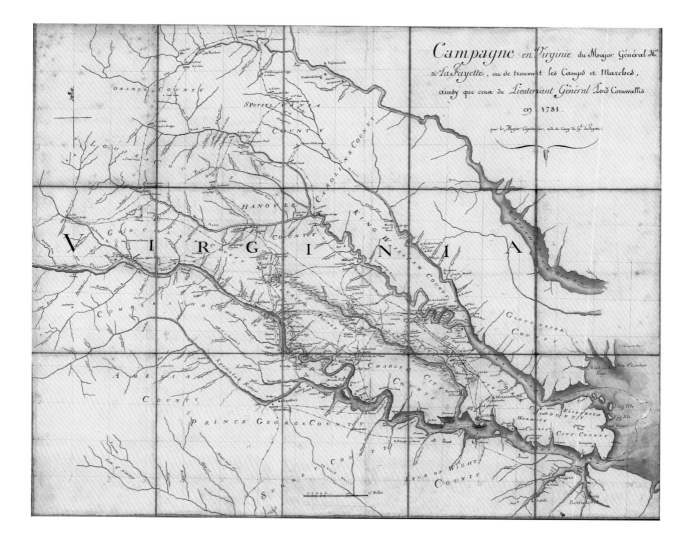

maneuvered around Cornwallis until the latter settled his troops in a defensive fortification in York and Gloucester. Lafayette then blocked Cornwallis by land until the French navy arrived with troops commanded by Washington and Rochambeau. In September the French navy defeated the British, which sealed Cornwallis's fate by preventing both British reinforcements and evacuation. In October, the combined forces of Rochambeau, Washington, and Lafayette defeated Cornwallis in the final major battle of the war.

While many maps document the battle between French and British fleets and the ensuing ground confrontation at Yorktown, few include the extended campaign precipitating that siege. Capitaine's detailed notes of each army's maneuvers allowed him to map these with precision. The map indicates how wildly unpredictable these skirmishes were: Lafayette showed great skill at engaging and distracting Cornwallis's superior forces, but he also enjoyed considerable luck and good timing at pinning the British by land while reinforce-

FIGURE 94.

Michele Capitaine du Chesnoy, "Campagne en Virginie du Major Général M'is de LaFayette" (Campaign of Major General Marquis de LaFayette in Virginia) (1781).

ments arrived by sea. Even more interesting is what the map reveals about the larger war. Patricia Van Ee argues that Capitaine probably drew the map in the winter of 1781–82 to help Lafayette convince the French crown to extend support for the American cause. If so, it reminds us that the battle of Yorktown— though now enshrined in textbooks and civic memory as the decisive victory in the war—was not understood as such for some time. Indeed, the colonists anticipated a battle over New York, and instead watched the British move to the Caribbean. Capitaine's map represents the contingency of history: what he saw as a hard-won campaign would only later be understood as a turning point in history (Cohen 1998; Van Ee 2006; Pritchard and Taliaferro 2002).

CARTOGRAPHY AND THE EMERGING NATION

Map production increased tremendously in the nineteenth century, primarily due to three trends. First, advancements in printing and transportation technology increased the output and lowered the cost of cartography. Second, the achievement of independence fed the demand for maps of the new United States. Finally, American exploration and acquisition of western territories stimulated the desire for more precise maps of the trans-Mississippi West. This was first and foremost a task of uncovering the topography of the interior, the major features of which were understood by the late 1840s. But after this had been achieved, geographers and cartographers designed emigrant guides, land and railroad surveys, and finally thematic and statistical maps of the nation as a whole. This territorial expansion, which in part fueled the sectional strife and subsequent war over slavery, explains the centrality of the nineteenth century to this chapter.

Even before Noah Webster articulated the need for a uniquely American language, Jedediah Morse asked for a "geography of our own country." For Morse, Webster, and many others, the Revolution marked not just a political but also an intellectual break from the past. Morse wanted *American* knowledge that grew from *American* minds, and this desire influenced the nation's emerging cartographic industry. Centered in Philadelphia, the fledgling community of map publishers and printers literally "represented the nation and the world to literate society" (Short 2001, 145). Many of their maps were designed to consolidate the new nation and articulate its identity, perhaps the single best example being John Melish's "Map of the United States with the contiguous British and Spanish Possessions" (fig. 95).

Melish was born in Scotland, and his anti-English sentiment translated easily into sympathy for the American cause. He traveled extensively through the young Republic before settling permanently in Philadelphia, where he made critical contributions to the city's map-printing industry. Melish's travels led him to publish a number of descriptive accounts of the country, and in

SUSAN SCHULTEN

the process he developed a lifelong passion for cartography. After publishing a number of atlases, Melish noticed an unmet demand for a large, national map of the United States, which he began to create in 1815, just as war with Britain was ending. He sent a prospectus of his map to publishers, and promised it would include the latest geographic information from well-known cartographers such as Aaron Arrowsmith as well as contemporary explorers such as Zebulon Pike (Ristow 1985, chapter 12).

The final map was drawn on six sheets with a scale of 1 inch (2.54 cm) to about 60 miles (96 km), and—measuring 36 × 57 inches (91.4 × 144.8 cm)—was designed to be publicly displayed. This unique, large national map went through seven editions in 1816 (including four more that were probably released the following year), and another five in 1818. Melish proudly insisted that only one hundred copies would be printed for each "state" of the map, since the rapid growth of geographic knowledge necessitated frequent updates. Five presidents are known to have owned a copy of Melish's map—John Adams, Thomas Jefferson, James Madison, James Monroe, and John Quincy Adams—and one of the 1818 editions was used to determine the border between the United States and Spain (later Mexico), codified by the Adams-Onis Treaty of 1819 (Ristow 1972, 162–82; Ristow 1985, 185–90).

Originally, Melish designed the map to end at the Rockies, then accepted as the country's western boundary despite its ambiguous topography. But in the process of creation, his ambition and imagination began to take over, and he realized the "propriety" of extending the map to the Pacific. This extension accounts for some of the map's enduring appeal: it anticipates western expansion, and therefore suggests a nation that later generations recognize as their own. Long before John O'Sullivan coined the term *manifest destiny*, Melish expressed his vision of national growth by describing the land west of the Rockies, which was then still foreign territory:

Part of this territory unquestionably belongs to the United States. To present a picture of it was desirable in every point of view. The map so constructed, shows at a glance the whole extent of the United States territory from sea to sea; and in tracing the probable expansion of the human race from east to west, the mind finds an agreeable resting place on its western limits. The view is complete, and leaves nothing to be wished for. It also adds to the beauty and symmetry of the map; which will, it is confidently believed, be found one of the most useful and ornamental works ever executed in this country. (Quoted in Short 2001, 145)

To look at Melish's map is to see the country as a transcontinental entity for the first time, decades before the Treaty of Guadalupe Hidalgo and the acquisition of Oregon. Perhaps this expansive view of the United States as a continental body even stimulated the territorial expectations of contemporary viewers. Melish celebrated his adopted country by touting the nation's

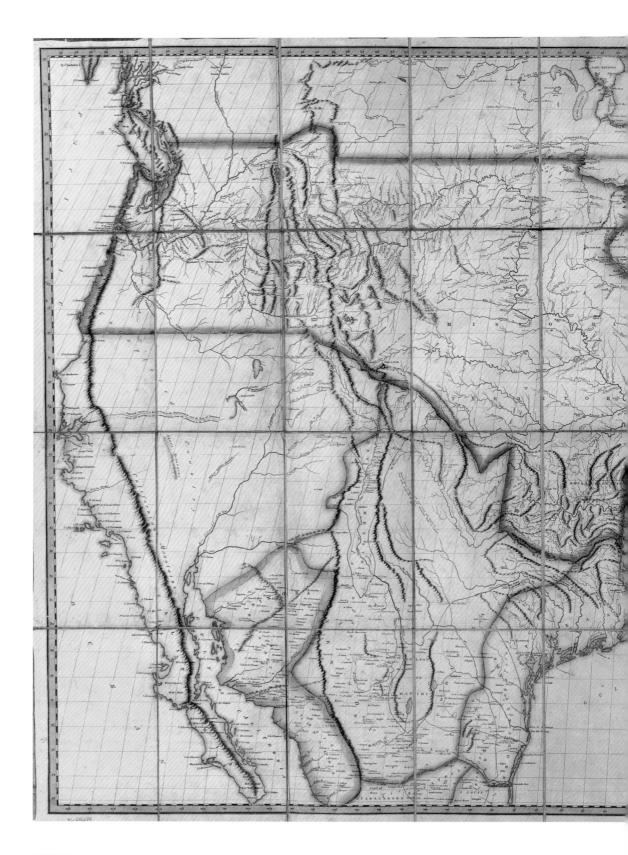

FIGURE 95.

John Melish, "Map of the United States with the Contiguous British and Spanish Possessions" (1819).

MAP
of the
United States
with the contiguous
BRITISH & SPANISH POSSESSIONS
Compiled from the latest & best Authorities
BY
John Melish

WEST INDIES.

economic potential and encouraging emigration. On his maps, he marked post offices as a way of endorsing federal power, and adopted the short-lived practice of marking longitude from the District of Columbia as well as London as a way to assert the country's cultural and scientific independence. Maps such as his were symbolic and synthetic, and as a result were often vague on details, particularly when it came to the minimally understood region west of the Mississippi. But in Melish we have an excellent example of cartography used for nationalistic ends (Short 2001, 136).

Melish published his map of the United States just as events in the West transformed the cartography of the country. At the time of the Louisiana Purchase in 1803, the geography of the West remained a mystery. As Carl Wheat has written, Americans knew next to nothing about the land they acquired, except that it was a

meagerly mapped wilderness, drained by a multitude of rivers most of whose sources remained unexplored, inhabited by numerous and warlike tribes of Indians, and vaguely bounded on the north by lands where British fur traders roamed, and on the west in part by possessions of Spain and in part by an undefined region later to become known as "the Oregon Country." (Wheat 1957–63, 2:1)

The West was known primarily through reports of Native Americans, traders, and explorers. This rough state of geographic knowledge is evident in Aaron Arrowsmith and Samuel Lewis's *New and Elegant General Atlas* for 1804. In the map of western North America, there is but one western mountain range—the Stony Mountains (later termed the Rocky Mountains)—and further west little information exists (fig. 96). The Missouri River is drawn far south of its course, while the courses of the southwestern rivers remain unclear. Stepping back, one notices that the entire continent is foreshortened, which might indicate the persistent hope that a waterway existed from the headwaters of the Missouri to the Pacific (Wheat 1957–63, 2:xii).

In this context, the expedition of Meriwether Lewis and William Clark was both a spectacular success and a limited failure. After more than two years of extraordinary endurance and remarkable luck, Lewis and Clark failed to discover a transcontinental waterway, but in the process they substantially advanced the state of western cartography. When their journals were finally published in 1814, they included a map adapted from Clark's original drawings. Clark worked on the map from the time of the expedition's return in 1806 until late in 1810. The next year, Clark's map was reduced to a smaller format (fig. 97) to guide the engraver, Samuel Lewis, whose printed rendition appeared in the journals of the expedition in 1814.

This map permanently altered efforts to understand the West, and is one of the nineteenth century's most important cartographic achievements. It clari-

SUSAN SCHULTEN

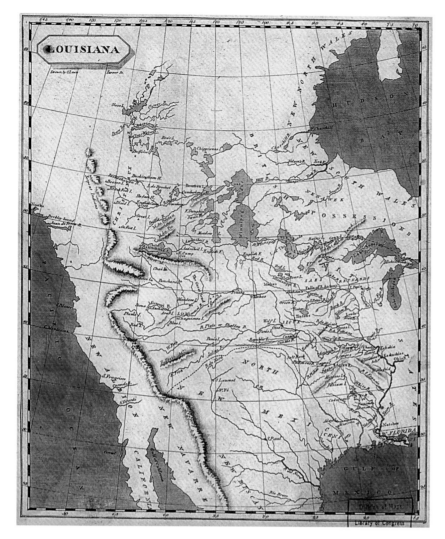

FIGURE 96.
Aaron Arrowsmith and Samuel Lewis, "Louisiana"
(1804).

fied and elaborated the Missouri and Columbia river systems, and raised the latitude of the Missouri's Great Bend well beyond that drawn by Arrowsmith and Lewis. Where earlier cartographers identified a single mountain range in the West, Lewis and Clark complicated the picture of the northern Rockies. As a result, the continent widened. Simply compare the 1804 to the 1811 map of the West to appreciate the contribution Lewis and Clark made to geographic knowledge. The latter remained the most detailed and accurate picture of the Far West from its publication in 1814 to the 1840s (Wheat 1957–63, 2: chaps. 13–14 and p. 57; Schwartz and Ehrenberg 1980, 227–28).

Lewis and Clark were followed by other important western explorers. Yet even after this rush of exploration, some myths remained, one of the most persistent being that the headwaters of the great western rivers were located in an area that could be covered in a single day's journey on horseback. Varia-

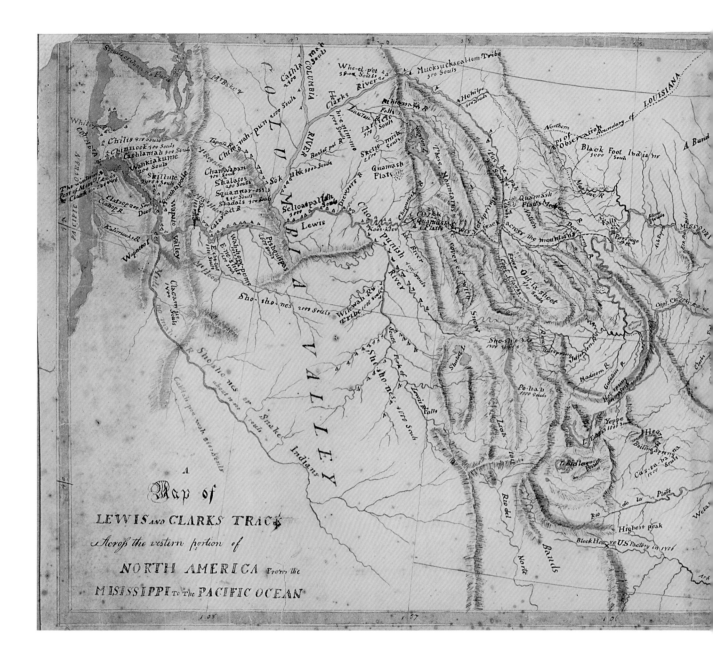

A Map of LEWIS AND CLARKS TRACK Across the western portion of NORTH AMERICA From the MISSISSIPPI To The PACIFIC OCEAN

tions on this "height of land" concept had grown for centuries in the minds of trappers, explorers, and cartographers, and was closely linked to the belief in a continental waterway. Jedediah Morse's *American Gazetteer* showed a link between the Missouri and the Rio Grande, based on an assumption that these rivers originated in the same region of the Rocky Mountains. Similarly, Zebulon Pike's 1810 map of the Southwest drew the headwaters of the Platte, the Missouri, and the Yellowstone in close proximity. This idealized core drainage area would persist in the minds of many cartographers until it was definitely disproved in the 1840s.

SUSAN SCHULTEN

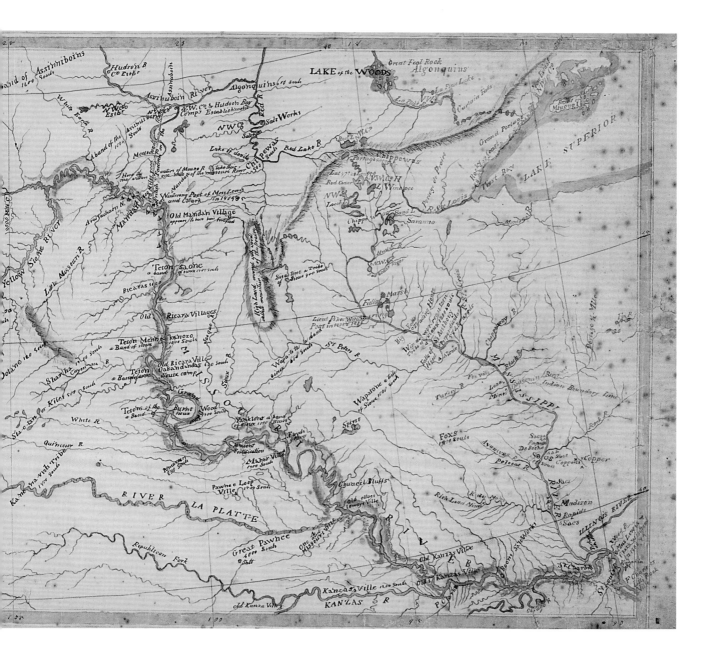

Part of the promise of the Louisiana Territory—at least for Jefferson—was its potential as a permanent settlement for Native Americans. When the expeditions of Pike and Stephen Long suggested that the land was unfit for cultivation in the 1810s and 1820s, it seemed even more appropriate to policy makers to turn the "Great American Desert" into Native American territory. While this idea conflated nomadic tribes (Comanche, western Sioux, Arapaho, and Kiowa) with agricultural ones (Pueblo to the south, and Osage, Kansa, Oto, Arikara, and Mandan to the east and north), the idea that all could be relocated to the trans-Mississippi West appealed to many Anglo Americans. In 1830 the federal government turned this geographic concept into policy by designating the plains as Indian Territory and outlining the possible contours of settlement. Along with its 1834 report came a "Map of the Western Territory" by the Topographic Bureau, which attempted to fulfill Jefferson's vision of an east-west axis of separation between Anglo Americans and Native Americans.

But the federal government had other reasons for attending to the western cartography, manifest in the founding of the Army Corps of Topographical Engineers in 1838. Simultaneously, exploration began to shift from the task of reconnaissance—such as locating landmarks and calculating distances—to asking new questions about the region. What were the resources? What were the western limits of the United States, and how would settlers travel overland? Americans were not even a step behind, and repeatedly pushed beyond the nation's borders: in the 1830s white settlers snapped up offers of cheap land in the foreign territory of Tejas; in the 1840s families migrated overland to Oregon Country, jointly administered by the United States and Great Britain; and in that same decade Mormons exiled from Missouri took refuge in the Mexican territory that would become Utah (Goetzmann 1959, 232).

Facilitating this early migration was John Frémont, a romantic hero to nineteenth-century Americans for his bold self-fashioning, his assertion of federal power in the West, and his sheer talent as an explorer. A member of the Corps of Topographical Engineers, Frémont was sponsored by his father-in-law, the powerful Missouri senator Thomas Hart Benton, perhaps the most vocal proponent of westward expansion. Benton had for years advocated a national road from Saint Louis to Santa Fe, and tirelessly promoted western exploration and surveying to this end. His support of Frémont gave the latter tremendous latitude and publicity. Upon returning from his first expedition to the Rockies in 1842, Frémont issued a report that electrified the public and secured his reputation as "the Great Pathfinder."

The Corps subsequently directed him to push beyond the Rocky Mountains to the Pacific so that the nation might understand the geographic relationship between the two. Frémont left Kansas in May 1843, passed through the Rockies in present-day Wyoming, and reached Fort Vancouver in November. He then proceeded south through Oregon and what is now Utah, crossed the Sierras into California, and returned east through what is now Nevada, Utah, Wyoming, and Colorado, then back to Kansas. His report of this expedition dramatically improved western cartography. Indeed, Brigham Young based his decision to take the Mormons to Utah in 1846 on Frémont's report. The power of the report rested in part on the accompanying map drawn by Charles Preuss, a surveyor for the Prussian government who immigrated to the United States in 1834 and worked for the U.S. Coast Survey before accompanying Frémont on his first, second, and fourth expeditions. The most noticeable feature of the map is that most of the page is left blank, for Preuss detailed only those areas he actually covered in order to heighten its credibility. With this "traverse"

SUSAN SCHULTEN

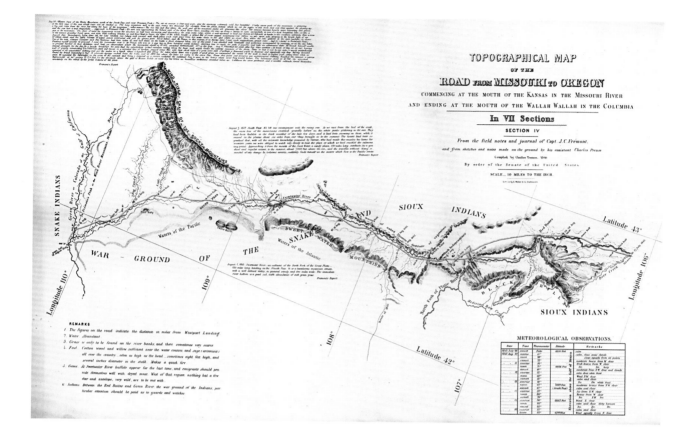

technique, the Kansas, Missouri, and Platte rivers are carefully mapped, as is the Salt Lake (Wheat 1957–63, 2:197–99).

This cartographic collaboration between Frémont and Preuss was critical in shaping perceptions of the West prior to the California gold rush, most importantly through their seven-sheet traverse map for westward migration (fig. 98). "Topographical Map of the Road from Missouri to Oregon" was drawn on the unprecedented scale of 10 miles (16 km) to the inch, which explains its utility and popularity with emigrants. Each sheet covered about 250 miles (400 km), and was oriented so as to make that portion of the route its major axis. The maps included information specifically for emigrants, such as weather patterns, vegetation and soils, animal life, water sources, the presence of Native Americans, and the lay of the land. Understood together, Frémont's achievements were impressive: he identified the Continental Divide, clarified the Great Basin, and documented the existence of two core drainage areas for the western rivers rather than one, thereby disproving the long-standing "height of land" theory (Ehrenberg 2005, 194–99).

The advances brought by Frémont and other explorers resulted in one of the first detailed maps of the entire trans-Mississippi West, Lieutenant Gou-

FIGURE 98.

Charles Preuss, "Topographical Map of the Road from Missouri to Oregon," section IV (1846).

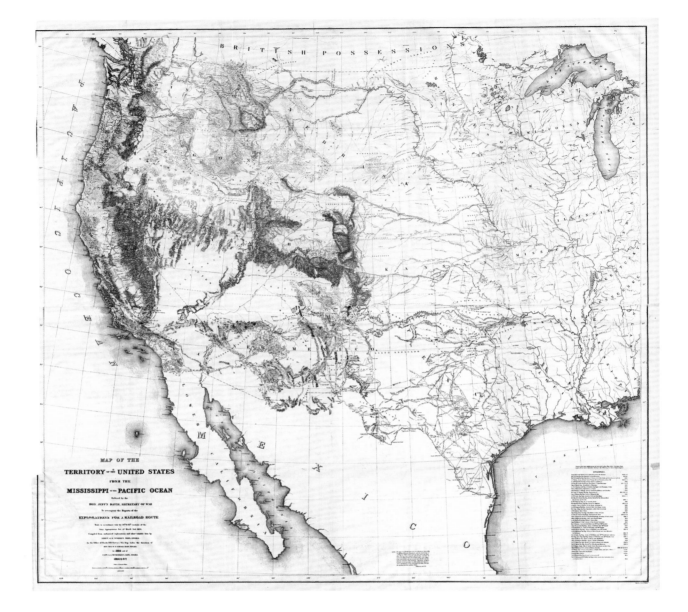

MAP OF THE
TERRITORY of the UNITED STATES
FROM THE
MISSISSIPPI to the PACIFIC OCEAN

FIGURE 99.

Gouverneur K. Warren, "Map of the Territory of the
United States from the Mississippi to the Pacific
Ocean" (1857).

verneur Kemble Warren's 1857 "Map of the Territory of the United States from
the Mississippi to the Pacific Ocean" (fig. 99). Warren worked for the Corps of
Topographical Engineers in the early 1850s, and was asked in 1854 to compile a
map of the land west of the Mississippi River as a step toward building a transcontinental railroad route. His task was to synthesize the disparate geographic
information present in existing regional maps into a comprehensive "view" of
the entire area, which called for reconciling conflicting cartographic accounts.
At the same time, he managed to standardize longitude—calculated variously
and erroneously on prior maps—which made this one of the most important

SUSAN SCHULTEN

foundation maps of the nineteenth-century West. Warren's map brought *precision* to the West: it carefully represented relief and drainage features, and left spaces blank where definitive information was lacking. In the process, he advanced railroad and commercial development, and anticipated an era of federally sponsored surveys in the 1860s and 1870s. Yet the scientific appearance of the map belies the fact that this was an essentially unstable region where tensions persisted between Americans and Native Americans, and the control of the federal government was limited. And though Warren's task was scientific, the map was submitted in 1857 to the Office of Pacific Railroad Surveys, whose aim was to propose a route that would—depending on its placement—privilege North or South in an era of strident sectionalism. Despite his achievement in creating the most comprehensive and important map of the West before the Civil War, Warren is chiefly recognized today for his topographic expertise, which proved critical to the Union victory at Gettysburg (Van Ee 2002).

These maps of the West reveal little of the division then gripping the nation. In fact, Warren served in the Union army, while Jefferson Davis—the secretary of war who authorized the work—would become president of the Confederacy. After the American victory over Mexico, the nationalist strain of manifest destiny became even stronger. Yet popular maps of the nation—designed in the tradition of Melish—were by the 1850s somewhat at odds with themselves. On the one hand they recognized the growing concentration of slave labor in the South, yet they were also designed to celebrate the country's epic growth westward and its implications for national identity. One contemporary national map reveals this conflicted state of affairs, its title itself suggestive: William C. Reynolds's 1856 "Political Map of the United States, designed to exhibit the comparative area of the free and slave states and the territory open to slavery or freedom by the repeal of the Missouri Compromise" (fig. 100). In all likelihood, this map was prompted by the Kansas-Nebraska Act, Senator Stephen Douglas's attempt to dampen the increasing sectional tensions over slavery in the West. Douglas used the act to introduce the concept of popular sovereignty to the territories carved out of the Louisiana Purchase, which by this time was no longer considered permanent Indian territory. Popular sovereignty proposed that the acceptance of slavery in a given territory be decided by the inhabitants or representatives of that territory. It was a politically disastrous proposal that Douglas would ultimately regret, for it alienated those who felt it gave too much power to proslavery forces while angering Southerners who considered it insufficient to protect and expand slavery in the West. Publisher William Reynolds made no effort to conceal his own position: "By the Democratic (?) legislation of 1854, in repealing the Missouri Compromise, the institution of Slavery *may be carried* into ALL of the Territories—the area of which is greater than that of all the States combined."

The map incorporated statistics from the census of 1850, including a table of revenues and expenditures favorably comparing free to slave states. In com-

FIGURE 100.

William C. Reynolds, "Reynolds's Political Map of the United States" (1856).

paring the sections, Reynolds noted: "Of the 6,222,418 white inhabitants of the South, only 347,525 are owners of slaves. And yet this faction controls every branch of the Federal Government, and wields its influence for the increase and perpetuation of Slavery." Reynolds used an array of statistics to document the disproportionate power of slaveholders and the drain they imposed on the national economy. Geographic area, racial breakdown, valuation of property, acres of improved land, miles of railroad, circulation of newspapers, and number of public libraries—nothing was overlooked in his drive to make the map a tool of sectionalism.

SUSAN SCHULTEN

All of these statistics were derived from the federal census, which since 1790 had given Americans an opportunity to see themselves in numeric terms. Originally, the census was a means of apportioning congressional representation, and thus was limited to calculating the domestic population. But by 1860 it had expanded to include inquiries about ethnicity, domestic wealth, economic structure, and manufacturing, all of which were of value to the Union war effort. Immediately following the Fort Sumter crisis, Census Superintendent Joseph Kennedy funneled these population figures directly to Lincoln while continuing to compile census tables for publication and maps for the military. This project included the use of postal maps of Southern counties overlaid with population figures for whites, free blacks, and slaves, along with data on livestock and crop production. This kind of cartography enabled Union troops to adapt and move quickly through the Confederacy (Anderson 1980, 63–64).

One of the first attempts to translate population data onto a map came out of the 1860 census, and illustrates the development of thematic cartography. Thematic maps are the subject of the next chapter; it is enough to say here that among them was a class of maps emerging in the nineteenth century that attempted to quantify—and therefore explain—the incidence of crime, illiteracy, disease, and other social statistics. By mapping a certain category of information through shading and symbols, thematic cartography "spoke to the eyes" and rapidly gained popularity in western Europe. By the 1850s, statistical maps began to proliferate in the United States, and bore a direct relationship to the census: thematic cartography depended on a significant body of data, and at the same time was uniquely able to organize and display this wide array of information (Anderson 1980, 68; Robinson 1982). Edwin Hergesheimer's map of Southern slavery was printed in September of 1861 and sold to raise money for sick and wounded Union soldiers (fig. 101). It identified the percentage of the population enslaved in each county, and the total number of slaves—four million, up from 700,000 in 1790—was a figure that could not have gone unnoticed by Americans living through such violent upheaval. By using this relatively new "choropleth" technique of shading, Hergesheimer showed Americans their country through the lens of slavery.

The "slave map" was of particular interest to Lincoln, as illustrated by Carpenter's *First Reading of the Emancipation Proclamation of President Lincoln* (fig. 87). The artist spent six months living at the White House in order to complete this work, and in that time repeatedly observed Lincoln studying the map. To master the detail on the map for his painting, Carpenter surreptitiously borrowed it; and when the president visited the artist in his White House studio a few days later he remarked, "*You* have appropriated my map, have you? I have been looking all around for it." According to Carpenter, Lincoln was once again instantly absorbed by the map and used it to trace the recent progress of Union troops through Virginia. It gave Lincoln happy news, for the areas conquered by the Union just that week were densely populated with slaves. Thus

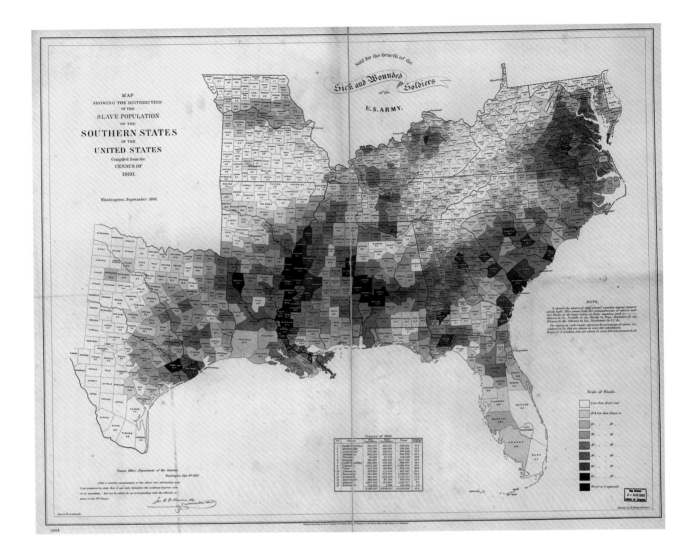

Hergesheimer's map appears in the corner of Carpenter's painting, a detail as meticulously chosen as the artist's arrangement of Lincoln's cabinet: those sympathetic to emancipation appear on the president's right, while the more conservative members are placed at his left. The map also appealed to Carpenter for its elegant organization of information. By just a glance, one could see the proportion of blacks to whites in the Southern states, which made it impossible to deny that slavery was at the heart of the rebellion.

CARTOGRAPHY IN THE MODERN AGE OF INFORMATION

The war allowed the Union to settle on a transcontinental railroad route, a decision that initiated a sustained period of rail construction that profoundly

SUSAN SCHULTEN

influenced the country, for the railroads have been called not only the most important force in industrial America but also its greatest result. Moreover, they represent the fluidity between public and private interests in that sector. The federal government granted enormous tracts of lands to the railroad corporations as an incentive to construct lines linking the Far West and Great Plains with the eastern United States. In return the railroads became as influential as the federal government over the lives of its citizens. Industrialization also bred a desire for maps that marked railroad routes and town and county divisions, and maps that contained political and economic information. This demand was met by new, commercial cartographic companies in the Midwest, most notably Chicago-based Rand McNally, which began as a ticket-printing company and only later tapped the growing domestic market for maps.

These midwestern companies used a newer, less-expensive printing technique known as wax engraving, which—unlike copperplate engraving and lithography—facilitated changes to the map as well as the inclusion of more place-names. This technique suited the public's interest in documenting the reach—both potential and real—of innumerable railroads, and allowed every possible town and station to be squeezed onto the map. As a result, maps were increasingly marketed not for their aesthetic beauty, but for the sheer amount of information they contained. Figure 102 is typical of this style, with extensive details of towns and railroads at the expense of topographic or relational information. It was precisely this style of map, one where geography meant primarily the identification of places, that many cartographers would challenge in the twentieth century (Woodward 1977).

An emphasis on maps as repositories of data extended to the federal government as well. Within a decade of Hergesheimer's map of slavery, the government brought the power of cartography to bear on the census in far more sophisticated ways. Consider the 1874 *Statistical Atlas of the United States*, the brainchild of Francis Amasa Walker. After serving in the Union army, Walker became the chief of the Bureau of Statistics in 1869 and Superintendent of the Census in 1870. As a professor of history and political economy, he had long been interested in a formalized and scientific approach to social reform, and the *Statistical Atlas* reflected this interest. It included fifty-four maps and graphs of the nation's physical geography, demography, and economy, along with extended discussions of national life. The *Statistical Atlas* translated these census figures into accessible pictures of the nation (Walker 1874).

The first ten brightly colored maps introduce viewers to the country as a geologic and physiographic entity (fig. 103). These appealing maps were among the first to profile the country's geology with color. These physical depictions were for many still unusual, accustomed as they were to seeing the country in political terms rather than according to statistics on rainfall, elevation, and geological divisions. Notice that just a few decades had passed since the Treaty of Guadalupe Hidalgo incorporated the Far West into the United States, yet

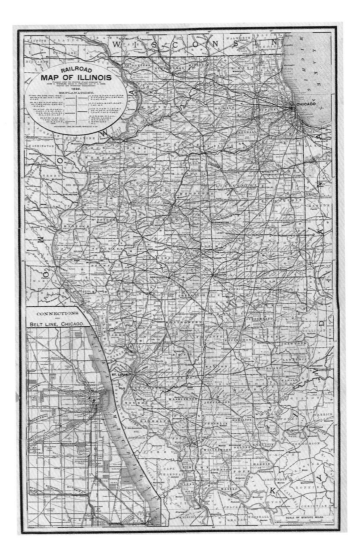

FIGURE 102.
"Railroad Map of Illinois" (1890).

this map suggests a nation of seamless unity. Yet while the *Statistical Atlas of the United States* was formally a national project, because it was a function of the census the maps and charts focused primarily on the eastern half of the country, where the vast majority of the population was still concentrated. Walker arranged the data using graphs and maps of race, ethnicity, economic statistics, vital statistics, gender, age, sex, and birth and mortality rates. For instance, the map in figure 104 uses shading to translate complex economic information into a snapshot of wealth distribution. Many of these thematic maps and graphs, like their forebears, focused on what Walker considered sources of social disorder, such as disease, immigration, blindness, and insanity. In part these were a function of his reformist sensibility and hope of understanding and controlling the effects of rapid modernization. For example, the maps

SUSAN SCHULTEN

FIGURE 103.

C. H. Hitchcock and W. P. Blake, "Geological Map of the United States" (1874).

of the Pacific Coast broke down the population along racial and ethnic lines, which communicated the tremendous diversity in the region.

One of the *Atlas*'s features is a map of population distribution, which views the nation as a whole but also strongly differentiates east from west (fig. 105). By 1900, the western half of the United States still contained only 4 million of the nation's 76 million inhabitants. So while the geographic expansion of the early nineteenth century was considerable, Walker's population map of 1874 reminds us of the disparity in population that persisted well into the twentieth century (note that many of the large areas of shading in the interior West were actually Indian reservations) (Meinig 1993, 32–35). One of Walker's most challenging tasks was to represent graphically the rapid growth of the nation over time. Chronological maps of population density did this well, and Walker used this technique to map other racial, national, and ethnic subgroups. The *Atlas* also illustrated the nation's shifting center of gravity by identifying on a map the geographic center of population in each of the prior nine censuses (fig. 106). This preoccupation with westward growth characterized the *Atlas* as

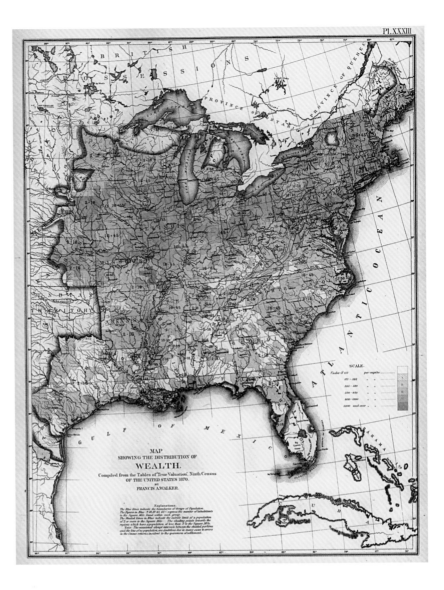

FIGURE 104.
Francis A. Walker, "Map Showing the Distribution
of Wealth" (1874).

well as the census figures it was based on. In 1890 the Census Office announced
the end of an unbroken "frontier" line in the West. The historian Frederick
Jackson Turner interpreted this moment as the end of the first period of American history, which suggests how much his interpretation was grounded in a
cartographic concept.

The *Statistical Atlas* was unprecedented: it conceptualized the nation as a
single entity, an aggregate of data that could be studied, organized, and potentially even reformed. That the federal government sponsored such an undertaking anticipates, among other things, the rising place of social statistics and
the role of experts in American life. And in terms of cultural importance, the
Atlas ranks alongside the map of John Melish as a statement of national unity.

SUSAN SCHULTEN

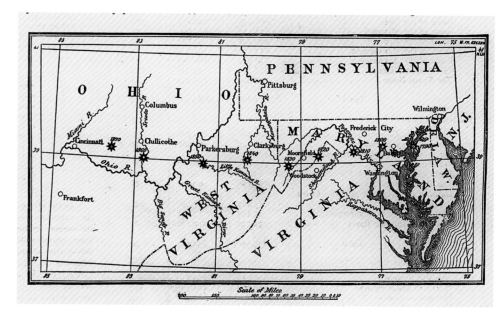

FIGURE 105.

(top) Francis A. Walker, "Map Showing the Degrees of Density, the Distribution, within the Territory of the United States of the Constitutional Population" (1874).

FIGURE 106.

(bottom) "Map Showing the Movement of the Geographical Center of Population" (1874).

(For more on Walker and the *Statistical Atlas*, see chapter 5 of the present text, pp. 237–38.)

Forty years later, Walker's contributions to thematic mapping influenced the U.S. approach to the peace negotiations after the First World War. After defeating the Austro-Hungarian, Ottoman, and German empires, Americans used these innovations in thematic cartography to redraw the boundaries of Europe. The United States had prepared for the Paris Peace Conference since entering the war in 1917, and relied extensively on thematic mapping to translate complex populations into simpler national units. Out of the defeated empires the victors drew new, more ethnically unified nations—including Yugoslavia, Czechoslovakia, Poland, and Iraq—with the goal of modernizing Europe.

According to Jeremy Crampton, the American delegates in Paris—known collectively as the Inquiry Committee—equated national identity with ethnic dominance, and then translated the territory under dominance into political sovereignty. Though the Inquiry was only a research body rather than a decision-making body, it shaped the Treaty of Versailles by privileging the idea of nation over clan, region, or other political units. Having witnessed the creation of Poland contributing to German resentment and aggression, and more recently the breakdown of Czechoslovakia, Iraq, and Yugoslavia, we see how consequential the work of the Inquiry was. Just as it treated race as a fixed category that could be mapped and bounded, it also understood nationalism as a natural and authentic source of identity. In stressing these categories on a map, the Inquiry reinforced their importance on the ground. In other words, the map itself not only reflected reality, but helped to create it (Crampton 2006).

Domestically, the turn of the century was a period of national reconciliation, but one that came at the cost of the rights and welfare of black Americans. Yet the Civil War continued to affect politics. The rising power of the Republican Party—born out of antislavery impulses in the 1850s—was in no small part aided by the willingness of Republicans to "wave the bloody shirt" and remind the nation of their party's leadership during the war and unbroken Unionism. This view is embodied in the 1888 map "Historical Geography," a vision of the nation very much at odds with the contemporary spirit of reconciliation (fig. 107). In this rendering the Civil War is only a symptom of a much deeper division traceable to the early days of colonial settlement and which turned on the decision to import slaves to Jamestown. From here, "history" brought forth two entirely different societies. To Plymouth came Liberty, "planted by Pilgrims upon the Bible . . . where it received God's blessing" in the form of intellectual, technical, and educational advantages unblemished by the sin of slavery. By contrast, "nearly every evil which exists in the political economy of our beloved country can be traced back to the pernicious teachings of the Jamestown settlers and their children."

In this view, history is the active engine of modern politics and geography, for though the slave colony of Jamestown no longer exists, "the colony still

SUSAN SCHULTEN

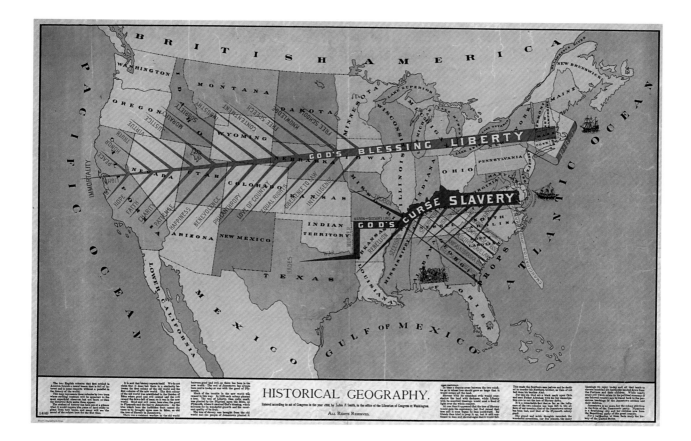

FIGURE 107.

John F. Smith, "Historical Geography" (1888).

lives in the form of the Democratic Party." Conversely, "Plymouth is a flourishing city and her children now form the Republican Party of this great country." This map is nearly a caricature of itself, and has more to say about regional and party identity than "historical geography" as such. Yet it also indicates the increased interest in national history at the end of the nineteenth century. Perhaps the war raised the public's sensitivity to contingency and the fragility of the Union, thereby initiating a deeper historical consciousness. In any event, Americans found the concept of national history especially compelling after the turn of the century, and we see this in the increased production of maps and charts of the past, which had prewar precedents but attained a new level of sophistication after the 1880s.

Behind this new historical sophistication was the professionalization of history as a discipline. By the 1880s the first generation of university-trained historians began to take control of the scholarship and teaching of the American past. Maps became tools to legitimate the aspirations of these new professionals, exemplified by Albert Bushnell Hart's *Epoch Maps Illustrating American History* (1891). Hart mapped the nation's past as a coherent narrative of evolution that began with the emergence of the colonies and came to fruition with westward expansion and the survival of Union. No longer was the past a series of chaotic events, but rather a story of national culmination. The pinnacle of this new tradition of American historical cartography was Charles Paullin's *Atlas of the Historical Geography of the United States*. In hundreds of maps—historical, thematic, statistical, topographical—Paullin created a historical and cartographic foundation that burnished the American past and elevated it as a subject of study (Paullin 1932). His volume also indicates the rising popularity of maps as tools of social science in the interwar period. Thematic mapping had widened the power of cartography, and scholars such as Robert Park began to use these techniques to study and reform contemporary American life.

Consider Ira De Augustine Reid's 1929 map of the Five Points neighborhood in Denver (fig. 108). A sociologist, Reid was asked by the Denver Inter-Racial Commission and the Urban League to study that city's black population, particularly its level of education, economic status, residential pattern, and employment profile (Reid 1929). In his report Reid noted that no legal segregation existed in Colorado, yet 5,500 of Denver's 7,000 black residents lived within the downtown neighborhood of Five Points. Their settlement in the first two decades of the century quickly prompted white out-migration, and as prospering blacks moved eastward in the 1920s they met with white resistance, through both violence and—according to oral histories—restrictive covenants forbidding home sales to blacks. Given that the Ku Klux Klan reached the height of its power in Denver in the early 1920s, such organized resistance is not hard to believe (Wasinger 2005, 28–39).

This background makes Reid's map interesting on several levels. First, a map drawn by an African American to advance the interests of blacks is unusual. We are far more familiar with maps that present African Americans as a *problem*—such as the Hergesheimer map of slavery—or that take the presence of blacks as a sign of instability, as in Walker's *Statistical Atlas*. More important are the multiple ways in which this map of Five Points could be read. For Reid, the map showed that "the distribution of Negro residences is wide and varied," and "no racial difficulties have appeared because of this intermingling." In fact, Reid's map suggests a high level of ethnic and racial *integration* in Five Points (Reid 1929, 3, 46). Yet the map also carried another meaning, highlighted by an exhibit of African-American history in Denver, where it was used to illustrate the confinement of blacks to Five Points through restrictive covenants in adjacent neighborhoods. Reid himself did not mention these restrictions, nor did he suggest the map be read this way. Yet with the census information before us, it is difficult to deny these restrictive residential patterns, or—for that matter—their gradual erosion after the mid-twentieth century. For African Americans sensitive to history, this reading of the map would be obvious. But what may be self-evident to one viewer may be lost on another, for maps are rarely transparent documents.

SUSAN SCHULTEN

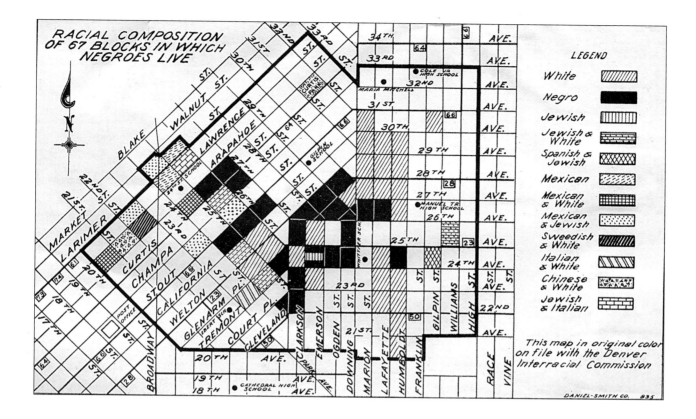

FIGURE 108.

Ira De Augustine Reid, "Racial Composition of 67 Blocks in which Negroes Live [in Denver's Five Points Neighborhood]" (1929).

Compare Reid's map of residential patterns in Denver to another map of—and created by—African Americans in Colorado in the 1920s, the plat map of Lincoln Hills (fig. 109 A). As Reid explained, Colorado state law prohibited residential segregation, yet custom, prejudice, and force had long kept African Americans out of certain parts of Denver. This pattern extended to Colorado's booming mountain towns, where blacks were prevented from owning land or homes. As a response, they founded the vacation community of Lincoln Hills on the Boulder Creek northwest of Denver (fig. 109B), a community that was advertised nationally to blacks through photographs, testimonials, and this property map. In the strong economic climate of the 1920s, Lincoln Hills grew briskly, as did the adjacent Camp Nizhoni for African-American girls. While the 1930s slowed its growth, it rebounded thereafter, in the process creating a vibrant community that continues to be very much a part of black culture and memory in Colorado.

The map of Lincoln Hills—included in its advertising—was visual evidence of the possibility of landownership for African Americans. In this respect the map and the community embody Thomas Jefferson's vision of a nation of landholders; and the rhetoric around the map stresses not race but rather the economic rewards of real estate investment—something white Americans

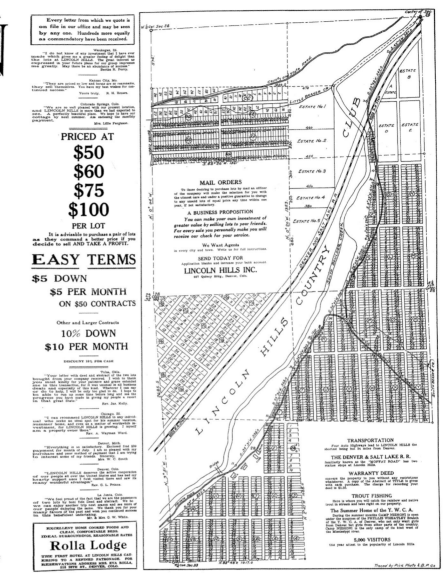

FIGURE 109A.

(right) Plat map of Lincoln Hills, Colorado (192–).

FIGURE 109B.

(left) Plat map of Lincoln Hills, Colorado, detail of title illustrations (192–).

could take for granted. This might explain the importance of the map today. In recent efforts to preserve Lincoln Hills, facsimiles of this map were distributed as a symbol of the local black experience. To white Americans, the map is like any other of its type: a utilitarian document devoid of explicit political messages. But to African Americans who faced clear obstacles to landownership, it represented not only the circumstances that forced the founding of an all-black community but also the pride in Lincoln Hills itself. We might also compare this map to Reid's and Hergesheimer's, each of which maps "race," but in different ways and on different scales. Hergesheimer mapped African Americans

SUSAN SCHULTEN

as slaves and placed them at the heart of the Southern insurrection. In Reid's map, blacks—though more or less confined to the Five Points neighborhood through restrictive covenants—also lived alongside other ethnic groups, a sign to Reid of limited social progress. In the Lincoln Hills map, blacks (ironically) do not appear as blacks, for the community was theirs entirely. Instead, they appear simply as landowners, while race disappears as a feature on the map even as it animated its creation.

One of the most irreverent and popular historians of the interwar era was Hendrik Van Loon, who immigrated to the United States in 1903. A successful career in journalism gave Van Loon intellectual breadth and fresh narrative skills, both keys to his wide and loyal audience. His *America* (1927)—filled with analogies, jokes, popular references, drawings, and maps—spoke of the nation's past in broad strokes and assertive tones, and in so doing challenged the style of contemporary academic history. Without the benefit of color, engraving, or topographic detail, his map of eighteenth-century North America enlivened the distant past for modern sensibilities (fig. 110). In straightforward language Van Loon described eighteenth-century struggles as imperial land grabbing. In New France, for instance, "All this looks very imposing on a map but it was really a neglected wilderness." Van Loon's was a political map, yet he did not draw the fixed boundaries that had become so popular on other maps of the same period, perhaps because they were not observed, agreed upon, or even in some cases *known* in any definitive way. With these techniques he managed to suggest both change over time and the instability of borders, both of which tend to be obscured by cartography.

Van Loon lived through the rise of aviation and mass communications, which prompted radical transformations in the representation of space. Maps—like modern sculpture and painting, as well as the newer genre of film—became a way to express these changes. Van Loon's images were equal parts art, cartoon, commentary, and information, and by mixing scale—notice the iconic palm trees and ships in the foreground, the celestial sky, and the exaggerated curvature of the earth—he conveyed an immediacy that traditional cartography could not approximate. His images took a modern approach to scale, proportion, and narrative, and in this respect anticipated the appearance and techniques of midcentury cartography.

The convergence of war and aviation in the 1930s and 1940s turned the nation's geographic attention abroad while also reconfiguring its understanding of a map. Indeed, it is difficult to underestimate—though equally difficult to calculate—the degree to which maps shaped American understandings of the Second World War. Such a role for cartography was not inevitable, but instead led by journalistic cartographers whose training enabled them to experiment with concepts of projection, perspective, and style in a way that captured the public's imagination. Among the most important of these individuals were Erwin Raisz, Charles Owen, and Richard Edes Harrison. Raisz was a lecturer

FIGURE 110.

Hendrik Van Loon, "The Royal and Imperial Game of Land Grabbing" (1927).

in cartography at Harvard during the war, and had previously established a reputation through his *General Cartography* (1938) and signature hand-drawn landform maps. The global scale of the war, together with the central role of aviation, prompted Raisz to "redraw" maps of the important regions, and to reconceptualize the world atlas as well. George Eisler, publisher of Raisz's *Atlas of Global Geography* (1944), introduced readers to this new world by explaining the contemporary revolution in technology and global politics, and their effect on cartography:

Our schools had long taught us that the world is round, but our knowledge remained theoretical and our maps remained deceptive. . . . We had no visual conception of the world as a globe, and many of us retained a mental picture of the world as a vast, sprawling cylindrical plain. Thus we almost forgot that the earth was round; our traditional concept became that of a cylinder in which neither North nor South Pole had more than a legendary place. . . . It was the airplane which finally made us relearn our geography. It was the airplane which gave the lie to century-old concepts which we had come to take for granted. (Raisz 1944, 5)

Simply put, new realities demanded new maps. Thus, Raisz drew a "smaller" and more interdependent world, and surrendered any claim to comprehensiveness in favor of relevance, topicality, and the ability to communicate complicated spatial concepts to lay readers.

The first part of the atlas included a series of maps profiling world climate, natural environment, the oceans, and transportation. Raisz then mapped world regions, approaching each as a piece in the larger puzzle of world strategy and economic interdependence. To bring the new "internationalist" sensibility home, Raisz drew these regional maps on an oblique orthographic projection, which resembles a picture of the territory from a fixed distance. His map of North America, for instance, gives the viewer an ability to see the territory from an aerial perspective (fig. 111). Through this technique Raisz managed to translate the three-dimensional realities of the air age onto a two-dimensional plane, and in the process created a sense of perspective that traditional map projections failed to convey. This relatively simple choice—repeated frequently in atlases and popular magazines during the 1940s—made it difficult to deny the power of aviation to reshape the world. Raisz conceded that some cartographers might object to this perspective for its approximation of a picture rather than a map drawn on a mathematical projection. Yet he defended it—as did many of his contemporaries—as a way of suggesting the influence of aviation, the *connection* between nations, and the importance of geographic regions.

Raisz tinkered not just with perspective but with the information featured on the map itself. In this instance Americans saw their home from a wider angle that highlighted economic resources, population distribution, and land

SUSAN SCHULTEN

The approximately 150 million people of North America—Canada, Alaska, the territorial United States—live under varied geographical conditions. In the north of the continent and also east of the Rocky Mountains the climate is "continental," characterized by extremely cold winters and hot summers; the growing season is very short in the far north and almost non-existent along the northern coasts of the Hudson Bay country, Labrador, and Greenland. In the south there is a gradation from a humid, almost subtropical climate to the Riviera character of southern California weather; in the Pacific North West of both Canada and the U. S. A. summers and winters are mild with much rainfall and dampness.

FIGURE 111.
Erwin Raisz, "North America" (1944).

cultivation. Political divisions and national boundaries were omitted—a significant departure from common American maps—both because aviation made them less important but also because, as Raisz put it, "who can know where the boundary lines of the future will lie?" (Raisz 1944, 8). He also stressed—even slightly exaggerated—landforms, and minimized the number of place-names in order to liberate the map from its role as a locational tool. Instead, this atlas used information to advance the understanding of political, economic, and cultural relationships.

The fact of war and the advance of aviation led many midcentury cartographers and illustrators to shift the look of the world map radically. Raisz's "The World Around Us" adopts an oblique azimuthal equidistant projection in order to demonstrate the distance from the central United States to cities around the world (fig. 112). The center of the map appears familiar, but the edges are dramatically distorted. Raisz accepted these distortions in order to convey to Americans that "the world is your neighbor. No inhabited place on the globe is more than 50 hours' flying distance from any point in the United States. This is fact, not a dream" (Raisz 1944, 45).

Raisz devoted the second part of the atlas to the geography of world problems, especially geopolitics, disease, hunger, poverty, and overpopulation.

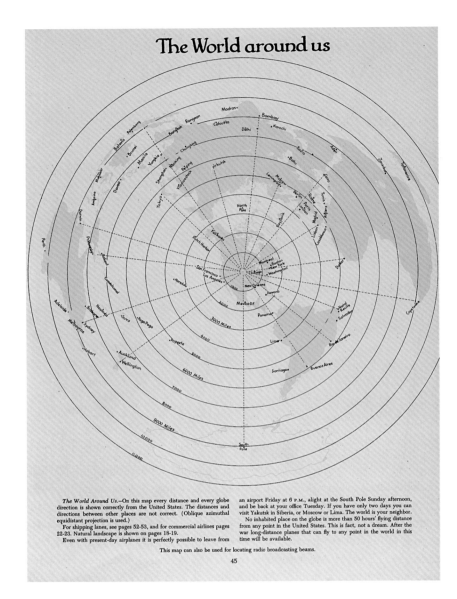

The World Around Us.—On this map every distance and every globe direction is shown correctly from the United States. The distances and directions between other places are not correct. (Oblique azimuthal equidistant projection is used.)

For shipping lanes, see pages 52-53, and for commercial airlines pages 22-23. Natural landscape is shown on pages 18-19.

Even with present-day airplanes it is perfectly possible to leave from an airport Friday at 6 P.M., alight at the South Pole Sunday afternoon, and be back at your office Tuesday. If you have only two days you can visit Yakutsk in Siberia, or Moscow or Lima. The world is your neighbor.

No inhabited place on the globe is more than 50 hours' flying distance from any point in the United States. This is fact, not a dream. After the war long-distance planes that can fly to any point in the world in this time will be available.

This map can also be used for locating radio broadcasting beams.

45

FIGURE 112.
Erwin Raisz, "The World around Us" (1944).

Here he reconceptualized cartography by drawing a new device to translate these problems into graphic images that Americans could quickly apprehend. Through "cartograms," Raisz combined the relational concept of a map with the statistical measurement of a diagram. The cartogram of the United States maintained a claim to geographic order while ranking size according to population (fig. 113 A). Similarly, the cartogram of world poverty "translated" data into geographic perspective (fig. 113 B). In many ways this was not unlike the approach of Walker's *Statistical Atlas* in the 1870s: both mapped "problems" such as disease, hunger, and population, and fused maps and graphs in order to communicate new concepts. Both were also products of their time: Walker was

SUSAN SCHULTEN

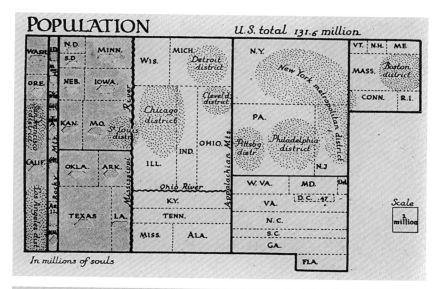

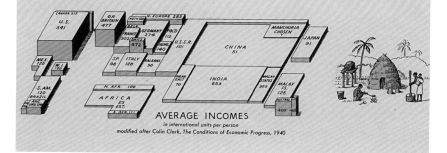

FIGURE 113A.

(top) Erwin Raisz, "Population [United States cartogram]" (1944).

FIGURE 113B.

(bottom) Erwin Raisz, "Average Incomes [world cartogram]" (1944).

part of a generation of experts determined to use the power of science to serve the nation, and Raisz was part of a generation of cartographers determined to demonstrate the radically altered state of international affairs wrought by global war and modern technology. Furthermore, while Walker stressed the interdependence within the nation and the importance of a continental view of domestic trends, Raisz stressed the interdependence of the nation and the wider world. Raisz made clear the humanitarian implications of this newly discovered interdependence by arguing that this was a world where "no one can live anywhere without being deeply affected by the lives of other people everywhere" (Raisz 1944, 7).

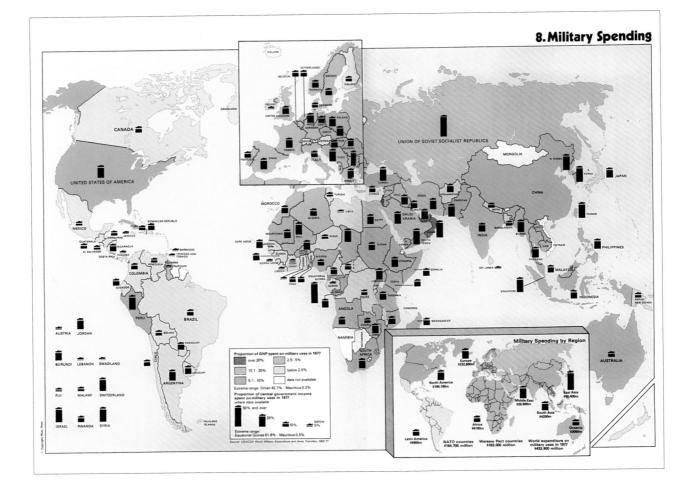

FIGURE 114.

Michael Kidron and Ronald Segal, "Military
Spending" (1981).

Raisz's work paved the way for subsequent political atlases, perhaps the
most well-known example being Michael Kidron and Ronald Segal's *State of the
World Atlas* (1981). As Kidron and Segal explain, their work evolved "from the
rich tradition of political atlases, born out of war or the threat of war." Theirs
was the protracted cold war, which generated interest in the geopolitical dimensions of international relations. The authors were admittedly—even self-
consciously—political, and brought their concerns about the environment,
the arms race, economic inequality, and human rights to the maps themselves.
Their bold, colorful techniques borrowed equally from graphic and cartographic traditions to create an unapologetically interpretive reference work.
"Military Spending" indicates the authors' concerns as well as their approach
to cartography (fig. 114). The colors of each country represent the proportion
of its gross national product spent on the military, while the symbols show the
proportion of national expenditures for the same. To the authors, the message

SUSAN SCHULTEN

is clear: far too many resources are devoted to the preparation for war (Kidron and Segal 1981, map 8).

Cartograms have only become more popular in recent decades, facilitated immensely by the Internet. Scholars at the University of Michigan received over a million visits to their Web site, which included a series of cartograms explaining the outcome of the 2004 presidential election (Gastner, Shalizi, and Newman 2004). Immediately after the election, a "blue state–red state" map of the nation circulated widely in the media (fig. 115). This map was misleading, because it generalized voting patterns on the state level, the equivalent of a blunt instrument that could not account for dramatic intrastate variations. Accounting for population distribution and other local differences could create a very different picture of the election, such as the one portrayed in figure 116. This cartogram, correcting for population distribution, amplified the "size" of Rhode Island (with 1.1 million inhabitants) so that it was far larger than Wyoming (500,000 residents). The authors went on to use these cartograms to show the limitations of cartography itself, which could not do justice to the complexity of the American electorate through the unit of states. In subsequent maps, the "red-blue" dichotomy became far more complex—even purple—when the maps accounted for representation in the Electoral College, population density, and other factors obscured by traditional political maps.

Of course, the number and variety of maps included here represent just a tiny part of the relationship between cartography and American history. Yet the larger themes prove enduring: the most powerful maps in the nation's history have been tools of exploration and discovery, statements and projections of national coherence and power, and instruments to explain the fundamental shift in spatial understanding brought by the modern era. Taken together with the other themes covered in this volume, they suggest the tremendously rich, complex, and varied traditions of cartography, and serve as a point of entry into the ongoing questions about the utility, power, and meaning of the map.

FIGURE 115.

(bottom, left) Michael Gastner, Cosma Shalizi, and Mark Newman, 2004 presidential election results by state, using a geographic scale (2006).

FIGURE 116.

(bottom, right) Michael Gastner, Cosma Shalizi, and Mark Newman, cartogram of 2004 presidential election results by state, scaling by population (2006).

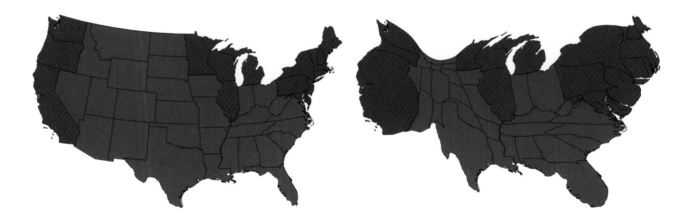

1234**5**67

VISUALIZING NATURE AND SOCIETY

Michael Friendly and Gilles Palsky

In addition to their standard uses—wayfinding and picturing geographic features and political boundaries—maps and related graphic methods have long played a significant role in scientific exploration, discovery, and explanation. In cartography, these latter uses are most well known under the rubric of "thematic cartography," but in fact, since the earliest such developments at the end of the seventeenth century, the histories of map-based and map-free information visualization have become increasingly intertwined. We thus construe our chapter title broadly to include various visual representations. We do not aim to give a comprehensive history of "first uses," but rather to explore the connections between image and scientific questioning through illustrative examples and their scientific context. In addition, we focus somewhat more on the side of cartography than of statistical graphics. More comprehensive historical treatments of data visualization may be found in Friendly (2006) and Friendly and Denis (2006), and of nineteenth-century thematic cartography in Palsky (1996, 2003).

The plan of this chapter is as follows: We first tell the story of two visual revelations that help us to understand how thematic maps and graphs can contribute to scientific explanation and discovery, and then try to identify some higher-level features shared by thematic maps and other data graphics in scientific inquiry and presentation. We then review the origin and development of thematic mapping, largely from a cartographic perspective. The fifth section illustrates some important developments in the history of statistical diagrams and graphs. Finally, we discuss the contributions of thematic maps and diagrams to the development of the social sciences.

VISUAL EXPLANATION AND DISCOVERY

Data and information visualization are fundamentally about showing quantitative and qualitative information so that a viewer can see patterns, trends, or anomalies, constancy or variation, in ways that other forms—text and tables—do not allow. Today, weather maps, maps of election results (for example, the "red" and "blue" states depicted in maps of U.S. presidential races; see figs. 115–16), and maps of disease incidence or outbreak (perhaps related visually to potential causes) are commonplace. But this was not always so; it took several small but significant conceptual leaps to move from showing purely geographic features (rivers, towns, terrain) that could be seen directly, to other things that might be measured locally, but could not be seen or understood globally without the aid of a map-based visual representation.

Thematic maps, overlaying a visual rendering of a spatial distribution of some data on the familiar forms of base maps, did not arrive until late in the seventeenth century, with meteorological charts and maps of magnetic declination at sea by Edmund Halley. Once invented, thematic maps provided a means for visual explanation and discovery that arguably could not have occurred otherwise.

In a parallel stream, the graphs and charts so widely used today had their origin in problems of physical, astronomical, and geodesic measurement also beginning in the seventeenth century. Throughout the eighteenth century, as new data-based problems were developing, graphic representation of scientific and economic data expanded to new domains and forms that were not primarily map based. To set the theme for this review, we briefly recount two stories of visual discovery.

The founding of geological cartography

If one can say loosely that geography is about seeing and understanding the distribution of things *above* the ground, one can also say that geology is about seeing and understanding what is *beneath* the ground. Geography is immedi-

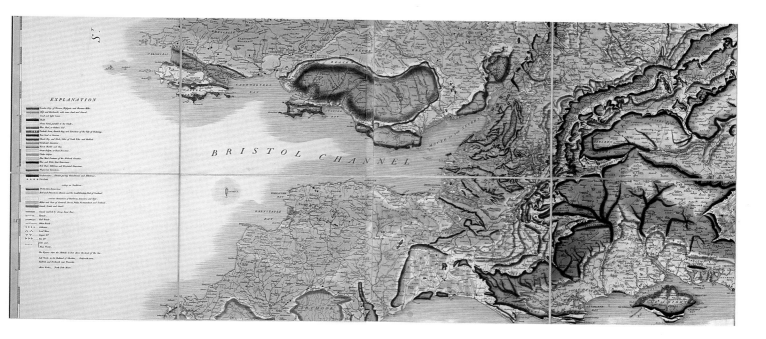

FIGURE 117.

William Smith, "A Delineation of the Strata of England and Wales with Part of Scotland," detail (1815).

ately visible, but geology is for the most part invisible, unless you happen on an exposed cliff or fault line. As we discuss below, mineralogical maps had been created as early as 1749. But the idea of mapping the invisible characteristics of the land beneath our feet and the first significant scientific breakthrough in what is now called "stratigraphic geology" came much later, in 1815–17, with William Smith's geological map of England and Wales (Smith 1815).

For more than thirty years, Smith traveled throughout the British country-side, surveying and recording the types of rocks and fossils that were found in places where various layers of soil and rock were exposed. Starting around 1799, he prepared a geologic table showing observed strata listed in order (from London clay and chalk at the top, to coal, slate, and granite at the bottom), together with the kinds of fossils found in each layer. He observed that the order of the strata *and* of the fossils were the same from place to place, a correspondence that compelled the view that geologic time and paleontologic time marched together in sequence through the layers of rock. Over time, he extended these observations and prepared maps showing the geographic variation of these geologic formations by colored shadings (fig. 117). The combination of paleontology, lithology, topography, and geographic location made visible by Smith's map and the final "Geological Table of British Organized Fossils" (Smith 1816) would lead to a deep understanding that geologic time and the history of the earth and its inhabitants could be studied by examining the layers in the ground beneath our feet.

The discovery of atomic number

The hallmark of good science is the discovery of laws which unify and simplify disparate findings and allow predictions of yet-unobserved events or phenomena. Mendeleev's periodic table, for example, allowed him to predict the physical and chemical characteristics of Gallium (Ga) and Germanium (Ge) before they were discovered decades later (Mendeleev 1889).

Mendeleev's table, however, arranged the elements only by a serial number, denoting an atom's position in a list arranged by increasing atomic mass. This changed in 1913–14, when Henry Moseley investigated the characteristic frequencies of X rays produced by bombarding samples of the elements from aluminum to gold with high-energy electrons, and measuring the wavelengths (and hence frequencies) of two specific peaks or spikes (called K and L) in their spectra (Moseley 1913). He discovered that, if the serial numbers of the elements were plotted against the *square root* of frequencies in the X-ray spectra emitted by these elements, all the points neatly fell on a series of straight lines (see fig. 118). This must mean that the atomic number is more than a serial number; that it has some physical basis. Moseley proposed that the atomic number is the number of electrons in the atom of the specific element.

Moseley's graph represents an outstanding piece of numerical and graphical detective work. Had he plotted *raw* wavelength or frequency itself, he would not have observed this remarkable linearity. In effect, Moseley had also predicted the existence of three new elements (without having observed them), corresponding to the gaps in the plot at atomic numbers 43 (Technetium), 61 (Promethium), and 75 (Rhenium). He also noted slight departures from linearity which he could not explain; nor could he explain the multiple lines at the top and bottom of the figure. The explanation came later with the discovery of the spin of the electron.

CONTENT, FORM, AND FUNCTION OF GRAPHIC DATA DISPLAY

Thematic cartography and data graphics share numerous visual features, general attitudes, and goals, but unfortunately do not share much in the way of a common language for analyzing the characteristics of each or explaining their historical development. However, both are ultimately concerned with the task of putting graphic marks on paper or some other medium to convey information to a viewer, and it may be of some use to consider them together from a wider perspective to discuss their development and contributions to visualizing nature and society.

In the abstract, maps and diagrams can be considered communication devices for conveying information from a source (author or creator) to a target (recipient or viewer) using visual signs and symbols. Various displays differ in terms of the information *content* (subject matter), the visual *forms* and attributes used to encode this information, and the *function* (task or communication goal) that the display is designed to serve. By analogy with language, visual form relates to the grammar and syntax of graphics (Wilkinson 2005),

MICHAEL FRIENDLY AND GILLES PALSKY

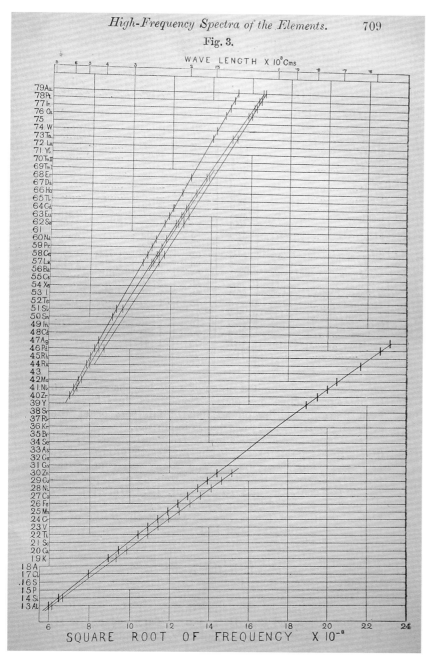

FIGURE 118.

Henry Moseley, graph of frequencies in X-ray spectra of chemical elements in relation to atomic number (1913).

content corresponds to meaning and semantics, while function relates to the pragmatic goals of communication.

At this level, both thematic cartography and data graphics are designed to convey quantitative or categorical information to a reader. They can use any of a wide range of visual attributes to map data items into graphic elements, and can be about similar subjects and topics. They differ, of course, in that carto-graphic visualization is primarily concerned with representation constrained to a spatial domain, but even spatially distributed data can be portrayed more or less usefully either in a map-based or graph-based form depending on the intended function or communication goal.

If we regard making maps and graphs from this perspective—as a form of communication, much like writing—we can distinguish at least three different functional roles served by both thematic cartography and data graphics, each with different design principles:

EXPLORATION. Exploratory data maps and graphs are designed to help reveal the pattern or structure of quantitative or qualitative information, to show variation across space, time, or circumstances. Most thematic maps fall into this category, as do data graphics used primarily to generate ideas and hypoth-eses. The use of dot or proportional symbol maps to show disease incidence (for example, Snow's dot map of cholera [fig. 140]) and of simple scatterplots to show relations between quantitative variables provide examples.

ANALYSIS. Analytic maps and graphs are driven more by goals of explana-tion, and are intended to aid in synthesizing, generalizing, or testing patterns and relations observed or suspected. To serve this end, the map or graphic design should provide for *direct* visual comparison. Examples include arrang-ing several maps or diagrams side by side (figs. 126 and 133); overlaying ad-ditional information regarding potential risk factors for disease on a map; or overlaying the predicted relation under some model on a scatterplot, as shown in Moseley's graph (fig. 118).

PRESENTATION. Maps and graphs designed for presentation intend to stimu-late thought or to convey conclusions. Their design principles include those of aesthetics, rhetoric, and exposition. Minard's well-known depiction of the fate of Napoleon's Grand Army is one example (Minard 1869); Playfair's chart of the national debt of England (fig. 130) is another.

THEMATIC MAPS

At first glance, the category of thematic maps, as distinct from general maps and even topographic maps, would seem both obvious and generally accepted. General maps attempt to show a variety of geographic features (waterways, roads, administrative boundaries, cities, and towns), while topographic maps add another layer of information to display something about an additional dimension, typically of elevation. These spatial features are all more or less concrete, fixed, and durable phenomena existing on the earth's surface. In con-trast, the thematic map displays the occurrence, spatial pattern, or variation of one or a small number of phenomena in the physical, biological, social, or economic world, such as climate, natural resources, population characteristics, and commerce.

The term *thematic* was first used by the German geographer Nikolaus Creutzburg in 1952, entered common cartographic usage in the 1960s, and merited a book-length historical treatment by Arthur Robinson in 1982 (Robinson 1982). Among theoreticians of cartography, there has been lively debate over precise definitions and attributes that distinguish thematic maps from other maps. To begin with, in terms of graphic form and content, there are no simple criteria for distinguishing between a "topographic object" and a "thematic phenomenon." Besides, topography could be considered an entirely separate theme. On the other hand, some maps appear to belong neither wholly to the thematic nor to the topographic category; there is no clear boundary between the two kinds, and mixed or intermediary forms exist.

One rule of thumb that can serve to distinguish thematic maps from other kinds of maps is their selective aspect. Thematic maps are designed to highlight the spatial distribution of a subject, an aspect, or a specific distribution, as opposed to the topographic map, a general map which represents various phenomena together. However, this characteristic is not enough, for some maps illustrate a theme emerging from topography such as administrative limits, roads, or hydrography. Thus, the difference between a topographic and a nontopographic object has to be specified. Thematic maps are also more abstract; because they attempt to show phenomena that are more or less invisible, they are more an intellectual construction than a straightforward depiction of land surface. For Elizabeth Clutton, "the thematic map presents a mental ordering of space, generalizing and arranging beyond the limitations of the original data to offer a visual image of more abstract truths" (1983, 42). Barbara Petchenik describes this as the difference between maps whose meaning relates to "being in place," compared with those whose meaning relates to "knowing about space" (1979). A more detailed discussion of the difference between the two categories of maps can be found in Palsky (2003).

Nevertheless, before becoming a recognizable, "nameable" category, thematic maps evolved slowly, starting in the last half of the seventeenth century. Historically, forms of transition are observed that could be called "para-thematic" or "pre-thematic." In the seventeenth and eighteenth centuries, several cartographers proposed novel cartographic representations, either because they broke with the customary cartographic synthesis, or because they testify to new geographic curiosities and an enriched nomenclature of the world. We designate these two categories of maps as special-purpose and hybrid maps.

Special-purpose maps

Cartography in the early modern period progressed according to a principle of accumulation: it used symbols expressing an increasing number of places and seemed all the more useful when it expressed a large quantity of information. Even so, in the beginning of the seventeenth century some maps would emphasize *one* element of the topographic inventory.

Road maps were among the first examples. Ogilby's renderings, assembled in *Britannia—a Geographical and Historical description of the Principal Roads thereof* (1675), portrayed the principal roads in England by vertical bands juxtaposed on the same section, echoing the tradition of medieval itineraries (see the discussion in chapter 1 of the present text, pp. 39, 42–45 and fig. 15). Moreover, true maps of networks were conceived by the beginning of the seventeenth century. Nicolas Sanson (1632) traced royal roads and the location of stages on a general map of France. The map, designed for travelers and traders, replaced the manuscript lists in use until then. In England, Ogilby's work influenced John Adams and George Willdey's ([1712]) maps of the road network (fig. 119).

The selective approach is also found in maps showing different administrative and jurisdictional divisions, reflecting an extraordinary concern for marking off sovereignties. Maps of the hydrographic network provide another example of special-purpose cartography, though these were rarer. The first one designed on the scale of France was Sanson's work (1634), republished in 1641 (fig. 120). Sanson showed only the outline of rivers, their toponyms, and their channels. The big hydrographic basins were shown by watercolor.

How can this kind of cartography be interpreted? Its *analytic* principle presented several advantages. First of all, the choice of one element of the inventory allowed a new precision. Thus, Sanson aimed his map of rivers at the curious amateur, priding himself on providing more information. To this was added a visual advantage: the map was less dense, and therefore more easily readable. All the same, its superiority also lay in the level of understanding. In 1697, the Abbé de Dangeau underlined this in his treatise of geography:

I realized by my own experience and those of others that all that prevents one from benefiting as much as one would like from maps and books, which until now were made to teach Geography, History and all that has a relation to it: mainly the multitude of objects that one sees simultaneously, in the poor order in which they are presented to the imagination. In order to improve this, I presented my work so as to show in sections on several different maps of the same country all that one sees in a single ordinary map. (Dangeau 1697, III)

The special-purpose map was in a way an echo of the Cartesian method of knowledge, which first invited a consideration of absolute things, that is to say singular, similar, equal, and independent, before going on to multiple and compound things. Moreover, Descartes proposed a comparison between his method of thought and the *view*, "for he who wishes to view several things simultaneously at a single glance sees none distinctly"; and similarly, "he who

is used to thinking of many things at the same time in one single act, is confused" (1996, 67–68).

Hybrid maps

It was not until 150 years after Sanson's river map that another effort to map the hydrographic network of France appeared. It was drawn by the geographer-engineer Dupain-Triel (1781) and emerged in a very different context: the new interest after 1750 in the development of internal navigable waterways.

FIGURE 119.

George Willdey, "The Roads of England according to Mr. Ogilby's Survey" (1712).

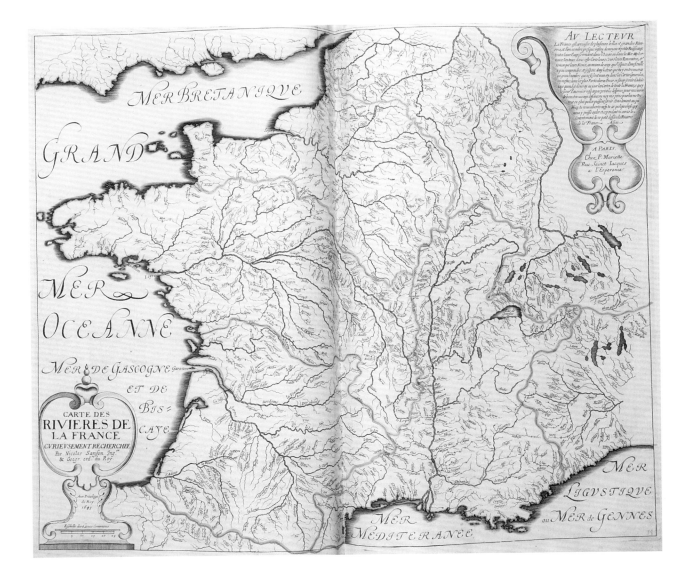

Selectivity remained rare in cartographic production in this period. Isolating an elementary geographic feature rather contradicted the natural tendency of cartographers, which was to add signs to complete the inventory.

At the beginning of the early modern period, many geographers drew attention to new objects, calling for their inclusion in maps. The Jesuit Antoine Lubin suggested, for instance, marking the outlines of big forests and distinguishing the types of trees, or indicating forges or mines, because of "the passion that men have always had for riches" (1678, 222). Initially, this advice had little effect. Knowledge and no doubt graphic language too was insufficient to supply these details.

MICHAEL FRIENDLY AND GILLES PALSKY

The evolution was nonetheless irreversible: space was being differentiated, the inventory completed, stemming as much from a more utilitarian vision of nature as from the progress of natural sciences, especially during the eighteenth century. But if the map drew attention to new objects, it was often by adding, and accumulating. Most of the special maps before 1800 were mixed or "hybrid" maps, as Robinson noted: "Occasional thematic additions had been entered on otherwise general maps, but the idea of making a map solely for the purpose of showing the geographical structure of one phenomenon seems not to have occurred to anyone" (1982, 17).

Hybrid cartography was based principally on developments in the natural sciences; thematic additions related to various fields, such as meteorological phenomena, plant species, formations, land use, or natural catastrophes. Mineralogical maps were typical examples. Before 1700, some maps were already pioneers in localizing resources of precious metals, such as Olaus Magnus's "Carta marina" (Mariner's Map; 1539). During the eighteenth century, these maps multiplied. One of the most famous was Christopher Packe's "New Philosophico-Chorographical Chart of East Kent" (1743). Conceived to show the system of valleys in this part of England, it contains several mineralogical symbols besides the usual information in a county map. In France, the naturalist J. E. Guettard wrote several treatises on the nature of soils between 1746 and 1786. The geographer Philippe Buache illustrated a few, and in particular drew the first mineralogical map of North America in 1752 (fig. 121).

The maps described here differ sufficiently from modern thematic ones to be considered a special category. They reflect a progress in geographic knowledge, but this improvement was obtained without breaking totally with the cartographic spirit of the time: early "para-thematic" maps look very much like traditional topographic maps, because they either concentrate on one feature of the topographic inventory (roads, rivers) or they add a new feature or two (plants, minerals) to an already detailed inventory.

These para-thematic maps represent a transitional solution that can be understood from a dual scientific and graphic point of view. Scientifically, the thematic additions were adapted to the discontinuous nature of observations. In order to communicate scientific conceptions through thematic maps, Helen Wallis recalled the necessity of following several preparatory steps, including the collection of a sufficient corpus of data and its subsequent organization into a coherent system (1973). Thus, mineralogical maps, precursor elements of geologic cartography, followed this schema of construction of knowledge. Before 1780, it was exceptional to see the representation of a zonal composition of rocks. Guettard clearly indicated a three-band arrangement on his "Carte minéralogique où l'on voit la nature et la situation des terreins qui traversent la France et l'Angleterre" (Mineralogical Map Showing the Nature and Situation of the Terrain across France and England)—"sandy," "marly," and "slate or

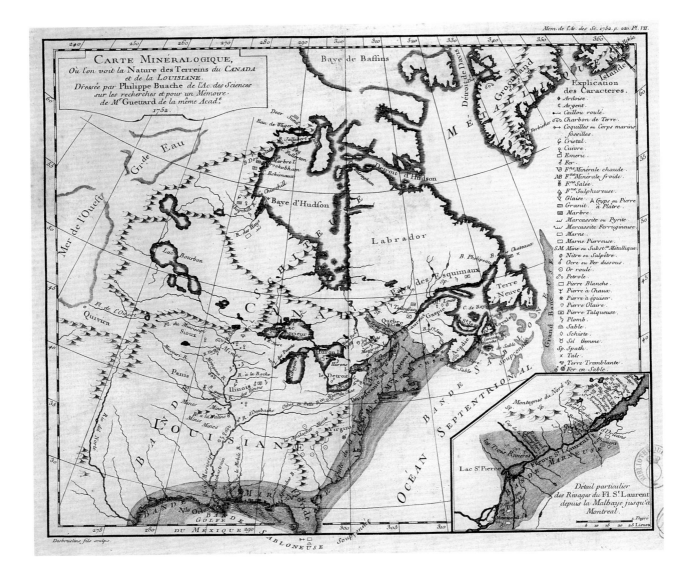

FIGURE 121.

Philippe Buache, "Carte minéralogique où l'on
voit la nature des terreins du Canada et de la
Louisiane" (Mineralogical Map Showing the Nature
of the Terrain of Canada and Louisiana) (1752).

metallic" (1746)—but his idea found little resonance, and pointwise additions
remained the most frequent medium for translating new phenomena.

From the graphic point of view, para-thematic cartography followed old,
familiar habits of presentation. As Umberto Eco put it, "The iconic representation produces real perception cramps and we are led to see things as they
have long been presented to us by stereotyped iconic signs" (1972, 183). The
situation is analogous to the usually slow pace of change in a society's art. Artists tend to paint the kinds of paintings their immediate predecessors painted,
and people looking at paintings are troubled when they don't see the types of
representations they are used to. The robust cartographic language that had
developed since the Renaissance was ill suited to the depiction of new, invisible

218

MICHAEL FRIENDLY AND GILLES PALSKY

phenomena, and its very iconic familiarity delayed the introduction of a more flexible, autonomous language of thematic cartography.

THEMATIC MAPS OF THE PHYSICAL WORLD

Thematic sea maps

There is no doubt that a real thematic cartography was born a little before 1700 in the realm of the sea chart, which linked the selective approach and the expression of abstract phenomena. The development of navigation, due to commercial and colonial activities, permitted the multiplication of observations on winds, sea currents, and terrestrial magnetism. Knowledge that earlier was simply observational combined with scientific knowledge in this period. But mariners' knowledge related to a portion of the surface of the globe that constituted a blank page and was truly experimental, a place on which symbols could be created without conflicting with topographic elements. In a space devoid of visible objects, cartographers were forced to invent signs to show the invisible.

The idea of these maps once again preceded their material realization by several decades. In France, for instance, J. François invited cartographers as early as 1652 to think of "the horizon with winds . . . currents and other sea movements . . . , magnetic variations in several places in the sea and on earth" (1652, 359). The idea was taken up some years later by Lubin: "It would then be necessary to chart each of the winds on the sea map and register all the observations I have just evoked"—that is to say, their direction, their "length," their "breadth," their season. Nonetheless, these maps appeared very complicated to produce. Thus, for the winds, Lubin underlined that "the enterprise would be very new and difficult, but not impossible and (that) it would bring more glory to those who could do it" (1678, 281).

The Englishman Edmund Halley imagined the first thematic maps at the end of the seventeenth century. He drew a first map of oceanic winds (untitled; it is known as "Halley's chart of the monsoons and trade winds" [1686]); then, after a scientific journey on the *Paramour*, the first map indicating the magnetic variations in the Atlantic Ocean: "A New and Correct Chart Shewing the Variations of the Compass in the Western and Southern Ocean" (Halley 1701), followed by a world map on the same subject (fig. 53). Halley used a method borrowed from topography: isometric lines. He was convinced of the novelty and effectiveness of the application of "curve lines," as he called them, to terrestrial magnetism.

Halley's maps were distributed on the Continent and quickly imitated. A little after 1710, the map of magnetism was copied in France, probably by Guillaume Delisle, under the title "Carte des variations de l'aiguille dans l'océan occidental et méridional suivant les observations faites en 1700 par Edmund Hallei" (Map of the Variations of the Compass in the Western and Southern

Ocean according to Observations Made in 1700 by Edmund Halley). Winds, currents, and magnetism became the object of numerous representations during the eighteenth century. Among them can be cited the "Chart of the Gulf Stream" (fig. 122), drawn by Benjamin Franklin after Timothy Folger (Franklin 1768), which seems to have adopted François' suggestion—showing a marine current as a river in the ocean. These maps sometimes combined several themes: winds and currents, or winds and magnetism, as on J. N. Bellin's "Carte des variations de la boussole et des vents généraux" (Map of the Variations of the Compass and of the General Winds; Bellin 1765; fig. 123).

A representational novelty of limited distribution and constituting an end in itself, the thematic map thus evolved toward becoming a tool of comprehension and management of territory, an instrumental map, "capable of shedding light on complex features closely incorporated into natural and human space" (Konvitz 1980, 304). At the same time, graphic language underwent its own particular evolution. It was above all in the nineteenth century that it acquired its autonomy in relation to the figurative or analogical code of topography. With

FIGURE 122.
Benjamin Franklin and Timothy Folger, "A Chart of the Gulf Stream" (1768), detail.

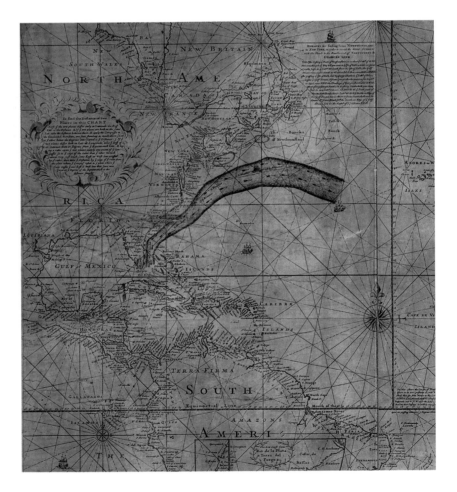

MICHAEL FRIENDLY AND GILLES PALSKY

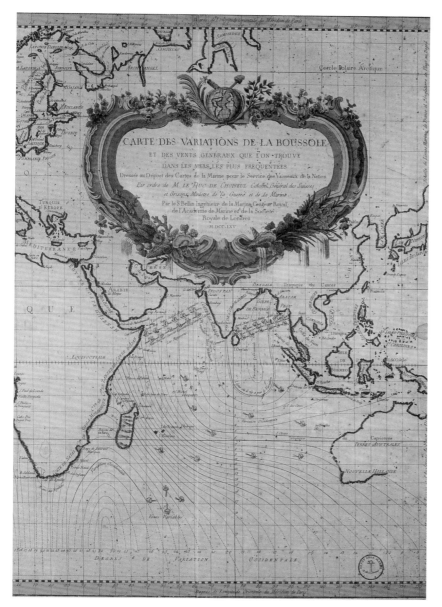

FIGURE 123.

Jacques-Nicolas Bellin, "Carte des variations de la boussole" (Map of Compass Variations), detail of the Indian Ocean (1765).

the development of an autonomous language of thematic mapping, the priority of representation was, in effect, reversed: topographic detail moved into the background and special themes into the foreground.

The influence of Carl Ritter and Alexander von Humboldt

The end of the eighteenth century saw an increasing number of maps featuring physical observations. Several colored maps recording the limits or extension of bedrock were produced in the framework of geognosy by Lommer (1768), Gläser

VISUALIZING NATURE AND SOCIETY

(1775), or Charpentier (1778). The first "zoographic" atlas, showing the distribution of animals on different continents with colored lines, appeared in 1800 (Jauffret 1800). These works, only a few examples of which are mentioned here, announced a flourishing cartography of physical phenomena in the nineteenth century. Carl Ritter and Alexander von Humboldt, the founders of scientific geography, played a fundamental role in the expansion of this cartographic field.

In 1806 Ritter published *Sechs Karten von Europa* (Six Maps of Europe), a small atlas that showed the distribution of cultivated plants, trees and shrubs, animals, and mountains, along with the limits of vegetation. The cartography remained very elementary: the essential information was expressed by text on the map and some colored highlights. However, the work enabled the correlation of natural data such as agricultural zones and climatic limits. It was the starting point of a rational and comprehensive physical cartography, itself an answer to Ritter's geographic project of a comparative knowledge of different parts of the globe (Ritter 1806).

Alexander von Humboldt's works appeared more original from the graphic point of view. As early as 1817, he drew the first "map" of isothermal curves, in reality a hybrid arrangement that was midway between a diagram and a map (fig. 124). This graphic was quite remarkable for several reasons. The top part shows smooth contours of constant temperature throughout the Northern Hemisphere; and although marked by scales of latitude and longitude, the map is nearly entirely implicit, save for a few place-names added for reference. Moreover, Humboldt recognized that temperature varies more with latitude and *altitude* than with longitude; to make visible this fourth dimension, he added the lower graph showing isothermal curves for altitude and latitude.

Humboldt described the results of his voyages in the New World in several volumes, often enriched by numerous illustrations. From the sum total of his observations, he wished to deduce laws of distribution of forms and considered graphic representations an essential tool for treating data: "The use of graphic means will shed light on phenomena of the greatest interest for agriculture and for the social state of the inhabitants. If instead of geographical maps, we only possessed tables . . . , a large number of curious connections that continents manifest in their forms and their surface inequalities would have remained unknown" (Humboldt 1817, 105). For example, Humboldt drew maps that show hydrographic, geobotanical, and geologic information, as well as vegetation sections, histograms, and other types of diagrams.

The observations of naturalists in the first half of the nineteenth century provided raw material for a diversified cartography. However, it must be stressed that the data were necessary, but not sufficient in themselves, for thematic cartography to flourish. In the present case, the use of diagrams and above all of maps was part of the scientific project of geography: spatializing in order to compare, classify, and explain. Humboldt's great scientific authority influenced the diffusion of graphic methods among several authors. His work

MICHAEL FRIENDLY AND GILLES PALSKY

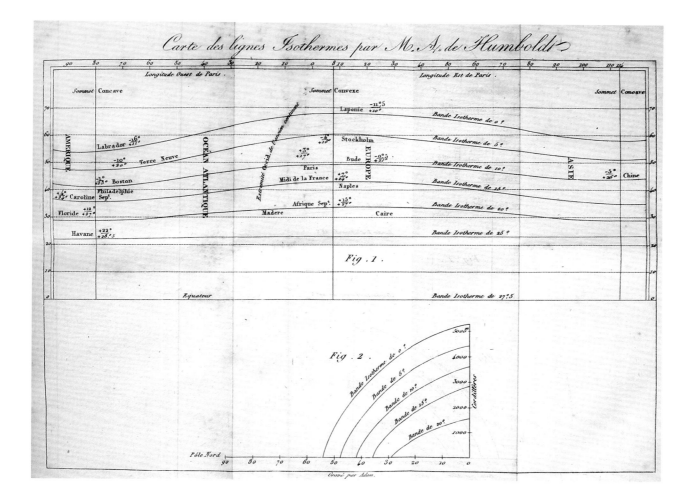

FIGURE 124.

Alexander von Humboldt, "Carte des Lignes Isothermes" (Map of Isothermal Lines) (1817).

on isothermal lines thus served as a reference for a veritable German school of cartographic treatment of meteorological data, marked by the works of Mahlmann and Kaemtz, the latter imagining "isobarometric" lines in 1832.

Nonetheless, the most important next step was Heinrich Berghaus's realization of the first big thematic atlas, the *Physikalischer Atlas* (Berghaus 1838–48; fig. 61). The work was undertaken on the suggestion of Humboldt, who saw it as the graphic counterpart of his project describing the "Kosmos." Borrowing from the best sources, it contained sixty maps, divided into six sections: Meteorology, Climatology, Hydrography, Geology, Magnetism, Phytogeography, and Zoogeography. The work was quickly copied and even plagiarized in Germany. It was also imitated abroad. In Scotland, A. K. Johnston (1848 and later editions) published the *Physical Atlas*, with plates (full-page illustrations) partly taken from Berghaus and partly original, among them the splendid, Humboldt-inspired planisphere of botanic geography decorated with vegetation profiles or sections (fig. 125). The German cartographer Augustus Petermann collabo-

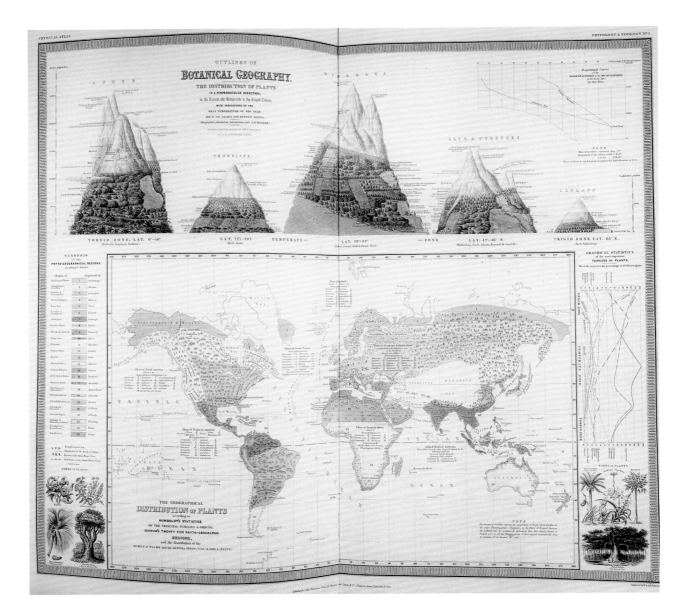

Alexander Keith Johnston, "The Geographical Distribution of Plants" (1850).

rated in this production and then published his own *Atlas of Physical Geography* (1850). In France, physical maps were included in later atlases, such as Hercule Nicolet's *Atlas de physique et de météorologie agricoles* (Atlas of Agricultural Physics and Meteorology; 1855), or J. A. Barral's *Atlas du Cosmos* (1861).

Geologic maps

Alongside these major works were more isolated ones. This was the case of geologic maps produced at the beginning of the nineteenth century. Modern

MICHAEL FRIENDLY AND GILLES PALSKY

geologic science took shape between 1800 and 1830 around a certain number of concepts that were new or had matured for a long time, including distinguishing among sedimentary soils in *chronological* units, dated from the fossils found there, rather than lithological units, identified by soil and rock types. Stratigraphic paleontology replaced litho-stratigraphy. The first modern geologic maps were linked to these developments by color coding soils by age. A pioneering work was the "Carte géognostique des environs de Paris" (Geological Map of the Environs of Paris) by Cuvier and Brongniart (1808). Based on Lamarck's research on fossils, the authors identified nine sedimentary formations around Paris. The colors of their map, laid in flat tones, were reproduced in an inset, vertically arranged in the very order of the deposits. The legend and the sections joined to the map (including an "ideal cross section," or stratigraphic scale) thus provided structural information: they allowed the extrapolation of the rocky volume in three dimensions.

As we noted at the beginning of the chapter, the grand project of early geologic mapping was accomplished by William Smith. Begun at the end of the eighteenth century, it culminated with the publication of "A Delineation of the Strata of England and Wales with Part of Scotland" (Smith 1815) on a scale of 5 miles (8 km) to the inch (1:316,800) (fig. 117). The huge map, measuring nearly 6 feet by 9 feet (1.8 × 2.7 m), is a real masterpiece. Different strata are identified by twenty-three colors, all applied by hand in this period before color lithography. Simon Winchester (2001) makes a good case for the practical and theoretical importance of Smith's map, calling it "the map that changed the world."

After 1820, maps based on stratigraphy multiplied. A second general map of England was published by Greenough (1820), then completed and republished in 1839. Leopold von Buch (1826) drew a geologic map of Germany and neighboring countries, as did Heinrich von Dechen (1838). In France, the inspector general of mines, Brochant de Villiers, was concerned about the absence of a sufficiently precise geologic map enabling the development of the mining industry. He elaborated a project in 1822, whose realization was given to two polytechnicians and mine engineers, Dufrénoy and Elie de Beaumont (1840). They stayed in England to learn the methods of work, and then made their observations in France from 1825 to 1835. Their map was engraved in 6 sheets at a scale of 1/500,000 and presented to the Academy of Sciences in 1841.

Such small-scale maps served as a basis for the detailed renderings at larger scales that were undertaken in the following decades. Geologists' work also extended to colonies and young republics of the New World, which were ideal fields for the observations of naturalists. Thus, the geologically colored "Carte générale de la république de Bolivia" (General Map of the Republic of Bolivia) testified to the results obtained by the Frenchman Alcide d'Orbigny during his trip to South America from 1826 to 1834, at the request of the Museum of Natural History in Paris (Orbigny 1842).

The development of charts, diagrams, and graphs in many ways runs parallel to that of thematic maps, and arose from similar problems and similar desires to move from observations and evidence to understanding and explanation. Among the most important problems of the seventeenth century were those concerned with physical measurement—of time, distance, and space—for astronomy, surveying, mapmaking, navigation, and territorial expansion. This century also saw great new growth in theory and the dawn of practical application of data to scientific problems: the rise of analytic geometry and coordinate systems (Descartes and Fermat), early theories of errors of measurement and estimation (J. Mayer and Gauss), the birth of probability theory (Pascal and Fermat), and the beginnings of demographic statistics and "political arithmetic" (Graunt). We use two early examples to illustrate these beginnings.

Scheiner's sunspots

In late 1609, Galileo constructed one of the first astronomical telescopes and almost immediately made a series of important discoveries—craters on the moon, a huge number of new stars, observations on the moons of Jupiter, sunspots—which he published in *Sidereus nuncius* (The Starry Messenger) in March 1610. News of Galileo's discoveries traveled quickly, and Christophe Scheiner (1573–1650), a Jesuit in Augsburg with talents in mathematics and instrument making, constructed a device to record the position and movement of sunspots over time. The substantive and philosophical issue was the Jesuit-assumed perfection of the sun (could sunspots be explained as shadows cast by moons?), but the evidentiary issue was how to display *changes* in configurations of sunspots over time.

Scheiner (1612) adopted a display form that appears quite modern today. He simplified the recording from each day from October 23 to December 19 of 1611 into a spot map showing the location and extent of sunspots, and then arranged these into a semigraphic[1] table (fig. 126), using a principle of graphic design that allows the eye to track changes across these thirty-seven images (Tufte [1983] calls this "small multiples"). Two main legends are used to identify the seven sunspots tracked (by the letters A–G) and the changing orientation of the sun over time. Part map, part graph, but more, Scheiner's visual representations of sunspots could win awards for graphic design today.

Scheiner's greatest work on sunspots, *Rosa Ursina* (The Rose of Orsini, after Scheiner's patron; 1626), pioneered new ways of representing the motions of spots across the sun's face by combining the daily configurations with movement of the sun across the ecliptic. Shortly thereafter, sunspot activity decreased drastically (the so-called Maunder Minimum, ca. 1645–1710) and interest in sunspots waned. A cyclic pattern in sunspots of approximately eleven

MICHAEL FRIENDLY AND GILLES PALSKY

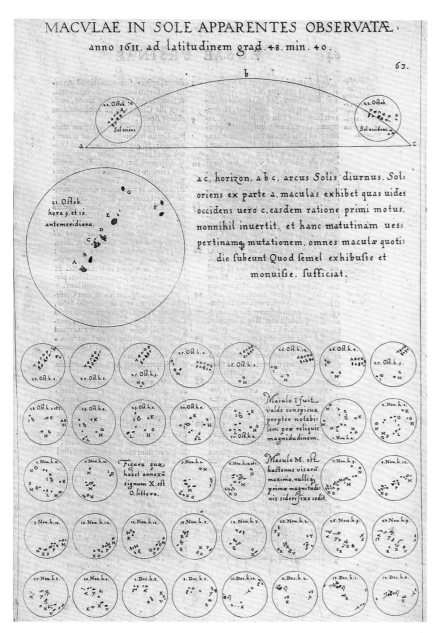

FIGURE 126.

Christophe Scheiner, "Maculae in sole apparentes observatae" (Spots Observed Appearing on the Sun) (1626; first published 1612).

years' duration would later be discovered by E. W. Maunder (1904) using a "butterfly diagram" to depict changes in sunspot location and intensity over time.[2]

Problems of longitude

Another remarkable example of the early interplay between maps and graphs is the attempt by Michael Florent van Langren (1600–1675), a Flemish cosmog-

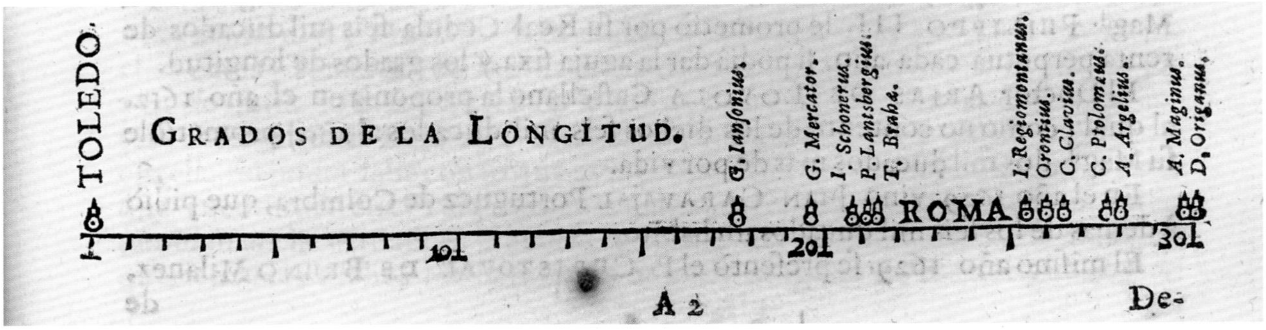

FIGURE 127.

Michael Florent van Langren, Variations in the estimates of the difference in longitude between Toledo and Rome (1644).

rapher to the court of Spain, to portray variations in the determination of longitude (fig. 127), believed to be the first visual representation of statistical data (Tufte 1997, 15). At that time, lack of a reliable means to determine longitude at sea hindered navigation and exploration (Sobel 1996).

This one-dimensional line graph (Langren 1644) shows all twelve known estimates of the difference in longitude between Toledo and Rome, and the name of the astronomer (Mercator, Tycho Brahe, Ptolemy, and so on) who provided each observation. What is notable is that van Langren could have presented this information more easily in various tables—ordered by author to show provenance, by date to show priority, or by longitude to show the range. However, only a graph shows simultaneously (1) the wide variation in the estimates (the range of values covers nearly half the length of the scale); (2) the central or estimated value (marked "ROMA"), along with (3) the names and values attached to the individual determinations.[3] Van Langren's graph is also notable as the earliest-known exemplar of the principle of "effect ordering for data display": "order information in graphs and tables according to what should be seen" (Friendly and Kwan 2003).

New graphic forms

With some rudiments of statistical theory, data of interest and importance, and the idea of graphic representation at least somewhat established, the eighteenth century witnessed the expansion of these aspects to new domains and new graphic forms. Abstract graphs and graphs of mathematical functions became widespread, and as economic and political data began to be collected, some novel visual forms were invented to portray them so the data could more easily "speak to the eyes."

Interest in a wider range of phenomena called for new abstractions and adaptations of visual forms. For example, in 1770, Philippe Buache published (with Guillaume Delisle) *Cartes et tables de la Géographie physique*, containing a chart and table of high and low water levels in the Seine over time, semiannually from 1732 to 1766 (fig. 128). Buache, a physical geographer, was quite used

MICHAEL FRIENDLY AND GILLES PALSKY

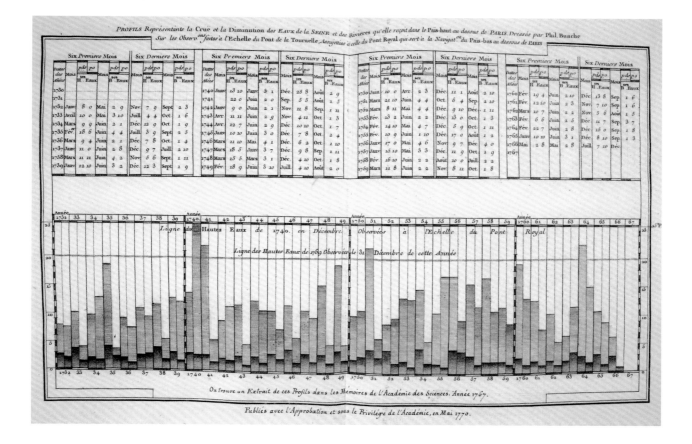

to profile maps of terrain elevation over space. To show changes over time, he substituted time for space, and used two levels of shading to distinguish high and low water levels (Buache 1770). A modest change visually, the substitution of concrete space by a more abstract dimension of time anticipates time series graphs and bar charts that would develop shortly.

As another example, geometric figures (squares or rectangles) and cartograms to compare areas or demographic quantities were introduced by Charles de Fourcroy (1782) and August F. W. Crome (1785). Figure 129 shows the English version of a "map" of the "statistical relations of Europe," produced in 1819 by Crome and engraved by Aaron Arrowsmith, notable for the combination of superimposed squares (showing area) and divided circles showing data on population and finances. Time lines, or "cartes chronologiques," were developed to portray people and events in history by Jacques Barbeu-Dubourg (on a 54-foot [16.2-m] scroll) and by Joseph Priestley (1765, 1769). Priestley's invention would shortly serve as an inspiration for William Playfair, whose work can be considered the origin of modern statistical graphics.

Playfair (1759–1823) is widely considered the inventor of most of the graphic forms used today—first the line graph (to portray economic data over time)

FIGURE 128.

Phillipe Buache, "Profils représentants la crüe et la diminution des eaux de la Seine et des rivières qu'elle reçoit dans le païs-haut au dessus de Paris" (Profiles Showing the High and Low Waters of the Seine and Its Inflowing Rivers in the High Lands above Paris) (ca. 1770).

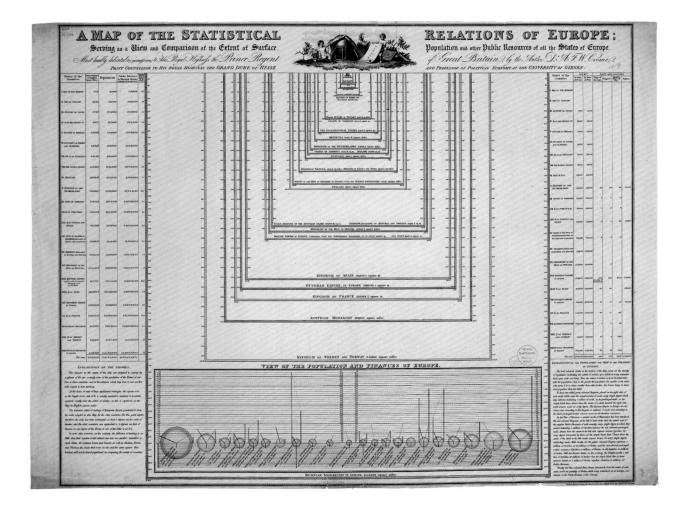

FIGURE 129.

August F. W. Crome, "A Map of the Statistical Relations of Europe" (1819).

and bar chart (1786), and later the pie chart and circle graph (1801). In the *Commercial and Political Atlas*, published in three editions in England (1786–1801) and one in France (1789) and the *Statistical Breviary* (1801; French translation 1802), Playfair adapted and invented an astonishing number of graphic constructions to convey economic data to the eye. "His genius was to realize that nonspatial quantities such as expenditures and historical time could be represented by physical space and that such representations offered advantages denied to tabular presentations" (Wainer and Spence 2005, 30). Such graphs were indeed novel; he referred to the idea to represent quantities by lengths of lines along a scale as "lineal arithmetic" and devoted several pages to description of how to read a graph.

Figure 130 shows his time series chart of the national debt of England. It is surprisingly modern in graphic design, with axis scales, major and minor grid lines, an aspect ratio (height to width) that enhances vertical differences, and text labels for significant historical events. Playfair's message is abundantly

MICHAEL FRIENDLY AND GILLES PALSKY

clear: the national debt has risen dramatically, and each sharp upward turn occurred in times of war.

Another chart (fig. 131) offers a creative combination of different visual forms: circles (used to show the area of nations), a pie chart (to show the divisions of the Turkish Empire), and lines (to show both population and taxes). In this figure the left axis and the left line on each circle shows population, while the right axis and line shows taxes in millions of pounds. Playfair intended that the slope of the line connecting the two would depict the *rate* of taxation, and argued that the British were overtaxed compared with the other nations. The graph is flawed, because the slope also depends on the diameter of the circle. It would also be considered sinful today, because separate *y* scales allow perceptions to be manipulated by rescaling one axis or the other. In Playfair's defense, the idea of calculating and graphing rates or other indirect measurements was still a half century away, and his main point is sustained because the line for Britain slopes in the opposite direction to most of the others.

Beginnings of modern statistical graphics

In the first half of the nineteenth century, all the modern forms of data display were invented: bar and pie charts, histograms, line graphs and time series plots, contour plots, scatterplots, and so forth. But another development—the widespread collection of population, economic, social, and medical data—spurred explosive growth in applications of statistical representations by graphic displays at a rate that would not be equaled until modern times.

As the modern states of Europe developed, it was seen that statistics (originally meaning "numbers of the state") were crucial for national planning, social legislation, and economic progress. Where should railroads and canals be built? What was the distribution of imports and exports? What should be done to control crime? Statistical bureaus that were established in many countries created an "avalanche of numbers" (Hacking 1990), but graphic methods often proved essential in deciphering them.

The use of diagrams and maps in understanding social or "moral" statistics will be detailed in the final section. Here we illustrate this period with a novel 1844 *tableau figuratif* (fig. 132) by Charles Joseph Minard, engineer for the École Nationale des Ponts et Chaussées (national school of bridges and roads) in Paris. Minard is, of course, best known for his depiction of the fate of Napoleon's Grand Army in what has been called the "best statistical graphic ever drawn" (Tufte 1983).

This inventive graph is related to the modern bar chart and mosaic plot (Friendly 1994, 2002), but Minard introduced two simultaneous innovations: the use of divided and proportional-width bars so that the *area* of each rectangle had a concrete visual interpretation. Through these variable-width, divided bars, the graph shows the transportation of commercial goods along one canal

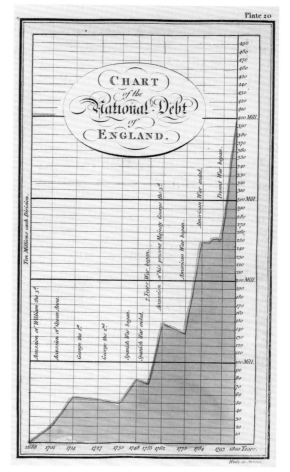

FIGURE 130.
William Playfair, "Chart of the National Debt of England" (1801).

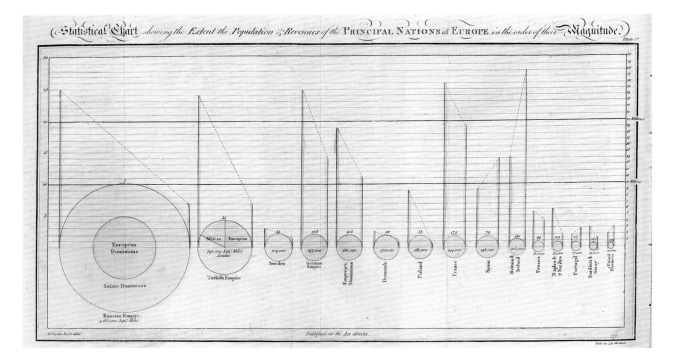

Statistical Chart showing the Extent the Population & Revenues of the PRINCIPAL NATIONS of EUROPE, in the order of their Magnitude.

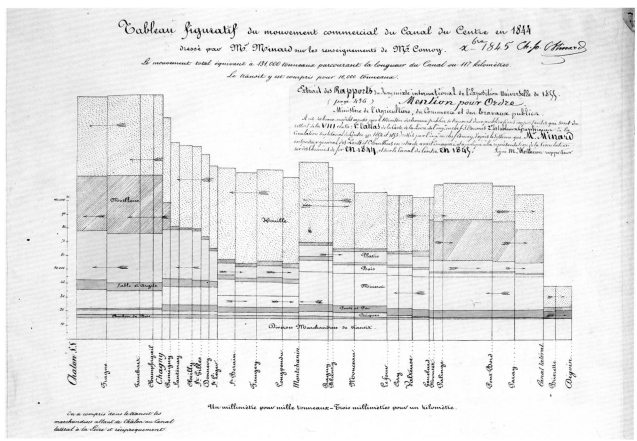

Tableau figuratif du mouvement commercial du Canal du Centre en 1844
dressé par Mr. Minard sur les renseignements de Mr. Comoy.

route in France (Minard 1845b). The questions at hand were how to plan and charge for transportation of various goods (coke, minerals, wood, and so forth) along various portions of the route (differential rates for partial versus complete runs; effect of direction of shipment). In this display the width of each vertical bar shows distance along this route; the divided bar segments have height proportional to amount of goods of various types (shown by shading), so the area of each rectangular segment is proportional to cost of transport. Direction of each type of shipment is indicated by arrows. Minard used this diagram to argue that differential rates should be set for various partial runs. Whereas Playfair had tried to make data "speak to the eyes," Minard wished to make them "calculer par l'œil" (calculate by eye) as well (Minard 1861, 4).

The golden age of statistical graphics

By the mid-1800s, the combination of abundant data of real importance, emerging statistical theory, and technological advances in reproduction provided the conditions for rapid growth—a "perfect storm" for advancements in data visualization. What started as the "Age of Enthusiasm" (Funkhouser 1937; Palsky 1996) for data graphics ended with what can be called a golden age, often with beauty difficult to find in modern graphics. So varied were these developments that it is even difficult to be representative; some highly selective examples must suffice to illustrate a few themes.

One important theme was the desire to display more complex phenomena and more than two variables simultaneously on a flat piece of paper. Earlier developments of isolines on thematic maps and contour diagrams of relatively simple and error-free physical data (variations of temperature or barometric pressure over time and space) were extended to three-dimensional surface plots of population by age and time (for example, by Gustav Zeuner in Germany and Luigi Perozzo in Italy) and, most important, to situations where the relations were statistical or only approximate, rather than functional ones or those measured with little error.

Among the scientific advances of this period that depended directly on insights gained from graphic analysis of statistical data in three or more dimensions, there are two that stand out, both due to Sir Francis Galton. The more well known is Galton's discovery of the bivariate normal correlation surface (Galton 1886), from data on the relation between heights of parents and their offspring. Galton constructed a table of grouped frequencies of heights of parents and children and drew smoothed isofrequency contours. He noticed that (1) the contours of equal frequency approximately formed a series of concentric ellipses and (2) the loci of the mean of y given x and of x given y were approximately the conjugate diameters of the ellipses. These relations would later form the basis for the theory of correlation and regression (Pearson 1901).

FIGURE 131.
(facing, top) William Playfair, "Statistical Chart Shewing the Extent, the Population and Revenues of the Principal Nations of Europe" (1801).

FIGURE 132.
(facing, bottom) Charles Joseph Minard, "Tableau figuratif du mouvement commercial du canal du Centre en 1844" (Figurative Chart of Trade on the Canal du Centre in 1844) (1845).

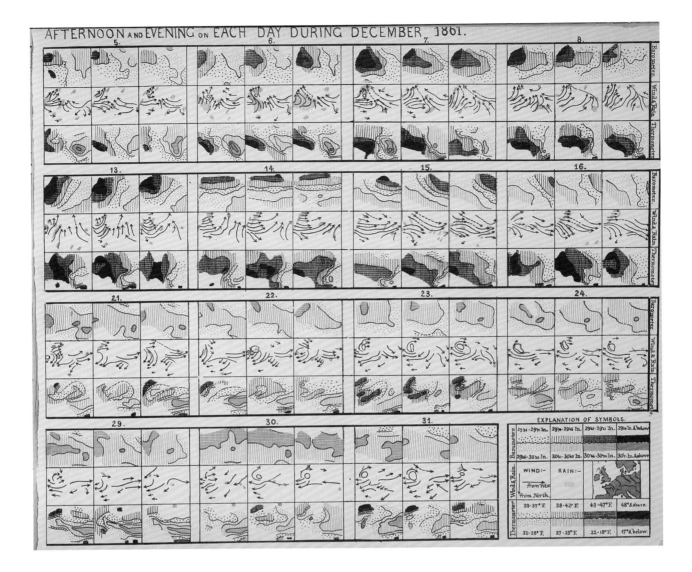

FIGURE 133.

Francis Galton, "Charts of the Thermometer, Wind, Rain, and Barometer on the Morning, Afternoon, and Evening on Each Day during December 1861," detail of right half (1863).

Less well known but perhaps the most notable nonstatistical graphic discovery of all time was that of the "anticyclonic" (counterclockwise) pattern of winds around low-pressure regions, combined with clockwise rotations around high-pressure zones. Galton's work on weather patterns began in 1861 and was summarized in *Meteorographica* (1863). It contained a variety of ingenious graphs and maps (over six hundred illustrations), one of which is shown in figure 133.

This remarkable chart, half of a two-page, multipanel display, shows observations on barometric pressure, wind direction, rain, and temperature from fifteen days in December 1861.[4] For each day, a panel of three rows and three columns shows nine schematic maps of Europe, mapping pressure (row 1), wind and rain (row 2), and temperature (row 3) in the morning, afternoon, and eve-

MICHAEL FRIENDLY AND GILLES PALSKY

ning (columns). One can clearly see the series of black areas (low pressure) on the barometric charts for about the first half of the month, corresponding to the counterclockwise arrows in the wind charts, followed by a shift to red areas (high pressure) and more clockwise arrows. Howard Wainer remarks, "Galton did for the collectors of weather data what Kepler did for Tycho Brahe. This is no small accomplishment" (2005, 56).

This chart was not the source of Galton's inspiration. Rather, it is the summary graphic he devised from ninety-three separate schematic maps of his data, each one using special iconic symbols he devised to show the weather measurements. As with Scheiner's graph of sunspots, the composition of many small figures into a single, "small-multiple" display permits visual comparisons of changes and patterns that could not be seen otherwise.

A second theme concerns transformations of data and maps to make relations simpler and enable their use for direct, visual calculation. Some examples from this period are the semilogarithmic graphs introduced by the economist William S. Jevons (1863) to show *percentage* changes in commodity prices over time; log-log plots to show multiplicative relations (Lalanne 1846) as linear graphs; anamorphic maps by Émile Cheysson (reproduced in Palsky 1996 as figs. 63–64) using deformations of spatial size to show a quantitative variable (for example, the decrease in travel time from Paris to various places in France over two hundred years); and alignment diagrams or nomograms using sets of parallel axes (Ocagne 1885, 1899) for calculating complex functions.

We illustrate this slice of the golden age with figure 134, a tour-de-force graphic by Charles Lallemand (1885) for determination of magnetic deviation of the compass at sea in relation to latitude and longitude without calculation. Lallemand was an engineer, best known as director general of the geodetic measurement of altitudes throughout France. This graphic combines many variables into a multifunction nomogram using three-dimensional figures, juxtaposition of anamorphic maps, parallel coordinates, and hexagonal grids. As with Galton's multivariate weather maps, those of us who do statistical graphing and mapping by computer would be hard pressed to create such displays today.

A final theme for the golden age is the production of impressive state-sponsored statistical atlases that began in the 1870s throughout many countries in Europe as well as the United States. This effort to present graphic views of population, trade and commerce, and social and political issues continued until the early part of the twentieth century, and was accompanied by international statistical congresses (begun in 1853 in Belgium) that attempted to develop standards for graphic presentation and were closely tied to the state statistical bureaus. The pinnacle of this period was undoubtedly the *Albums de Statistique Graphique* published annually by the French Ministry of Public Works between 1879 and 1899 under the direction of Émile Cheysson (see Dainville 1972). They were published as large-format books (about 11 × 15 inches [27.9

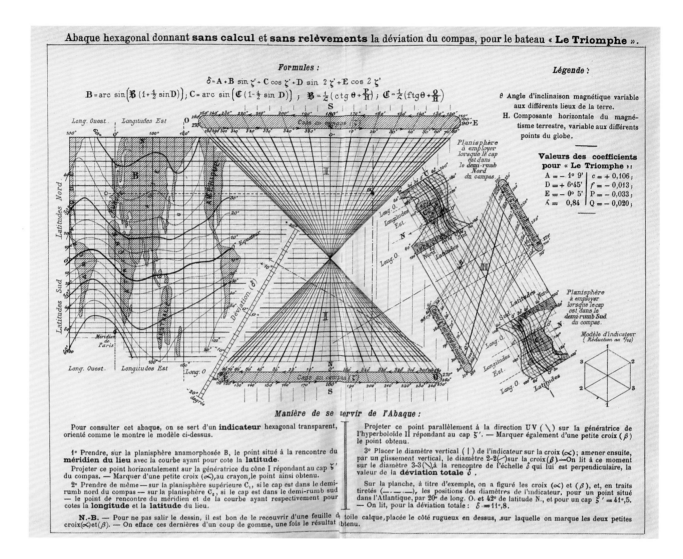

FIGURE 134.

Charles Lallemand, "Abaque hexagonal donnant sans calcul et sans relèvements la déviation du compass, pour le bateau 'Le Triomphe'" (Hexagonal Abacus Giving the Variation of the Compass without Calculation or Plotting, for the Ship "Le Triomphe") (1885).

× 38.1 cm]), and many of the plates folded out to four or six times that size, all printed in color and with great attention to layout and composition. We concur with Funkhouser (1937, 336) that "the *Albums* present the finest specimens of French graphic work in the century and considerable pride was taken in them by the French people, statisticians and laymen alike."

Many of these albums were designed to show temporal changes or provide graphic comparisons of related quantities for population, trade and commerce, and transportation in relation to the geography of France and the world. To do this, many forms of graphic symbols were appropriated and adapted to the problem at hand (flow lines, proportional and divided circles, divided rectangles, planetary diagrams, and so forth). The collection of these images can be regarded as an exquisite sampler of the graphic methods then known.

236

MICHAEL FRIENDLY AND GILLES PALSKY

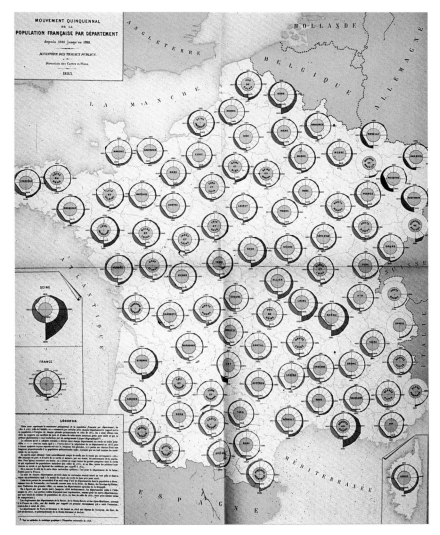

FIGURE 135.

"Mouvement quinquennal de la population par département depuis 1801 jusqu'en 1881" (Quinquennial Change of the Population by Department from 1801 to 1881) (1881).

Figure 135, for example, uses spiral symbols to show the population of each department (administrative district) over the five-year periods from 1801 to 1881. The radius of each circle corresponds to the population in 1841, and the area between this reference circle and the curve for actual population is shaded red or blue, respectively, according to whether the population was less than or greater than that in 1841.

At about the same time, other statistical albums and atlases were prepared in Europe and the United States, and among these, those from the U.S. Census Office deserve special mention. The *Statistical Atlas of the Ninth Census*, produced in 1874 under the direction of Francis Walker, contains sixty plates, including several novel graphic forms (United States Census Office 1874). It had the ambitious goal of presenting a graphic portrait of the nation, and covered a

VISUALIZING NATURE AND SOCIETY

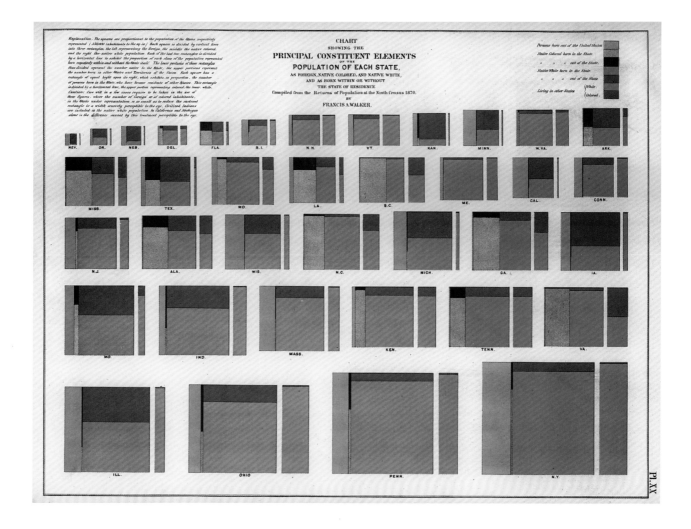

FIGURE 136.

Francis A. Walker, "Chart Showing the Principal Constituent Elements of the Population of Each State" (1874).

wide range of physical and human topics: geology, minerals, weather, population, wealth, literacy, rates of death and disease, and so forth. Figure 136, for example, uses mosaiclike divided rectangles to show state populations classed as foreign, native colored, and native white, and as born within or outside the state of residence. Walker is also credited with the invention of "age pyramids" (back-to-back bilateral frequency histograms and polygons), which he used to compare age distributions for two classes (male/female, married/single, and so forth). (For more on Walker and the *Statistical Atlas*, see chapter 4 of the present text, pp. 189–92 and figs. 103–106).

Following each subsequent decennial census for 1880 to 1900, reports and statistical atlases were produced with more numerous and varied graphic illustrations. The 1898 volume from the census of 1890, under the direction of Henry Gannett, contained over four hundred graphs, cartograms, and statistical diagrams (United States Census Office 1898). Figure 137 is an example

MICHAEL FRIENDLY AND GILLES PALSKY

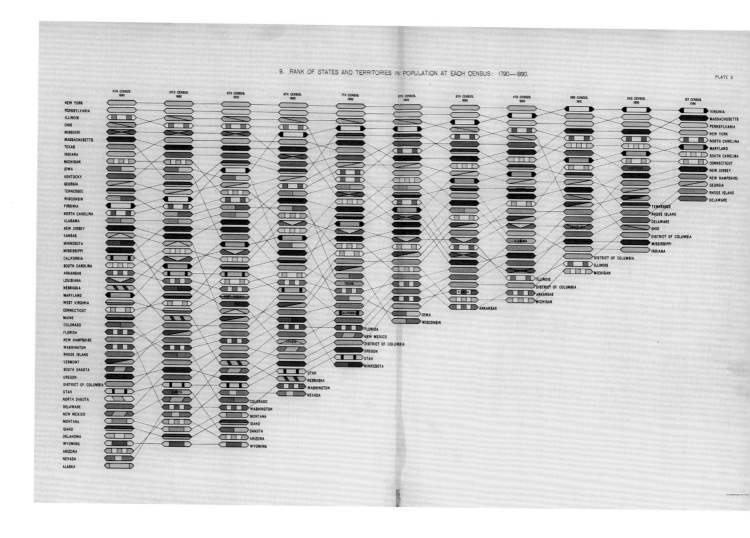

FIGURE 137.

Henry Gannett, "Rank of States and Territories in Population at Each Census, 1790–1890" (1898).

of a ranked-list, *semigraphic* display of a form similar to what is now called a parallel coordinate plot. The presentation goal was to show the ranking of populations of the states from all censuses from 1790 to 1890 and allow increase or decrease to be tracked by state. To achieve this, the symbol for each state was colored and shaded distinctively and connected across adjacent census periods by lines.

MAPS AND DIAGRAMS IN SOCIAL SCIENCE

If the early modern period (before about 1800) of thematic maps and graphs can be characterized as driven by concerns from the physical sciences, many innovations in the nineteenth century stemmed from a new concern with understanding human conditions and activities that would give rise to modern

VISUALIZING NATURE AND SOCIETY

social science, including economics, sociology, and epidemiology. There were several forces at play here, among which an explosive growth in data collection on social and economic topics figured prominently, as we noted earlier. Another theme that arose was the possibility of formulating "laws" or relations of the social order, akin to those that had been developed to understand relationships of the physical order. The principal methods of present-day quantitative cartography were imagined in this context between 1826 and 1845; the graphic methods invented by Playfair for economic data were extended, modified, and appropriated to comprehend a wider range of social phenomena and their relations.[5] In this endeavor, the natural sciences offered a model to follow in the search for patterns of constancy and variation in social data, as can be inferred from early statisticians' frequent references to Humboldt, Berghaus, and others.

The rise of moral statistics

To express these new social phenomena, cartographers had to reform radically the graphic language for the portrayal of statistical data in order to create new relationships between graphic forms and numbers as well as establish a new logic and a new syntax, much as Playfair had done with his "lineal arithmetic." The first modern statistical map is credited to Baron Charles Dupin in France. His "Carte figurative de l'instruction populaire de la France" (Figurative Map of Popular Education in France) was the starting point of a graphic revolution, whose consequences can still be felt in contemporary mapping. Trained as an engineer at the École Polytechnique starting in 1801, Dupin turned his attention in the 1820s to "statistics, an entirely new science [that] had never been usefully applied. Mr. Charles Dupin resolved to make it serve to observe our country's progress in the path of moral and material interests" (Hoefer 1858–78, vol. 5, p. 320). Dupin presented some of his results in lectures at the Conservatoire National des Arts et Métiers. In 1826, he submitted his map (fig. 138) in a paper on the "effects of popular education on France's prosperity" (Dupin 1826). The map was printed one year later in his major work: *Forces productives et commerciales de la France* (Productive and Commercial Power of France; Dupin 1827). It illustrated primary education, considered a sign of general development. This topic allowed him to represent a basic opposition between northern and southern France through an original graphic method: "To render the most important of these differences visible," he wrote, "I hit upon the idea of giving those departments that sent fewer pupils to schools the darker shades" (Dupin 1827, 249).

The method, now called the choropleth map, had no known antecedent. However, a plausible hypothesis about its conception can be advanced. The map became famous as the "map of enlightened and obscure France"[6]; Dupin himself used the expression "enlightened France" when he presented it at the Conservatoire. It would be reasonable to conclude that a scale of moral values

MICHAEL FRIENDLY AND GILLES PALSKY

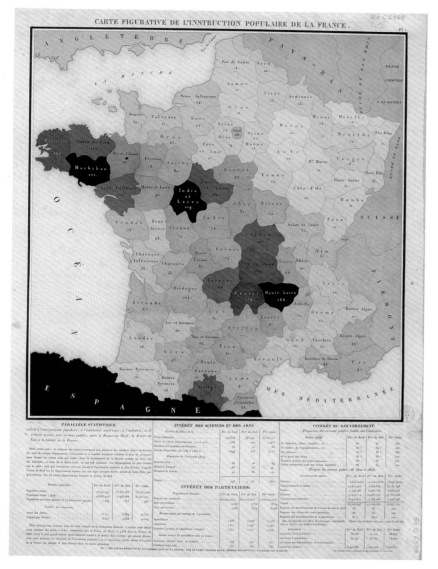

FIGURE 138.

Charles Dupin, "Carte figurative de l'instruction populaire de la France" (Figurative Map of Popular Education in France) (1826).

directly inspired the gradual shadings of the map. The shading gave the impression of a light cast on the map, comparable to the light of knowledge. Moreover, we will see that most of the maps that immediately succeeded Dupin's were based on the same principle: the scale of shadings always transcribed a scale of values, always with sense that darker meant worse.

Very quickly, Dupin's invention served a wider role in the origin of social science as "moral" and social statistics became more widely available. In 1825, the French Ministry of Justice instituted the first centralized, national system of crime reporting, recording the details of every charge laid before the courts. In 1829, André-Michel Guerry, a Parisian lawyer with a penchant for numbers, joined with the Venetian geographer Adriano Balbi to produce the first *comparative* maps of crime and education, to examine the *relation* between these moral variables (Balbi and Guerry 1829). By 1833, Guerry produced the first comprehensive study of such social data, also including data on suicides, donations to the poor, illegitimate births, and so forth. Using shaded maps and graphic tables, he showed that rates of crime, suicide, and other variables remained remarkably stable over time, yet varied systematically from place to place. He said, "We are forced to conclude that the facts of the moral order are subject, like those of the physical order, to invariable laws" (Guerry 1833, 14). This, along with the contemporaneous work of Adolphe Quetelet in Belgium (1831, 1835), may be regarded as the foundational study of criminology, sociology, and modern social science. Guerry's last work (1864) proposed a new form of *analytical statistics* presented visually with both maps and graphs, and comparing data from France and England over a thirty-year period. Figure 139 shows one of seventeen plates from that volume, here for crimes against persons in France. Recognizing that shading on the map is an overall summary, various graphs around the periphery were designed to dissect these by time or other factors or to highlight noteworthy patterns and trends.

Proportional symbols

Sometime after Dupin, A. Frère de Montizon, an officer who became a professor of sciences after the fall of the First Empire, conceived a second graphic method employing dot symbols. He published several educational works on morals or history at the beginning of the century. Later, his "Carte philosophique figurant la population de la France" (Philosophical Map Showing the Population of France; Frère de Montizon 1830) represented population distribution in absolute values. The population was indicated by departments, using a number of dots proportional to the number of inhabitants, 1 dot to 10,000 persons. The map was "philosophical" because Frère de Montizon wished to relate the population to "the physical, intellectual and moral state of the country." He thus traced a line *AB* on his map, going from Saint-Nazaire to Maubeuge—a "thermometric line" dividing the territory into two climatic zones of differing agricultural production. In his view, this fact explained the general distribution of population. The map was visually not very effective, for the procedure was difficult to master. The image presented was more of a uniform distribution, because the very small dots and the observation scale (by department) made the spatial contrasts less visible. Nonetheless, the "Carte philosophique" initiated an important method and revealed the new curiosity toward demography. This was shown yet again a few years later by George P. Scrope's publication of a map of the world's population according to three bands of density (Scrope 1833).

MICHAEL FRIENDLY AND GILLES PALSKY

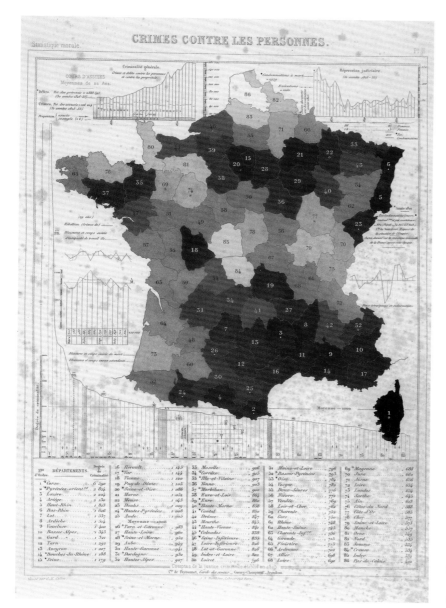

FIGURE 139.

André-Michel Guerry, "Crimes contres les personnes" (Crimes against Persons) (1864).

Innovations in the sphere of cartographic statistics in the 1830s were due above all to engineers. In the context of the industrial revolution, their major task was to establish the main railway lines. Laying out tracks called for the integration of varied data: physical and technical conditions, political or strategic contingencies, but also the distribution and mobility of population, resources, and wealth. Therefore, engineers were among the first to take an interest in demographic and economic data and sometimes turned to graphic

VISUALIZING NATURE AND SOCIETY

language in order to better exploit them. The atlas presented by the railway engineer Henry Harness to accompany a report of the Irish Railway Commissioners provides a first example (Harness 1838). It included six very original maps on the distribution and circulation of goods and passengers: three of them applied the system of proportional circles to urban populations, and the other two were the first maps depicting the flow of passenger and freight road traffic. The cartography of population and transport then became a tool for objectively determining the main railway tracks.

Diffusion of procedures

The methods imagined by Dupin, Frère de Montizon, and Harness spread unevenly. Statistical maps with dots (or other discrete symbols) remained rare in the nineteenth century. The procedures reappeared in medicine, but quite independently of Frère de Montizon's map, which remained little known; cases or deaths linked to epidemics were localized on a large scale, often in an urban context (Palsky 1995). A well-known example was Dr. John Snow's 1855 map (fig. 140), which showed the effects of cholera in a London neighborhood and the link between the deaths and the probable source of infection from the public water pump on Broad Street (Snow 1855). Snow's map figures prominently in the history of epidemiology as a graphic argument linking cases of cholera to a probable cause: the Broad street pump around which the majority of deaths clustered (Koch 2005, 75–155).

This was not the first map of disease seeking to understand causation through its spatial distribution. In 1798, Valentine Seaman published two dot maps to illustrate the distribution of yellow fever in New York (Seaman 1798). After the first outbreak of Asiatic cholera in Great Britain, Dr. Robert Baker constructed a map showing the distribution of the 1,800 cholera cases in Leeds in the particularly severe outbreak in 1832 (Baker 1833). But Seaman's and Baker's graphic techniques, combined with their limited scientific understanding of the causes of disease, were insufficient to lead to Snow's eureka experience. For example, Baker simply used uniform hatching in red to denote "the districts in Leeds in which cholera had prevailed." He could only note "how exceeding the disease has prevailed in those parts of the town where there is a deficiency, often an entire want of sewage, drainage and paving" (Baker, 1833, 10).

Maps with proportional symbols soon found other applications. Several engineers drew flow maps after 1845 in France, Belgium, Austria-Hungary, and Russia. Thus, the French engineer Minard transposed to cartography the idea he had in 1844, when he drew diagrams of passenger and freight traffic (fig. 132 above). In March 1845, he presented his "Carte de la circulation des voyageurs par voitures publiques sur les routes de la contrée où sera placé le chemin de fer de Dijon à Mulhouse" (Map of the Circulation of Travelers by Public Conveyances over the Region through Which the Dijon to Mulhouse Railroad Will

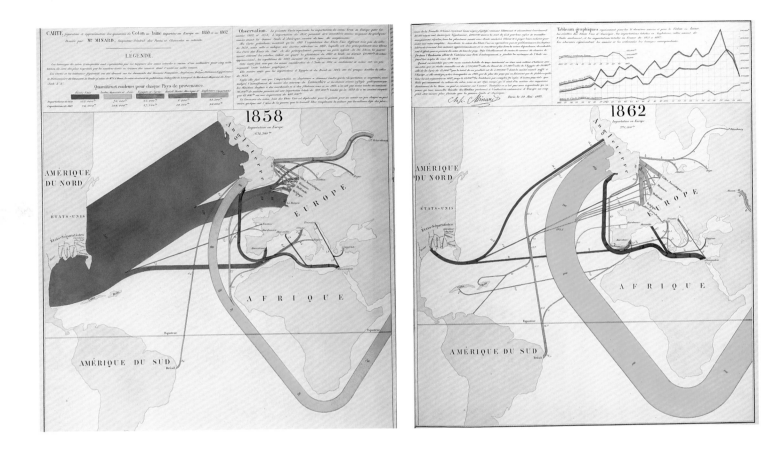

CARTE *figurative et approximative des quantités de* Coton en laine *importées en Europe en* 1858 *et en* 1862

Dressée par Mr MINARD, *Inspecteur Général des Ponts et Chaussées en retraite.*

LÉGENDE.

Tableaux graphiques

1858

1862

AMÉRIQUE
DU NORD

AMÉRIQUE
DU NORD

ÉTATS−UNIS

ÉTATS−UNIS

Angleterre

Angleterre

EUROPE

EUROPE

AFRIQUE

AFRIQUE

AMÉRIQUE DU SUD

AMÉRIQUE DU SUD

FIGURE 141.

Charles Joseph Minard, "Carte figurative et approximative des quantités de coton en laine importées en Europe en 1858 et 1862" (Figurative and Approximate Map of Quantities of Raw Cotton Imported to Europe in 1858 and 1862) (1863).

Run; Minard 1845a), in which he drew bands of thickness in proportion to the annual passenger traffic. This map was part of a running debate on two different routes for a railway line between Dijon and Mulhouse since 1841. It argued in favor of one of the routes (through Besancon and the Doubs valley), on the grounds of a clear distinction in the existing road traffic, which had more to do with visual evidence than with partisan interests.

Between 1845 and 1871, Minard drew several other "figurative maps," which often backed analysis of political economy and sometimes testified eloquently to the irregularity of the traffic. His maps of imports of raw cotton to Europe before and during the American Civil War (fig. 141) were among the most striking visual demonstrations (Minard 1863). In particular, the visual message of the comparison is clear: before the war, most imports came from the southern United States; by 1862, substantial amounts had been replaced by Indian and Egyptian cotton, but the clothing makers in Europe were probably still wanting.

Further, Minard, perhaps inspired by Playfair, was one of the first to draw proportional circles divided in sectors on maps, for example in order to translate harbor circulation, or show both the total and proportions of kinds of meat shipped to Paris. These proportional symbols were equally popularized by the

MICHAEL FRIENDLY AND GILLES PALSKY

famous German cartographer Augustus Petermann, who toward midcentury constructed several population maps according to this procedure (fig. 142), but also, more surprisingly, a map showing the quantity of rainfall (around 1852).

In the realm of statistical graphics, Florence Nightingale (1857) introduced a polar-area chart, or "coxcomb diagram," showing mortality in the British army in the Crimean War over time by angular sectors whose area was pro-

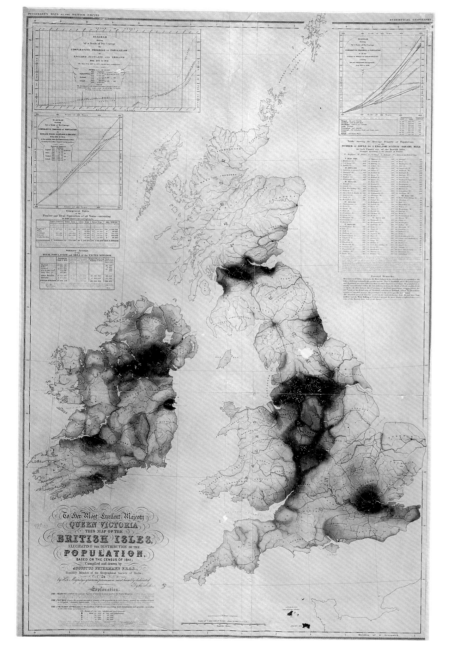

FIGURE 142.

Augustus Petermann, "Map of the British Isles, Elucidating the Distribution of the Population" (1850).

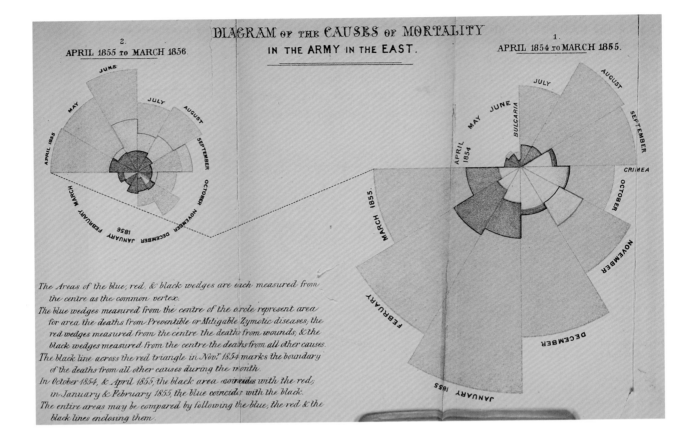

DIAGRAM OF THE CAUSES OF MORTALITY
IN THE ARMY IN THE EAST.

2.
APRIL 1855 TO MARCH 1856.

1.
APRIL 1854 TO MARCH 1855.

The Areas of the blue, red, & black wedges are each measured from
the centre as the common vertex.
The blue wedges measured from the centre of the circle represent area
for area the deaths from Preventible or Mitigable Zymotic diseases, the
red wedges measured from the centre the deaths from wounds, & the
black wedges measured from the centre the deaths from all other causes.
The black line across the red triangle in Nov.r 1854 marks the boundary
of the deaths from all other causes during the month.
In October 1854, & April 1855, the black area coincides with the red,
in January & February 1855, the blue coincides with the black.
The entire areas may be compared by following the blue, the red & the
black lines enclosing them.

FIGURE 143.

Florence Nightingale, "Diagram of the Causes of
Mortality in the Army in the East" (1857).

portional to the number of deaths (fig. 143). The dark inner portions for each month show direct deaths on the fields of battle; the larger, outer portions show deaths due to "other causes," meaning largely preventable death due to wounds and disease. She used such diagrams to argue persuasively to the British public and parliament for the institution of better battlefield nursing (the modern mobile army surgical hospital, or MASH unit). She became known as the Lady with the Lamp, the mother of modern nursing. If Playfair was the father of statistical graphics, we might well consider Florence Nightingale its mother.

But one graphic method of cartography appears to have spread most widely outside the restricted circle of cartographic specialists: the map with graduated shadings introduced by Dupin and elevated to a tool for the scientific study of moral and social statistics by Guerry. Other applications of this method in the social realm appeared in France (Angeville 1836; Parent-Duchâtelet 1836), Holland (Somerhausen 1827), England (J. Fletcher 1847–49; Mayhew 1851–62), and elsewhere in Europe as the general study of "moral statistics" took shape. These shaded maps dealt with education, criminality, begging, prostitution, poverty (called "pauperism," a peculiar English term likening destitution to a

MICHAEL FRIENDLY AND GILLES PALSKY

disease), and other social topics. Like Nightingale's coxcomb diagrams but on a much wider scale, maps, with their striking and persuasive aspect, figured prominently as arguments in scientific or political debates. For example, Guerry (1833) argued from his maps that levels of crime are not related to the level of instruction, as many in his time believed, and that crimes against persons and against property show different relations to other variables.

Progressively, the connection between shadings and moral values seemed to fade as the method was applied to other topics, now with dark representing "more" and light representing "less."[7] However, the association between colors and moral values remained, perhaps unconsciously, for a long time, as shown by the maps on poverty that Charles Booth attached to his monumental social inquiry of London (1889–91). Booth applied a color to each street that corresponded to a social category. He used dark and cold colors for the poor (black for "the lowest, most vicious, semi-delinquent class"; dark blue for the "very poor, occasional workers. Chronic misery," etc.) and hot colors like pink, red, and orange for middle and higher categories (fig. 144). By combining the codifications of moral cartography from the beginning of the nineteenth century, Booth stigmatized the most ignorant, the most criminal, or the most miserable parts of the urban territory by dark shades.

Near the turn of the twentieth century, several social surveys were modeled after Booth's study. Residents of Hull-House, a social settlement on Chicago's West Side, published the results of their investigation of the neighborhood, a poor immigrant district (Addams 1895). They created a series of detailed maps, directly inspired by those of Booth, and retaining in particular his use of thematic coloring. As with previous "moral" maps, the Hull-House renderings were intended as graphic evidence of social problems, in order to promote reform.

CONCLUDING REMARKS

Over the last decades of the nineteenth century, state-sponsored statistical bureaus flourished throughout Europe and the United States, and the discipline of social statistics became solidified in national and international organizations, which began to consider the adoption of international standards for graphic methods in both maps and diagrams.[8] Between 1869 and 1901, the congresses held by the International Institute of Statistics in The Hague, Vienna, St. Petersburg, and Budapest debated the variety of visual encodings used in maps and diagrams in an attempt to define a universal grammar, with much of the discussion concerning scales of values and coloring schemes for choropleth maps. By 1914, a set of more general recommendations for statistical diagrams was published by the American Statistical Association (Joint Committee on Standards for Graphic Presentation 1914), and by this time graphic methods

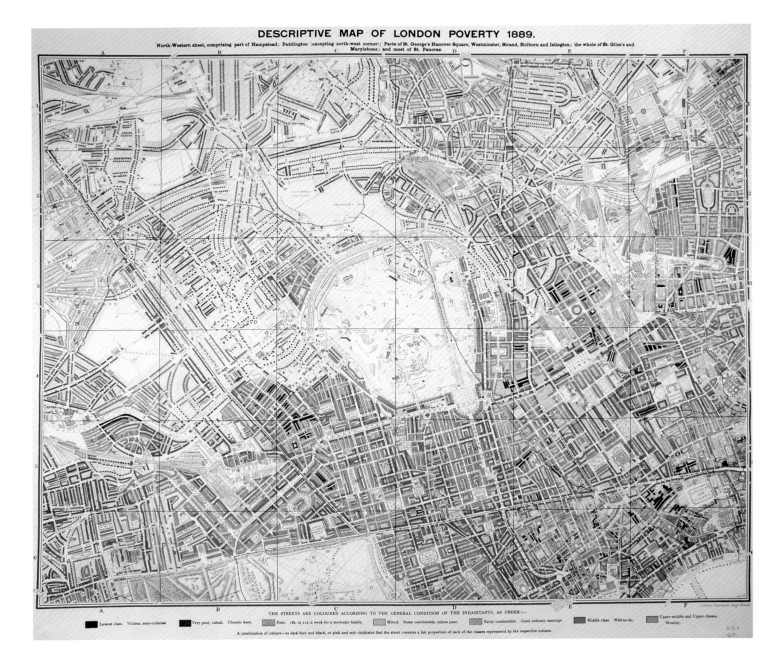

FIGURE 144.

Charles Booth, "Descriptive Map of London Poverty," northwestern sheet (1889).

became mainstream, entering textbooks, the curriculum, and standard use in government, commerce, and social and natural science.

Paradoxically, the golden age of innovation, enthusiasm, and graphic excellence in statistical graphics and thematic cartography had drawn to a close in the early 1900s, being supplanted by the "modern dark ages" of visualization, the rise of quantification and formal models in the social sciences (Friendly and Denis 2000). Statistical models and parameter estimates were precise; graphs and maps, on the other hand, were just pictures: pretty or evocative, perhaps, but incapable of stating a "fact" to three or more decimal places. So it seemed to many statisticians.

The Zeitgeist for statistical mapping and graphics changed again beginning in the 1950s, with several significant developments. At this time, postwar rebuilding and economic planning spurred huge developments in the application of thematic cartography, and the framework for a modern theoretical appraisal of visual symbolism was laid by Arthur Robinson (1952). In France, Jacques Bertin published the monumental *Sémiologie Graphique* (1967) that appeared to some to do for graphics what Mendeleev had done for the organization of the chemical elements. In the United States, John Tukey (1962) issued a call for the recognition of informal, robust, and graphically based data analysis distinct from mathematical statistics. Finally, initial steps in the computer processing of statistical data and computer-generated visual displays offered the possibility to construct new graphic forms (or at least construct them more quickly) and interactive graphic applications.

The present landscape for visualizing nature and society includes many new specialties, including geographic information systems (GIS), an emerging science of geovisualization (for example, Dykes, MacEachren, and Kraak 2005), volumetric brain and medical imaging, and visualization of high-dimensional and massive data sets. But it also faces new and globally important challenges: biotechnology, threat detection, global warming and environmental change, and patterns of propagation of AIDS and influenza are just a few current topics of application. Coupled with these forces, new technology now provides the means for dynamic interaction with statistical data and graphs. As new graphic methods are developed to help comprehend such phenomena, it is useful to understand also the deep roots that these methods have in the history of thematic cartography and statistical graphics. And although we think of these maps and graphics as the products of science, their makers' desire to communicate information effectively produced work that is often elegant and even beautiful.

Notes

This work was supported by Grant 8150 from the Natural Sciences and Engineering Research Council of Canada to M. Friendly. We are grateful to the members of Les Chevaliers des Albums de Statistique Graphique for historical information, images, and helpful suggestions.

1.   The term *semigraphic* was invented by John Tukey (1972) to refer to visual displays that were a mix of graph, table, and numbers. In French, Charles Joseph Minard and others referred to *tableaux figuratifs* and *cartes figuratives* to denote combined forms of tables, maps, and graphs.

2.   Maunder plotted latitude of sunspot activity on the sun versus time and observed a cyclic pattern that resembled the wings of a butterfly (see http://www.windows.ucar.edu/tour/link=/sun/activity/butterfly.html). The explanation for this recurrent migration is still unknown.

3.   The estimates were all biased upward from the true distance (16" 30'), likely due to underestimation of the earth's circumference (Tufte 1997).

4.   In July 1861, Galton distributed a circular to meteorologists throughout Europe, asking them to record these data synchronously, three times a day for the entire month of December, 1861. About fifty weather stations supplied the data; see Pearson (1914, 37–39).

5.   Not everyone concerned with statistical data and social problems was enthusiastic about graphic methods. In France, the influential early statistician Jacques Peuchet (1805) considered Playfair's figures childish games, irrelevant to science. In England, early statisticians ("statists," as they called themselves) who formed the basis for the Royal Statistical Society were more concerned with the presentation of statistical "facts" in tables, leaving the generalization to laws and theories to others.

6.   It showed a relatively clear-cut separation between the north and south of France along a line from Geneva to Saint-Malo in Brittany. This sharp cleavage between "la France éclairée" (enlightened France) and "la France obscure" would become reified as the "Saint-Malo–Geneva line" and generate much debate about causes and circumstances through the end of the nineteenth century.

7.   The direction of the visual encoding of "more" and "less" is also crucial in understanding the relations between different data shown on maps. Thus, Guerry (1833) was careful to arrange the scales of data on crime, instruction, suicide, and so forth, so that darker always meant "worse" (more crime, less education—or more illiteracy).

8.   Initially, at the 1832 founding of the Statistical Section of the British Association for the Advancement of Science (later to become the Royal Statistical Society), the view was expressed that its concerns were restricted "to facts relating to communities of men which are capable of being expressed by numbers, and which promise when sufficiently multiplied to indicate general laws." That might sound good, until it is realized that the "facts" were to be expressed simply as numbers in tables, and the "general laws" were to be inferred by others: "the sciences of morals and politics are far above the spec-

MICHAEL FRIENDLY AND GILLES PALSKY

ulations of our philosophy" (Mouat 185, 15). By 1885, at the Silver Jubilee of the Royal Statistical Society, graphic methods and reasoning to conclusions had become mainstream, as Alfred Marshall addressed the attendees on the benefits of the graphic method, and Émile Levasseur presented a survey of the wide variety of graphs and statistical maps then in use (see Marshall 1885; Levasseur 1885).

MAPPING IMAGINARY WORLDS

Ricardo Padrón

Anyone who has read J. R. R. Tolkien's *The Lord of the Rings* knows very well just how helpful a map can be when we're trying to follow the story line of a travel narrative. The adventures of Tolkien's characters range across a geography just as detailed as it is imaginary, making the maps that accompany each of his volumes an indispensable tool for finding our way through his stories. But this hardly exhausts the interest that Tolkien's imaginary world inspires. At least two atlases have been published providing myriad maps of Tolkien's imaginary universe at all different scales, dedicated to different themes, and charting the course of different events (fig. 145). Their authors combine painstaking attention to Tolkien's writing with considerable technical and artistic skill as mapmakers—as well as hefty doses of imagination—to round out, fill in, and extrapolate from the geographic information provided by the texts themselves (Fonstad 1991; Strachey and Tolkien 1981).

The sheer effort expended in producing these atlases speaks to the intrinsic fascination of such projects. Far from serving merely as reader's aids, maps of imaginary worlds, be they Tolkien's Middle Earth, Dante's Inferno, C. S. Lewis's

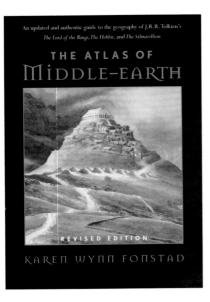

FIGURE 145.

Edoras, by Alan Lee; cover of Karen Wynn Fonstad, *The Atlas of Middle Earth* (1991).

Narnia, Frank Baum's Oz, or any other, have value as creative endeavors in their own right. Their appeal is not unlike that of other art forms, whether verbal or plastic. Maps of imaginary worlds delight. They distract. They reveal truths. They whisper secrets. They unsettle. They reassure. Yet, like any artifact of the human imagination, maps of imaginary worlds do all these things in particular ways, and for particular purposes. This chapter explores some of these maps, in the hope of understanding at least some of the possibilities involved in this vast but peculiar branch of mapmaking.

But what makes a world "imaginary"? And what makes a representation of such a world a "map"? When our example is the map of Middle Earth, then these questions have answers so obvious they don't bear repeating; but when we cast our nets more widely, we soon run into problems. As the literary scholar Thomas Pavel reminds us, the fictionality of fictional worlds often lies in the eye of the beholder (1986, 80). Sacred geographies represent the most obvious example. To the believer, maps of the afterlife or of the worlds of the gods are anything but imaginary, while to the nonbeliever, they are anything but real (D'Aulaire and D'Aulaire 2005) (fig. 146). What about maps of lost continents like Lemuria or Atlantis, which are mere myths for most of us, but have at different times by different people been believed to be real?[1] The worlds of many literary works have been subject to similar changes in perception (Pavel 1986, 81). There is a whole genre of literary maps dedicated to tracing the real-world settings of fictional events, or the location of events once thought to have been real but now recognized as fiction (see Hopkins, Buscher, and Library of Congress 1999) (fig. 147). Do these maps count? How about disproportionate maps that tell us how certain people perceive certain spaces (fig. 148), or maps that show how territory would have worked out had some war been won by the losing side (Post 1973, vii)? What about allegorical maps, which render moral lessons or philosophical ideas as territories (see Swaaij and Klare 2000; fig. 149)?

And why limit ourselves to iconographic maps? After all, as Denis Cosgrove reminds us in chapter 2 of this volume, "'World' is a social concept." There is a difference, in other words, between a physical geography, however vast, and a "world," conceived as a structured set of social, political, or cultural affairs, or a more or less systematic collection of ideas. For this reason, many imaginary worlds, from those of ancient mythology to the fanciful creations of contemporary literature, are known to us primarily, if not exclusively, through stories and verbal descriptions. Whether or not these worlds are mappable, the fact remains that very many of them remain unmapped. Sometimes these worlds are explicitly described, as is the case of the labyrinthine library of Jorge Luis Borges. Sometimes these worlds are constructed piecemeal by the story, as in the case of Macondo, the fictional town that serves as the setting for *One Hundred Years of Solitude*, by Gabriel García Márquez. And so it goes with many other fictional worlds that we can identify in the writing of Italo Calvino, Franz

RICARDO PADRÓN

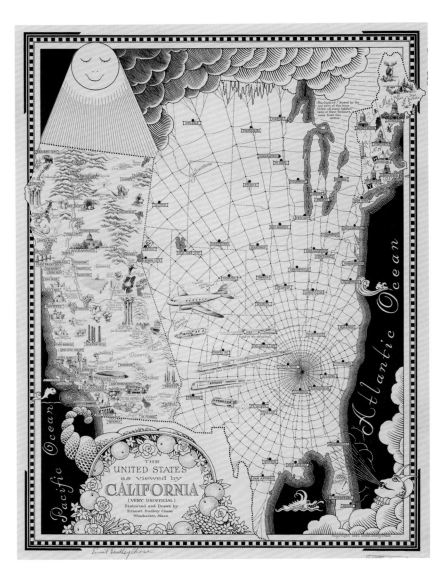

FIGURE 148.
Ernest Dudley Chase, "The United States
as Viewed by California (Very Unofficial)" (1940).

Kafka, Margaret Cavendish, Edgar Rice Burroughs, Jules Verne, Jan Amos Ko-
mensky, Henri Michaux, Julien Gracq, and so many others: they become real
to us in the reading.

Couldn't we say that these texts themselves, therefore, constitute some
sort of map, even if they do not come accompanied by illustrations? After all,
they allow us to create mental images of the places they describe, even in the
absence of actual illustrations. Not only do they allow us to picture places and
spaces, but by telling stories that take place in them, or by sculpting characters
associated with them, they give those places life and meaning. Indeed, any
iconographic map of the worlds imagined by these texts might even miss the

RICARDO PADRÓN

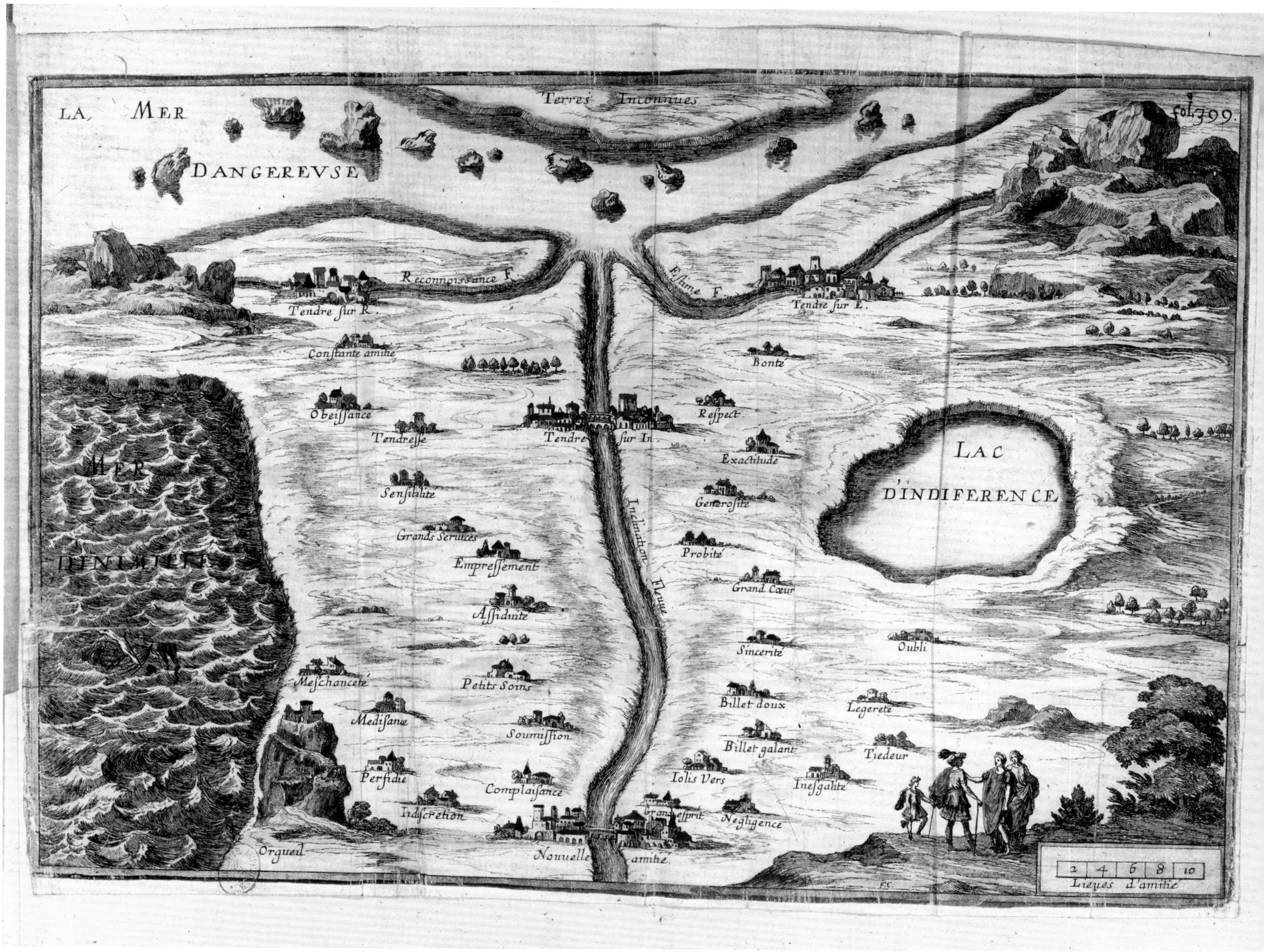
LA MER DANGEREVSE
Terres Inconnues
Reconnoissance F.
Estime F.
Tendre sur R.
Tendre sur E.
Constante amitie
Bonte
Obeissance
Tendresse
Tendre sur In
Respect
Exactitude
LAC D'INDIFERENCE
Sensibilite
Generosite
Grands Services
Probite
Empressement
Grand Cœur
Inclination Fleuve
Assiduite
Sincerite
Oubli
Meschancete
Petits Soins
Billet doux
Legerete
Medisance
Soumission
Billet galant
Tiedeur
Perfidie
Iolis Vers
Inegalite
Complaisance
Indiscretion
Grand esprit
Negligence
Orgueil
Nouvelle amitie
2 4 6 8 10
Lieues d'amitie

FIGURE 149.
"La Carte du Tendre" (Map of Tenderness). From Madeleine de Scudéry, *Clélie* (1654).

point, by reducing their rich engagement with space and place to the fixity of a cartographic image. It is even possible that the words themselves can be considered a map of sorts, when their arrangement on the page suggests cartographic meanings in the manner of a concrete poem (Conley 1996).

But to cast the net so widely would blow this discussion out of all manageable proportion, and would lose sight of some important distinctions. There is something unique that happens when we take pen in hand and attempt to draw a map of an imaginary world, or when we have the opportunity to explore such a map drawn by someone else. However effective words may be at helping us imagine spaces, at allowing us to enter them and inhabit them, however common and even powerful verbal mapping may be, words do not have the same impact, do not provide quite the same experience. That impact has everything to do with the seductions of *seeing* a world that is not our own. Like all seductions, this one offers both pleasures and perils.

And so I will make remarks about verbal worlds only in conjunction with the maps that depict them, and I will limit my discussion primarily to iconographic maps of imaginary worlds. I will also limit myself to those imaginary worlds that are depicted as some sort of physical geography invented at a particular time by a particular person, a geography that all readers in their right minds understand to be fictional.[2] Furthermore, I will favor those worlds that have some kind of connection with literature and storytelling. To some extent this decision stems from the limitations of my own expertise, but it also represents an attempt to turn that limitation into an opportunity. Maps of imaginary worlds that are linked to literature and storytelling come with a prepackaged point of comparison, the text that they purportedly illustrate. By comparing the map with the text, however generally, I hope to get at some of the things that the maps do. After a voyage through some interesting maps of fictional worlds that will regrettably leave out many fine, charming examples, I will have some things to say about another category of creative maps that I have no choice but to leave out: the maps and mapping practices of the visual arts.

THE RENAISSANCE GOES TO HELL

We learn something crucial about maps of imaginary worlds by looking at some of the earliest efforts to map one of the most famous literary worlds in Western culture, Dante's Hell. Most people are familiar with the general outline of the *Divine Comedy*, the long narrative poem in which Dante tells the story of an imaginary journey through Hell, Purgatory, and Heaven. Along the way Dante witnesses the fates of dozens of individuals, past and present, general and specific, who count among the damned or the saved. The poem constitutes a rich and complex meditation on the philosophical, theological, moral, and political issues central to Dante's time, and indeed to other times as well. It also maps the afterlife in a very rough manner, by telling the story of the journey through it. Along the way, we encounter a smattering of descriptive passages like this one, from the eleventh of the *Inferno*'s thirty-four cantos:

"My son, within these rocks," he then began, "are three lesser circles, one below another, like those you are leaving. All are full of accursed spirits; but in order that hereafter the sight alone may suffice you, hear how and why they are impounded. . . . All the first circle is for the violent: but because violence is done to three persons, it is divided and constructed in three rings. To God, to one's self, and to one's neighbor may violence be done." (Dante 1980, 11.16–33)

RICARDO PADRÓN

Virgil addresses Dante as he leads him from one part of Hell to another, and maps out what they will be seeing as they descend to its lowest levels.

During the fifteenth century, a Florentine architect by the name of Antonio Manetti decided that one could gather the information presented in these passages and extrapolate from it to map out precisely the size, shape, and location of Dante's Hell. Manetti's work would not make it into print for some time, but his ideas would be popularized in summary form by others, fueling what John Kleiner (1994, 24) has called "the heyday of infernal cartography," stretching, roughly, from 1450 to 1600.[3] Italian intellectuals, particularly Florentines, debated, questioned, and refined Manetti's "Dantean cosmography," and even converted his argument to maps that accompanied their own editions of Dante's poem and their commentaries on it.[4] Dantean cosmography became an intellectual fad that attracted the attention of some leading thinkers, including no less a figure than Galileo Galilei (Kleiner 1994, 24–26).

One of the earliest and most important contributions to the cartography of Hell was made by none other than Sandro Botticelli, one of the leading artists of the Italian Renaissance. His detailed and colorful chart of Hell formed part of an ambitious series of illustrations for the *Divine Comedy* that he drew during the 1480s and 1490s at the behest of Lorenzo di Pierfrancesco de' Medici, a cousin of Lorenzo the Magnificent (fig. 150).[5] Unlike previous illustrators, who had mapped Hell as a schematic vertical cross section, Botticelli rendered Dante's Hell in three-dimensional space, and populated it with minute figures derived from scenes in the poem itself (Morello 2000, 323). Hell appears as a large funnel tapering down from the top of the parchment page to the bottom-most ninth circle, where we find Satan trapped in eternal ice. In the upper left-hand corner, we can spot the figure of Dante, dressed in a red tunic, as well as that of his guide, Virgil, dressed in blue. The two figures then reappear at other places in the chart, marking different encounters between them and the damned. Specific scenes from the *Inferno* can be made out quite clearly. At the upper left, we see Dante and Virgil entering Hell, and approaching the ferry of Charon, the boatman of the Underworld. At the upper right, the city of Limbo houses the virtuous pagans who have died without baptism. Although Botticelli's image never made it into print during the Renaissance, other maps of Hell, like the one printed in Venice in 1515 (fig. 151), clearly follow the model he established (Morello 2000, 324).

One might well ask why anyone would go to so much trouble. After all, in our own day, this sort of project has been thoroughly discredited. The most influential modern Dante critics insist that there is no coherence to be found in the descriptive passages that punctuate the *Inferno*, that passages like the one cited above have a primarily stylistic purpose, that of enhancing the *Inferno*'s realism.[6] In the *Divine Comedy*, Dante feigns that his story of a journey through Hell, Purgatory, and Heaven is actually the account of a dream vision that he has experienced, and so takes measures to make his story as convincing as pos-

sible. The descriptive passages contribute to the illusion he is trying to create, and nothing else. They cannot be made to yield Hell's overall design, at least not with the precision sought by the Dantean cosmography of the Renaissance, without making any number of unwarranted assumptions and questionable interpretations. According to these critics, Manetti and the rest did not discover the infernal architecture that underpinned Dante's poem, but instead projected onto the poem their own aspirations for it.

In their defense, they were not alone in their attempts to picture Hell. Dante's poetic depiction of the kingdom of the damned is so vivid, so compelling, that it seems to cry out for illustration. Artists as diverse as Jean Fautrier, Salvador Dalí, Renato Guttuso, Tom Phillips, Robert Rauschenberg, William Blake, and Gustave Doré have responded to the challenge with drawings, paintings, or engravings (Nassar 1994). Auguste Rodin cast a series of scenes from the *Inferno* in bronze in his monumental *Gates of Hell*. Yet there were specific reasons why Renaissance intellectuals may have wanted to map Hell the way they did, with greater precision than the poem seems to allow. John Kleiner

RICARDO PADRÓN

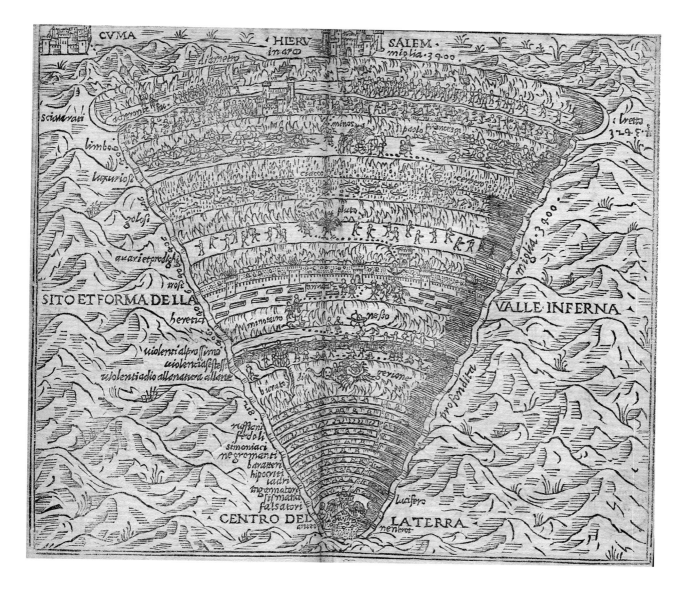

FIGURE 151.

"Sito et forma della valle inferna" (Position and Form of the Valley of Hell) (1515).

points out that much of the work that was done in Dantean cosmography bore the stamp of civic pride, that Florentine intellectuals were eager to reclaim Dante as one of their own. He also reminds us that Renaissance Florence was abuzz with the excitement of the new scientific cartography that issued from the rediscovery of Ptolemy, and that this excitement was itself related to a strong taste for an aesthetics of order and geometric rationality (Kleiner 1994, 26–33). These men who were so committed to mapping Hell, in other words, were trying to find in Dante a mirror of their own modernity. And so when they read the *Inferno* and tried to extrapolate the size, shape, and site of Hell from its meager descriptive details, they found the sort of symmetry and geometric order that shows up not only on all Renaissance maps of Inferno, but on so many

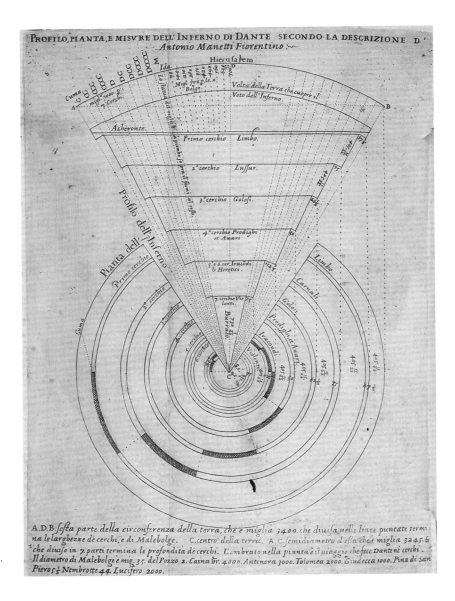

PROFILO, PIANTA, E MISVRE DELL' INFERNO DI DANTE SECONDO LA DESCRIZIONE D'
Antonio Manetti Fiorentino :~

Hierufalem

Volta della Terra che cuopre il
Voto dell'Inferno.

Acheronte.

Primo cerchio   Limbo.

2.º cerchio   Luffur.

3.º cerchio   Goloſi.

4.º cerchio Prodighi
et Auari.

5.e 6.cer. Iracôdi,
& Heretici.

7. cerchio Vio-
lenti.

Profilo dell'Inferno

Pianta dell'

Primo cerchio

Limbo

Carnali

Goloſi

Prodighi, e Auari

Iracondi

Violenti

A.D.B. feſta parte della circonferenza della terra, che è miglia 3400. che diuiſa nelle linee puntate termina le larghezze dé cerchi, e di Malebolge.   C. centro della terrá.   A. C. ſemidiametro d'eſſa che è miglia 3245.⅘
che diuiſo in 7. parti termina le profondità dé cerchi. L'ombrato nella pianta è il uiaggio che fece Dante né cerchi.
Il diametro di Malebolge è mig. 35. del Pozzo 2. Caina br. 4000. Antenora 3000. Tolomea 2000. Giudecca 1000. Pina di San
Pietro ⅓ Nembrotte 44. Lucifero 2000.

FIGURE 152.

"Profilo, pianta, e misure dell'Inferno" (Profile, Plan,
and Dimensions of Hell) (1595).

Renaissance maps of real places (fig. 58). This interest in symmetry and order
can be readily seen in Botticelli's map, and even more so in the very abstract
and geometric map that accompanied the 1595 edition of Dante prepared by
Florence's Crusca Academy (fig. 152). In this way, Renaissance Florence made
Dante its own, just as every place and time makes literary classics its own by
reading them in the light of its own concerns. It sought and found in the poet's
vision of Hell the sort of order and symmetry that it hoped to find in its own
world.

What is true of these maps of Hell is true of many maps of imaginary
worlds. Rarely do literary texts provide enough of the right kind of detail to al-

RICARDO PADRÓN

low us to map their worlds in conclusive, indisputable ways. Mapping involves interpretation, and interpretation always contains an irreducibly subjective component. Sometimes that interpretation may take unwarranted liberties, particularly when it ignores contradictions among descriptive passages that make the world of the text impossible to map without ignoring certain details. In fact, Dante's *Inferno* may be precisely just such a text. Sometimes mapping these worlds may be beside the point. Literature of all kinds has a great deal to tell us about space and place, but the things it has to communicate are not necessarily of the sort that lends itself to cartographic representation. Mapping involves visibility, stasis, hierarchy, and control. Literature often works to subvert these things. It has us experience space and place in myriad ways that have little to do with mapping it, just as it has us experience time in many ways that cannot be measured by a clock.[7] Yet people make and enjoy these maps. In doing so, they make the text and its world their own, and in doing so may be mapping themselves just as much as anything else.

ISLANDS AND INSULARITY

One of the reasons that Dante's Hell lends itself toward the sort of speculation I have just discussed is because the *Inferno* clearly suggests that Hell is an orderly, well-bounded space, however vague and contradictory it may be about the details of size, proportion, distance, and location. Spaces of this sort seem like they should be visible, and therefore suggest that they might be mappable. In this way, Hell resembles an island, one of the favorite spaces of the cartographers of the imaginary. An island is clearly bounded and set off from the rest of the world. It has no *terrae incognitae*, no feisty neighbors, no disputable borders, no porous frontiers. Unlike a continent, with its vast spaces, islands can be taken in at a glance, giving us the impression that we can know them completely (Lestringant 1980). Many imaginary islands even come conveniently equipped with a centrally located mountain or hill from which to do so.

Islands are also the perfect setting for an adventure story. Georg Simmel notes that adventures, like islands, are set apart from life and have clear-cut boundaries, that is, clear-cut beginnings and endings (1959, 244–45). Arrival on an island, particularly by ship or shipwreck, is one of the very definitions of embarking upon an adventure. Departure from that island clearly marks the end of the ride. It is no surprise, then, that we can point to any number of adventure stories, including *Robinson Crusoe* (1719–22) and *Treasure Island* (1881), that are set on imaginary islands.

In the case of *Treasure Island*, the island and its map play an illuminating role. In this instance, the map of the island came before the adventure story (fig. 153). Robert Louis Stevenson drew it with his father and stepson, and

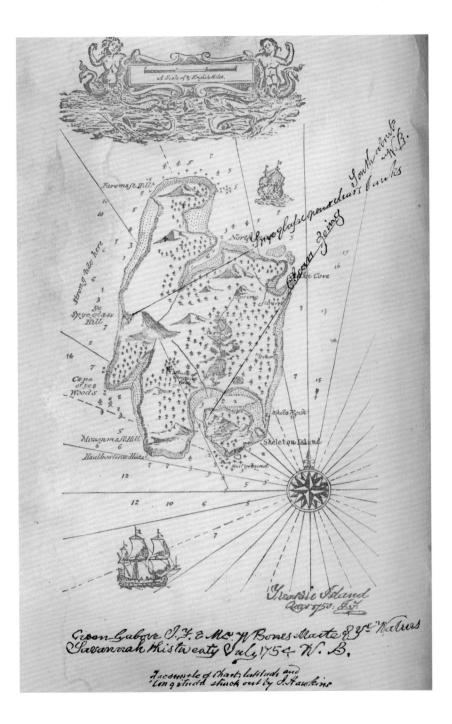

only afterward thought to write a pirate story to go with his treasure map. The combination of the insularity of the island with the possibility of discovering treasure there must have proved irresistibly seductive for Stevenson, as it does for readers of his book and as it has for real-life explorers. We discover the map

RICARDO PADRÓN

with Jim Hawkins, and we have to get to the island, find out what's there, see what happens. The island itself, that perfectly possessable geographic object, displaces the treasure as the reader's object of desire, and the map becomes the symbol that draws us to it. Angus Fletcher even suggests that "the map seduces the reader into an expectant state of unfulfilled desire. . . . One wants to say that in its magically seductive power the map is the most dangerous symbol in the world" (2005, xxxvi).

With their clear geographic distinctiveness, islands are also perfect settings for utopias and dystopias (as are planets, their science-fiction cousins). This has been true since the origins of Western utopian literature in the writings of Sir Thomas More. His *Utopia*, originally written in Latin and entitled *Concerning the Best State of a Commonwealth and the New Island of Utopia*, first appeared in 1516. It packages a program for social and moral reform in a learned and witty literary device. More tells the story of a fictional meeting with a fictional traveler, Raphael Hythloday, who claims to have visited a hitherto undiscovered country on the far side of the globe. During the conversation, Raphael describes this country, Utopia, an ideal state set on an island, in which a wise government keeps its people on the straight and narrow by implementing a sober constitution based on a small number of reasonable principles. The result is a microcosm removed from the tumult of history, blessed with order, peace, prosperity, and virtue (see Schaer, Claeys, and Sargent 2000).

Both the first (1516) and the third (1518) editions of *Utopia* came equipped with maps (figs. 154 and 155). Like the Renaissance maps of Inferno, these renderings take some liberties with the geographic descriptions provided by the text in order to produce a coherent, mappable geography that may not, in the end, be supported by every interpretation (More 1992; Bony 1977; Goodey 1970; McClung 1994). Unlike those Inferno maps, however, the first of these seems to have been authorized by More himself, and the second seems to have been produced, if not with his consent, then at least without his interference (More 1965, 276–77). It thus becomes more difficult to dismiss these maps as external impositions by others seeking to advance a particular critical agenda. But what is their relationship with the text? They so simplify or distort Utopia's geography that they are useless as reader's aids. For this reason, they have been subjected to various symbolic interpretations. Scholars have speculated about the meaning of Utopia's crescent shape; about the way it encloses the central bay as if it were a womb of some kind; about its insularity; about its relative closure to the outside world (Jourde 1991, 110–11; Marin 1984). I add that these maps also extend what *Utopia* has to tell us about the links we make among bounded geographies, mapping, visibility, and knowledge.

More's description of Utopia is shot through with all sorts of ironies. Even its name is built out of an erudite pun with opposites. *Utopia* is a coinage of More's that puns on the Greek *ou-topos*, meaning "no place," and the Greek *eu-topos*, meaning "best place." It is the best place, but it exists nowhere. Uto-

UTOPIAE INSVLAE FIGVRA

FIGURE 154.
(above, left) "Utopiae insulae figura" (Map of the Island of Utopia) (1516).

FIGURE 155.
(above, right) Ambrosius Holbein, map of Utopia (1518).

pia is presented to us, moreover, in the form of a questionable traveler's tale, drawing on a long tradition of dubious travel accounts describing fantastical places, to be believed only by the gullible. And what is presented to us is a country built upon principles so alien to sixteenth-century European society that it could hardly be reproduced back home without changes so radical as to be unrealizable. For all these reasons, scholars have debated how to understand *Utopia*'s message regarding politics and ethics: whether to interpret it as a literal blueprint, as an impractical ideal, as a yardstick of progress, or something else.

The key to understanding the maps is to recognize that they do not help us understand Utopia or its meaning: they *just* help us visualize it, and in doing so, they play a subtle game with us. It is a game quite common to the sort of playful, literary travel narratives that More draws upon. Even as they describe absurd and fantastical locales, they insist upon their own veracity, and offer different kinds of evidence to "prove" that they are telling us the truth.[8] The map of Utopia can be interpreted as just this sort of "evidence," a real map of an

RICARDO PADRÓN

imaginary place, meant to blur the line between the fictive and the real, meant to play with our credulity when it comes to images.[9]

Our suspicions only intensify when we turn to the map that Ambrosius Holbein prepared for the 1518 edition. Like the 1516 version, Holbein's image follows the general outline of Raphael's description, but it alters some aspects of the composition of its predecessor, and adds some crucial elements of its own (Bony 1977). These include three signs naming Utopia's capital city, and identifying the source and the mouth of its principal river, the Anydrus. The signs hang from a garland that is itself suspended from the frame of the map. In this way, signs and garland contribute to the creation of an illusory third dimension. They hang in the frame through which we look out over Utopia, and even block part of the view. We share the view with three figures in the foreground. One of them, identified as Raphael Hythloday, points at the island while he speaks with one of the other two men. These men, we can presume, are More himself and his companion Peter Giles, Hythloday's interlocutors in the text. The 1518 map, therefore, does not just allow us to picture the island, but to picture the scene in which Raphael describes the island to his friends.

A profound irony immediately leaps out at us. The image has us *see* Raphael *showing* the island to More and Giles, and it makes it possible for us to *see* Utopia along with them. But all of this looking and seeing stands in marked contrast with what is actually happening. Utopia cannot be seen, by us or by anyone else, because it does not exist. Raphael is a dubious fictional traveler who describes Utopia to More and Giles, and we read More's imaginary recreation of this fictional conversation. Utopia, in other words, is an idea that comes to us through a series of narrative filters of questionable reliability. The 1518 map erases these filters, staging an illusion of immediate visual access to the country that Raphael describes. It invites us to forget the mediating role of words, and the possibility that those words might be distorting the message. In this way, ironically, it underscores the caginess of More's text.

That text, the critics tell us, is deeply ironic. It is just as much about the problems involved in knowing the truth and putting it into practice in the here and now as it is about painting an ideal image of a utopian world (Wooden and Wall 1985, after Berger 1965, 40; and Greenblatt 1980, 23). This is the reason why fully half the book is dedicated to a conversation among More, Giles, and Hythloday in which they lament the sorry state of sixteenth-century society (Baker-Smith 1991). This is the part that sets Utopia at a distance that is not just geographic but philosophical as well. It sets it behind the mediating role of representations, and underscores our place in the here and now, where these questionable representations are all we have. It invites us to think about how far we are from the ideal, and about what we could do, if not to make Utopia real, then at least to make our own world better. The maps, therefore, are not just maps of an imaginary island, made available so we can see and know it.

They are emblems of our *desire* to know and possess that island, itself a symbol for the true, the beautiful, and the good, those treasures that so often elude us in real life. Our job as readers is to recognize these maps for what they are: fantasies about not only Utopia, but our quest for knowledge itself.

Something similar happens two hundred years later, in the *Travels into Several Remote Nations of the World* (1726), by the master of English satire, Jonathan Swift. In this early protonovel, better known as *Gulliver's Travels*, the Englishman Gulliver visits a series of imaginary places, all of them set on islands or, in one case, an isolated peninsula. Each island features a distinctive, self-contained culture that serves to satirize some aspect of English or European society, or of the human condition more generally. The original edition of *Gulliver's Travels* included five maps of these countries: Lilliput (fig. 156), Brobdingnag, Laputa, Balnibarbi, and the Land of the Houyhnhnms. Like the maps of Utopia, these work in tandem with the text they illustrate to trigger a critical reflection upon the nature and value of certain kinds of knowledge.

On one level, the object of that reflection is cartography itself, with all its claim to scientific authority. Swift disdained geography and cartography as one of those modern sciences that provided knowledge of the physical world, but added nothing to the moral improvement of human beings (Bracher 1944, 73). In his poetry, he ridiculed it memorably:

> So Geographers in *Afric*-Maps
> With Savage-Pictures fill their gaps;
> And o'er unhabitable Downs
> Place Elephants for want of Towns. (Ibid., 65)

Like the geographers of these lines, Gulliver is keen to fill in the blank spaces on existing maps. He reports that he has shared his geographic discoveries with the geographer Herman Moll, a real person, for incorporation into his world map. But Moll, he tells us, has rebuffed him, and "hath rather chosen to follow other authors" (Swift 2004, 269). This eagerness to round out the map of the world, however, is just an aspect of Gulliver's own foolishness. Swift's protagonist, it has often been said, acquires knowledge but not wisdom: he accumulates information but undergoes no real transformation (Hunter 2003, 224). He has learned where these countries lie, but has not assimilated any of the real lessons they have to teach him. His eagerness to fill in the map, like all cartography, turns out to be nothing but empty pedantry.

One of the ways the satire works is by mapping the islands Gulliver has discovered into real-world spaces traced from a 1719 map of the world by Herman Moll himself (Bracher 1944). Lilliput, we see, lies not too far from Sumatra, while Brobdingnag constitutes a peninsula on the Pacific coast of North America, and so forth. Certain inconsistencies with the text of *Gulliver's Travels* suggest that Swift did not have a hand in producing these maps, but neither

RICARDO PADRÓN

did he come out against their addition to the text by his printer (Bracher 1944, 64). Why would he? As a parody of a real travel narrative, it is only fitting that *Gulliver's Travels* come equipped with parodic maps. Their addition only extends the many ways in which Swift blurs the boundaries between fact and fiction in order to satirize the value his contemporaries attached to certain kinds of writing and knowledge.

But the relationship between the maps and Swift's satire may run deeper than that. Carole Fabricant (1995) points out that one of the fundamental characteristics of *Gulliver's Travels* is its constant transgression of boundaries that should be stable. The text blurs the line not only between fact and fiction, but also between novel and travel narrative, understanding and ignorance, the human and the nonhuman, wisdom and foolishness. In book 4, Gulliver finds himself among the hyper-rational Houyhnhnms and the basely passionate Yahoos, and it is clear that his own humanity constitutes a blend of these two thoroughly distinct and segregated creatures, or lies somewhere in between them. Gulliver, however, never understands how he blends the two, or where he lies in relationship to them. Speaking metaphorically, we can say that the text can map (that is, identify, locate, describe) the reasonable Houyhnhnms and the passionate Yahoos, but not the human being.

The maps, therefore, in their enthusiastic attempt to locate these new discoveries, miss the point just as thoroughly as does Gulliver himself. No man, in *Gulliver's Travels*, is an island—but not in the sense that we usually understand this expression. Human nature is contradictory, slippery, difficult to pin down. Unlike an island, the boundaries of the human are difficult to delineate, and its location relative to other beings, difficult to pin down. In this way, it is fundamentally "unmappable." The maps that accompany *Gulliver's Travels*, by contrast, confidently map the islands inhabited by the settled, well-bounded, mappable, and therefore phony cultures that Gulliver discovers. Like the maps of Utopia, the maps of Gulliver's discoveries need to be handled with caution. They masquerade as reader's aids, but should be understood as traps. They invite reverie, but their invitation needs to be resisted. In order to learn their wisdom, we cannot just look at them, we cannot inhabit them imaginatively. We must learn how to interpret them, how to recognize them for the deceits they are.

FIGURE 156.

Herman Moll, map of Lilliput (1726).

FANTASY, FUN, AND FAIRIES

Not so with what are certainly the most famous maps of imaginary worlds of our own day, the ones that accompany twentieth-century works of fantasy and science fiction (Post 1973). These literary genres have roots deep in the past of literary history, and draw on Swift, More, Dante as much as on anything else; but they do not emerge in their modern forms until the mid-nineteenth century (see Mathews 1997; Roberts 2006). According to Richard Mathews, fantasy as we know it—an adventure story set in an imaginary world constructed from fragments of medieval literature and history, mythology, folk legends, and the like—develops primarily in the English-speaking world as a reaction to both literary and historical trends. It rejects the realism that was then dominant in the novel, and reacts in ways that are sometimes quite conservative and at other times quite radical to the profound changes wracking the Western world at the time, such as industrialization and secularization (Mathews 1997, 16–20). From early on, landscape, geography, and even maps figure prominently in fantasy literature, and through the influence of C. S. Lewis and John Ronald Reuel Tolkien, they become standard fixtures of the genre.

The maps that accompany the fantasy novels of J. R. R. Tolkien are undoubtedly the best known of all, and they are by far the most interesting.[10] Part of their appeal must have to do with the way in which they help to make Middle Earth so palpable. Not only do they help us picture Tolkien's world, but like the map from *Treasure Island*, they allow us to imagine that they themselves constitute artifacts from Middle Earth. This is especially true of Thror's map, which appears as one of the endpapers to *The Hobbit* (1937). The map depicts the Lonely Mountain, where the dragon Smaug guards the treasure that the dwarves in the story want to recover (fig. 157). Like the Marauder's Map in J. K. Rowling's *Harry Potter and the Prisoner of Azkaban* (1999), this is a map used by the characters themselves to understand their world and plan their adventures. The narrator in *The Hobbit* even invites the reader to join the characters in doing so. As Gandalf, Bilbo Baggins, and the dwarves decipher the runes on the map, the narrator addresses us directly, in a parenthetical remark: "Look at the map at the beginning of this book, and you will see there the runes in red" (Tolkien 1996a, 19).

The other maps that come bound with *The Hobbit*, *The Lord of the Rings* (1954–55), and *The Silmarillion* (1977) are not presented the same way; they are clearly authorial or editorial impositions, rather than feigned documents from the imaginary world itself (fig. 158). But they too work to make Middle Earth palpable, primarily by providing a surfeit of geographic information. All are rich with place-names that figure very slightly in Tolkien's stories. Their presence on the maps invites us to explore. Traveling across the map of Middle Earth, we encounter places like the Mountains of Angmar, or the *terrae incognitae* of Rhûn and Hardwaith. In various ways, Tolkien links these places to his archvillain Sauron, but tells us little about them. Like the young narrator in Joseph Conrad's *Heart of Darkness* (1899) gazing at the blank spot in the map of Africa, we find these places where the known meets the unknown and wonder what they might contain (Jourde 1991, 127). There, the world created for us by the author recedes into the mists of what Tolkien has left unsaid, allowing us to imagine and speculate.

This invitation to daydream arises not only from the wealth of information on the Tolkien maps, but also from the way in which most of them are drawn.

RICARDO PADRÓN

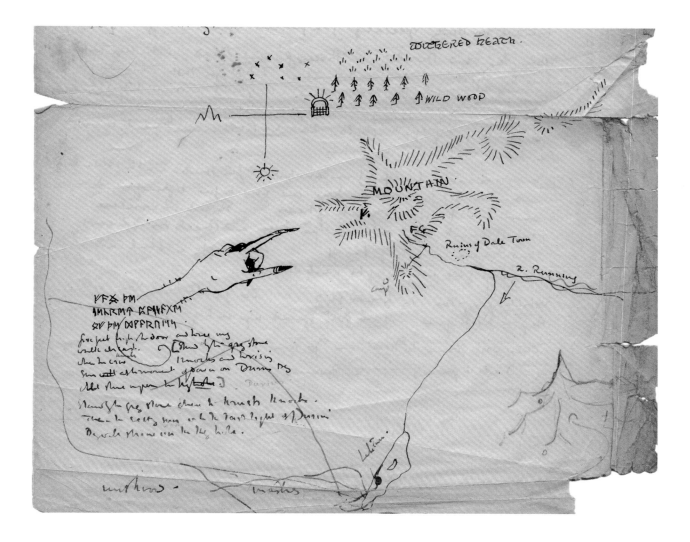

The relatively large-scale map of the kingdoms of Gondor and Mordor is exceptional in this regard. It depicts the mountain ranges surrounding the kingdoms with modern contour lines. The other maps, by contrast, demonstrate more consistency of style. Although they bear a scale and sport a compass rose, on the whole they resist the abstraction of modern cartography, preferring a deliberate, stylized archaism that echoes Tolkien's writing. Mountains, hills, and forests are depicted iconographically, in profile, rather than with contour lines or abstract symbols. The lettering looks vaguely old-fashioned. We get the impression that these are maps from times long gone by, of a world that has vanished. But we are also drawn into the maps. The places drawn in profile do not allow us to remain aloft, looking down on Middle Earth from that imaginary point of view way on high that maps usually assign to us (see Wood and Fels

FIGURE 157.

J. R. R. Tolkien, "Thror's map of the Lonely Mountain" (author's manuscript draft, ca. 1935–36). © The J. R. R. Tolkien Copyright Trust 1937.

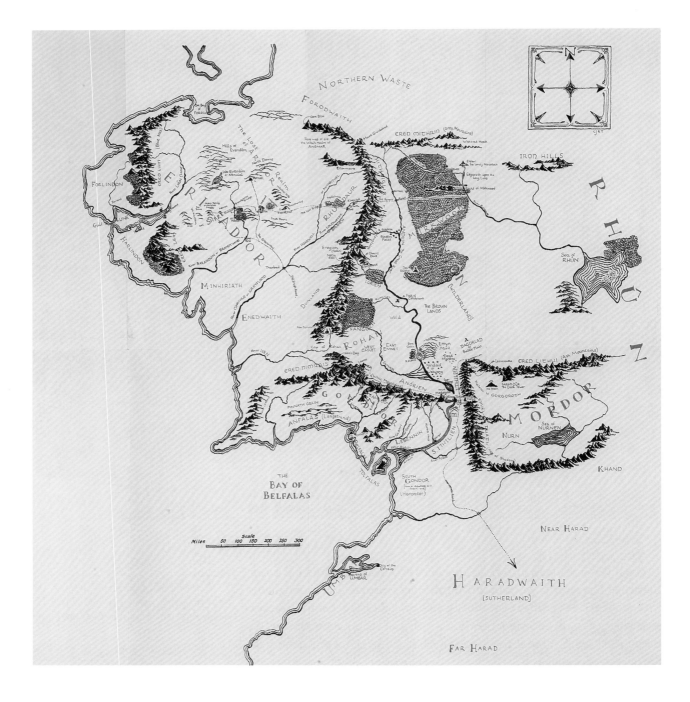

FIGURE 158.

Christopher Tolkien, "The West of Middle-Earth
at the End of the Third Age" (1954). © Christopher
Reuel Tolkien 1954.

1992, 152–76). They pull us down to earth (to Middle Earth, that is), inviting us to consider the landscape from the perspective of someone traveling through it. We follow the roads through the forests, across the mountains, along the rivers, sometimes tracing the paths of Frodo and the others, and sometimes forging our own way.

RICARDO PADRÓN

The world we move in, moreover, is fraught with danger. Pierre Jourde outlines this for us by way of a contrast with More's Utopia (1991, 126–29). While Utopia is orderly, balanced, proper, and unchanging, Tolkien's Middle Earth is shot through with a menacing opposition between Good and Evil. The west of Middle Earth is divided into two well-bounded regions, Eriador and Gondor, the lands of civilization and of Goodness, while the east is divided into two regions of wildness and Evil, Rhovanion and Mordor, regions that themselves open up to the mysterious *terrae incognitae* allied with Sauron. In this way, the map sandwiches the lands of the good guys between Evil and the sea. While its many place-names speak of a rich past, in the form of stories half forgotten or entirely unknown, its structure attests to the uncertainty of the future. Middle Earth lies suspended between a deep past and an impending apocalypse, and this is encoded on the map.

Many readers have associated that apocalypse with the real-world troubles that plagued Tolkien himself. It has often been said that his fiction reacts against many of the horrors of the modern age, just as nineteenth-century fantasy novels had reacted against the changes wrought by industrialization and secularization. As a veteran of the trenches of the First World War, Tolkien had seen firsthand the destructive potential of the machine age, and longed for a simpler, quieter time more closely linked to the rhythm and beauty of the natural world. The Shire, in effect, is his utopia. It is a fantasy recreation of an arcadian Britain that never was, one that finds itself threatened by evil forces that themselves represent the destructive forces of the machine age. It is no accident, then, that the landscapes of the Shire, on Tolkien's map, features rolling hills and folksy names. It is also no accident that Saruman, the lesser of the two evil wizards in *The Lord of the Rings* but the one that has a direct impact on the Shire, is associated with the destructive possibilities of wheels, machines, and gunpowder (Tolkien 1996b, 525).

Danger of this kind explains much of the appeal of fantasy literature, and of some of its more recent successors. In many ways, such novels are less about the triumph of Good over Evil than they are about the restoration of safety in the face of menace. The same is true for many of the imaginary worlds of massive multiplayer online games (MMOGs) like EverQuest, Westward Journey, and World of Warcraft. In such games, players with a subscription and the appropriate software use their computer to enter an online virtual world in the guise of a fictional character. There they interact with other characters, some controlled by the program and some by other players the world around. The games can accommodate thousands of people at once, and are available for play twenty-four hours a day. One of their attractions is their vast and varied physical environment, which is rendered graphically, and even breathtakingly so, in three dimensions on the player's computer screen. Another is the sense of adventure, and even of danger, that the games offer. Players take on various tasks, develop exotic skills, and, depending on the game, battle myriad foes.

None of the other imaginary worlds I discuss here pose the difficulties presented by these virtual worlds. As Edward Castronova (2005) has amply demonstrated, these games tend to blur the psychological, emotional, and economic boundaries between the real and the imaginary in ways so striking as to force us to rethink the distinction. But what interests us here is their cartography. Maps of the virtual world can be indispensable for game play, particularly when the game involves exploration and quests. When the makers of EverQuest, a fantasy-world MMOG descended from Tolkien by way of the tabletop role-playing game Dungeons and Dragons, failed to provide their world with a map, players assembled an online atlas of their own (*EQ Atlas* 2006). Players of another such fantasy game, World of Warcraft, have done the same. The makers of the game provide a base map depicting the two continents that constitute the world of the game, Kalimdor and The Eastern Kingdoms, along with some accompanying islands (fig. 159). World of Warcraft players have collectively turned this world map into the base map of an online atlas (World of War 2006). There are maps of individual regions and of cities. There are thematic maps depicting the resources available in different areas, and the quests one can undertake there.

But just as the fascination of Tolkien's maps does not end with their usefulness as reader's aids, so the attraction extends far beyond their role in facilitating game play. According to Richard Bartle, a pioneer in the design of such virtual worlds, people come to online-gaming because they are interested in social interaction, achievement, conquest and control, or exploration. The explorers "come to see what there is and to map it for others. They are happiest with challenges that involve the gradual revelation of the world. They want the world to be very big, and filled with hidden beauty that can only be unlocked through persistence and creativity" (Bartle 2004, 130; see also Castronova 2005, 72). These are the players for whom the game map holds a special appeal. What sort of appeal? As one avid player explained to me, he feels that he knows the world of World of Warcraft better, in some ways, than he knows the suburban area where he lives. After all, his journeys through that world are limited to a repetitive commute: he does not explore, and even if he did, what would he find? World of Warcraft offers things suburbia does not: novelty, adventure, and danger. It offers a rare commodity: a blank map and the opportunity to discover new worlds, all without having to abandon the security of suburban life (Scarborough 2006). While Tolkien's maps speak nostalgically of a preindustrial landscape, the maps of the World of Warcraft offer an escape from the drudgeries of a postindustrial one.

Imaginative travel is indeed one of the functions of all of these maps. While More and Swift warn us of the perils of such travel, Tolkien, Stevenson, the MMOGs, and others invite us to revel in it. Those who have played Risk know the thrill of embarking upon the conquest of Asia, and those who have

RICARDO PADRÓN

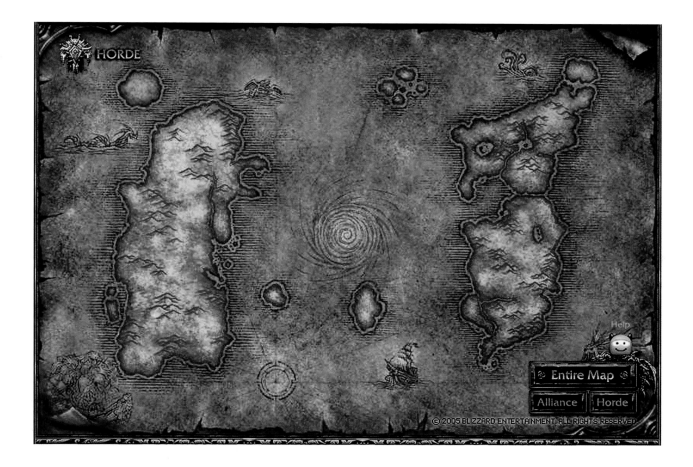

shared Candyland with a small child know how much fun it can be for them to make their way to the Candy Castle. One map, in my opinion, stands out for the pleasures of this kind that it has to offer children, Jaro Hess's *The Land of Make Believe*. Jaro Hess was a Czechoslovakian artist who immigrated to the United States, where he eventually entered his oil painting, *The Land of Make Believe*, into a contest at the Century of Progress Exposition, the world's fair hosted by Chicago in 1933.[11] The picture was enthusiastically received, and has gone into print as a poster various times since (Rosen-Ducat Imaging 2006). It collects various characters and scenes from well-known children's stories into a single landscape view. Among these we find the wall that Humpty Dumpty fell from, the Emerald City of Oz, and The City of Many Towers Where the Beautiful Princess Lives. Everything is depicted pictorially: nothing is reduced to cartographic abstraction.

There are other such maps, like Bernard Sleigh's "Anciente Mappe of Fairyland" (fig. 160), that do much the same thing (Post 1973, 90). The collection of these disparate characters and scenes into a single space is itself quite charm-

FIGURE 159.

Map from the online fantasy game World of Warcraft (2006).

ing, but what makes these images truly captivating is the way that space is organized. Both maps set the individual elements into their respective landscapes quite carefully, placing some in the foreground, others in the background, and connecting them all with routes that invite the eye to wander from one to the next. Both maps provide commanding heights that the viewer can occupy imaginatively, creating a sense of mastery over the landscape. The Sleigh map, however, puts everything on an enormous island, producing the sort of coherent and well-bounded geography we saw earlier. The Hess rendering, by contrast, opens out in various directions through a particularly powerful use of horizons and vanishing points. This feature is what makes the Hess work stand out.

At the center of the image, a strange tower of earth rises precipitously above the castle of Old King Cole. Atop the tower sits the house of Grandfather Know-All. Here, it seems, is Hess's answer to the Spyeglass Hill that, in Stevenson, allows one to take in the whole of Treasure Island and the surrounding sea. And it is only one such site. The background features a row of towers and cities and mountains from which one could enjoy a similar perspective. And if one does not want to remain earthbound in order to survey the Land of Make

RICARDO PADRÓN

Believe, there are the magic carpet, the Moo-Moo Bird, and the Moon. All of them look down over this enchanting world, taking it all in.

But not quite all of it. Along the left-hand side of the image, a river winds its way into a fjord or canyon of sorts. A ship makes its way up that river, under a bridge, toward the place where the north wind lives, just past the point where the walls of the canyon seem almost to converge. In the center background, a placid sea recedes into the horizon. At the foot of Grandfather Know-All's strange tower is a dark Bottomless Lake. Along the right-hand side of the image, a waterfall drops from an equally dark ravine that snakes into the background. The blackness of the lake and the ravine stands in marked contrast with the bright colors that characterize the printing. We know the river leads upstream somewhere, but do not know what we will find there. Neither do we know what mysteries, perhaps dangers, lurk in the crevice, the lake, or the sea. As in the map of Middle Earth, the eye runs up against these vanishing points that articulate the known with the unknown. They create openings for our imaginations to travel beyond the geographies that the map provides, beyond the familiar world of fairy tales and nursery rhymes, into the uncharted spaces of our own imaginations.

FIGURE 160.

Bernard Sleigh, "An Anciente Mappe of Fairyland" (1920?).

MAPPING IMAGINARY WORLDS

We have seen maps that project onto stories the desires that certain readers have for them. We have seen maps that challenge readers to resist those desires, to recognize their dangers and deceits. We have seen maps that invite readers to extend their stay in imaginary worlds, courting make-believe dangers and charting the unknown reaches of their own daydreams. Yet another kind of map stands out from the many maps of imaginary worlds tied to literature and storytelling that we can explore. These are interesting because of the kind of literature they accompany, a kind of novel that would seem to eschew any sort of cartography. Yet both of the maps I have in mind were drawn by the novelists themselves.

One of these is very well known—or at least its author is. It is the map of Yoknapatawpha County drawn by William Faulkner in two separate versions: one to accompany the original edition of *Absalom, Absalom!* (1936) and the other the original edition of *The Portable Faulkner* (1946). The two maps depict the imaginary Mississippi county that provides the setting for some of Faulkner's most famous novels and stories. Yoknapatawpha's history and geography are derived from that of Lafayette County, Mississippi, Faulkner's adoptive home; but they differ in many ways from the real thing, and thus depict a place usually understood as imaginary and not merely "fictional," in the way that we might speak of the London of Charles Dickens, a real London inhabited by imaginary characters (Aiken 1977, 1979). Faulkner's depiction has been analyzed in its own right and has been pressed into service as a symbol for his entire body of work. Some have seen in it, for example, "a spiritual geography of Christendom," while others have found in it a microcosm of the American South (Aiken 1979, 331).

The other map is surely unknown to most English-speaking readers. It depicts a place called Región (fig. 161), drawn by the Spanish novelist Juan Benet and published separately as a supplement to his groundbreaking novel, *Volverás a Región* ([1968] 1996), which appeared in English translation as *Return to Región* (1985).[12] Región is the imaginary world for this novel and several others by Benet that reflect upon the trauma of the Spanish Civil War and its aftermath. His "Mapa de Región" (Map of Región; incorporated in Benet 1983) provides a dramatic contrast to Faulkner's maps of Yoknapatawpha, not to mention others. While Faulkner's maps exhibit the graphic simplicity of a novelist's sketch map, Benet's map, with its abstract cartographic signs, its contour lines, and its distance scale, is indistinguishable from a real topographic map of an actual place. It represents, as far as I know, the most ambitious attempt ever undertaken by a writer to map his or her imaginary world in the idiom of modern scientific cartography.

Yet neither Faulkner nor Benet is the sort of author from whom we might expect a map. Both are famous for their complex, modernist narrative style, a

RICARDO PADRÓN

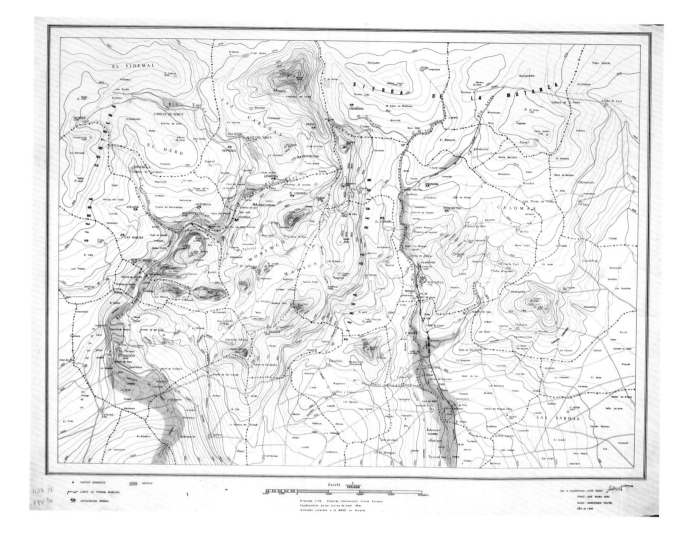

FIGURE 161.

Juan Benet, "Mapa de Región" (Map of Región)
(1968).

style that belies the clarity and mastery of their respective cartographies. Take, for example, the crucial matter of point of view. Throughout this chapter, I have discussed the ways that maps of imaginary worlds comment upon the relationship between seeing and knowing, or wishing we knew, or thinking we know. At every turn, I have been able to assume that the maps in question, much like a painting constructed in linear perspective, imply a single ideal reader or onlooker eager to see and know the world depicted. Whatever the map may be saying to that onlooker about the world it depicts, we can always assume that he or she is there, implied by the map.[13]

But it is precisely this sort of singular, imperious onlooker that novels like *Absalom, Absalom!*, *The Sound and the Fury* (1929), or *Return to Región* do their best to undermine. These and other works by Faulkner and Benet tell their stories from different, incommensurable points of view, alter the natural chronology of the tales, put them in the mouths of unreliable and even incompetent narrators, omit crucial episodes from any of the fragmentary narrative threads that together allow us to piece together the whole. They thereby frustrate the reader's efforts to master the characters, events, and even the settings of their stories. A paradox emerges: the maps provide precisely the sort of mastery that the novels undermine. They imply the sort of commanding perspective that the narrative techniques of the fiction abandon. We have seen tensions between stories and maps before, but here it is taken to an unprecedented extreme.

It is tempting, then, to understand Faulkner's and Benet's maps, quite simply, as very necessary reader's aids that help us make our way through the muddled worlds of their novels. As we saw, however, the fact that a map functions as a reader's aid does not mean it that it does not lend itself to other purposes as well. In the case of Faulkner, about whom so much has been written, we have no lack of attempts to interpret the maps. This is because space and place clearly matter, and matter deeply, in his fiction (Brooks 1990). J. Hillis Miller goes so far as to say that "Faulkner has a strongly topographical imagination" (1995, 272). His tales are spun in an elaborately mapped space in which people are associated with places, and places are invested with meanings. Those meanings, in turn, often unfold under the multiple curses of Civil War, slavery, racism, and incest.

The most interesting aspect of Faulkner's two maps of Yoknapatawpha, therefore, is not that they map the physical geography of Yoknapatawpha County, but that they map the characters and events of Faulkner's stories into that geography. We see the location of places like Sutpen's Hundred, the plantation at the heart of *Asbalom, Absalom!*, or of Compson's Mile, home of the once-aristocratic Compson family whose tale is told in *The Sound and the Fury*. Faulkner's rendering does not do so, however, merely for the purpose of helping us understand these texts. In fact, the map is probably too simple, not informative enough, to provide much help at all. Instead, it marks the profound connection that Faulkner makes between place, identity, and fate. It also identifies Yoknapatawpha County, Hillis Miller (1995, 272–73) argues, as a place built by human action over time, and in which human beings ultimately find themselves ensnared (see also Baldwin 1991–92; C. Brown 1962; Duvert 1986).

In Faulkner, then, it seems that a more or less realistic geography lurks beneath the modernist style. He can map Yoknapatawpha because he imagines its geography as something mappable, no matter what his writing style does to the stories that unfold there. Not so with Benet's Región: the relationship between the map and the novels is quite different, because the novels themselves handle space and place differently. *Return to Región* reads like something of a nightmare. Like a nightmare, it has moments of terrifying lucidity. Some of these moments can even be described as "cartographic" for the way they clearly describe places and spaces, going so far as to cite precise distances and elevations. But also like a nightmare, the story is radically incoherent. Its various narrative threads return to many of the same scenes of violence as different characters dredge up their memories of the horrors of civil war. The reader struggles to piece together the whole, but finds him- or herself frustrated by Benet's "miswritings," the deliberate silences and contradictions that he builds in to the various accounts. In this way, his prose becomes what one critic has called "a study in the essential inapprehensibility of reality" (Margenot 1991, 38).

What a relief, then, to have such an accurate and detailed map! As I mentioned above, the map charts Región as if it were a real space, with all of the visual rhetoric of modern cartographic science. It doesn't even mark the boundaries of Región, suggesting that the place is perfectly contiguous with the rest of Spain, that its map is a slice of the map of the nation. Yet we soon discover that, on the map as in the novels, something has gone awry. As John Margenot (1991, 32–40; 1988) has demonstrated, there are various contradictions between the geography of the novels and that of the map, contradictions that make it very difficult to follow the narrative threads of Benet's novels. It would seem, then, that just as Benet miswrites the story of Región, he also mismaps it. His map promises to capture his imaginary world for us and guide us through it, but with its complex web of roads, rivers, and contour lines, it is probably better understood, following Margenot, as the graphic representation of a labyrinthine cosmos ready to bedevil and ensnare us.

VISUAL ARTS: WILL THE REAL MAP PLEASE STAND UP?

We could go on exploring maps of imaginary worlds. It would be tempting, especially, to look at maps that have nothing to do with literature and storytelling, particularly maps that appear in the plastic arts, or maps that are drawn as a form of artwork. Maps have appeared in paintings since at least the Renaissance, and some early modern art has even been analyzed for the "mapping impulse" at the heart of the way it renders space (Alpers 1983; Woodward 1987). The twentieth century, particularly its latter half, witnessed an explosion of interest in maps and cartography on the part of visual artists. Early twentieth-century avant-garde movements sometimes incorporated maps into their collages, or drew parodic maps of various kinds. From the 1960s on, increasing numbers of conceptual, performance, and installation artists executed pieces that in various ways engaged questions of maps and mapping, often critically and in the service of political ends (Cosgrove 2005; Curnow 1999; Wood 2006a, 2006b). Very few of these can be said to "map an imaginary world," if by an imaginary world we mean the geography of a nonexistent place, country, or planet. Many of them, indeed, do not even look anything like what we are accustomed to calling maps. All of them, however, engage imaginatively with maps and mapping, and many of them could be said to map geographies that are in some sense "imaginary." Even when it is our own world that they map, they invariably do so from a unique perspective, converting that world to its imaginary double.

Take, for example, Jasper Johns's *Map* (1963), a picture that maps the lower forty-eight states in oil on canvas (fig. 162). Obviously, the geography depicted is a real one, but the picture's rich textures, thick brushstrokes, and playful colors steer our attention toward the aesthetics of its surface, rather than to the geography it represents. In this sense, *Map* is the opposite of a map, a counter-cartographic composition that reminds us of something that we often forget: that the map is a kind of image, made of color, line, and letter, and that it can be appreciated as such, rather than an objective representation of a real space. In this way, like much of the map art produced since the 1960s, and like some of the maps we have seen in this chapter, Johns's composition questions the claims that modern maps make to truth and authority (Yau 1996, 34–40). But it also does something else. Yes, the piece can lead us to question the representational authority of maps, but it can also remind us—in a world where maps have become so commonplace that we take them for granted—that a cartographic image can be just as much about the beauty of its colors, lines, shapes, and lettering as anything else. That much is clear to a map artist at the other end of the spectrum of fame and recognition, a man in Michigan who finds this sort of beauty can even be found in a common road map. Adrian Leskiw (2006) creates road maps of imaginary places whose color and composition speaks of the aesthetic charm of even this most practical and commonplace sort of map (fig. 163).

REAL WORLDS, IMAGINATIVE MAPPING

Like Mr. Leskiw's artwork, all of the maps mentioned in this chapter eventually refer us to maps of real worlds, although perhaps with a heightened awareness of what those maps have to tell us, or of what we can do with them. Some of them teach us to doubt the sort of knowledge we find in maps, suggesting that wisdom lies elsewhere. Some of them help us understand our interest in maps, with their promise of intellectual mastery over places and spaces. Some of them help us escape from our all-too-real world into fantasy realms that either speak of lost ideals or provide compensatory thrills. In doing so, however, all of them help us cultivate an imaginative relationship with maps and mapping that transfers quite well to maps of the real world.

Such a relationship is possible because all maps are, in some sense, maps of imaginary worlds in that all of them involve a process of selection, representation, and conceptualization that inevitably falsifies the territory they represent, even as they communicate valuable information about that territory. In this way, the production of a map introduces the values and the prejudices, the perceptions and the misconceptions, the insights and the blind spots, the ideology and the culture, of the mapmaker into the representation of the territory.

RICARDO PADRÓN

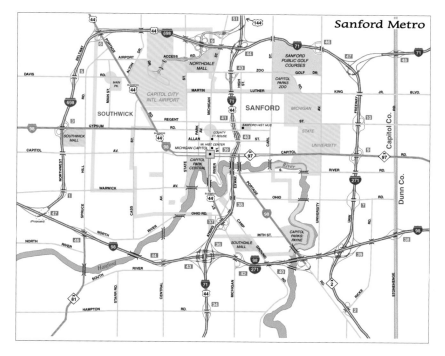

FIGURE 162.

Jasper Johns, *Map* (1963). Art © Jasper Johns / Licensed by VAGA, New York, NY.

FIGURE 163.

Adrian Leskiw, "Sanford Metro" (2006).

Truly, there are no maps of real worlds, only maps of not-so-imaginary worlds that we take for reality. All maps, moreover, mediate an imaginary relationship with that real territory. We do not make maps of spaces that we can, without effort, take in with our eyes. We only map what we cannot see, in order to be able to see it. Every map, therefore, provides for us the perspective that has been so important to my discussion here, the privileged point of view ordinarily enjoyed only by birds, pilots, astronauts, angels, and gods.

And any maps, when we engage them thoughtfully and sensibly, can provide many of the experiences that those depicting imaginary worlds are made to provide. What happens when Jasper Johns converts the map of the United States to a textured collage? Don't the function and the meaning of that map change? In this way, it becomes a worthy successor to the satire of cartography we saw in *Gulliver's Travels*. What happens when we sit in front of a map of our hometown or of a country whose history we know very well? We travel. We revisit memories. We march in the company of armies, explorers, or peaceful protesters. We recall childhood games and personal tragedies. Put simply, we recover the stories and meanings that we associate with those places and spaces (Harley 1987). What happens, then, when we come across maps of unfamiliar places, whether full of blank spots or dense with names? We explore, we imagine, we give play to our fantasies and desires.

And so, while maps of imaginary worlds do indeed delight, distract, reveal truths, whisper secrets, unsettle, reassure, perhaps they do not do so because they are maps of imaginary worlds, but because they are maps. We allow them to trigger our imaginations because the world they depict is imaginary, or because we encounter them in the context of dealing with works of art, such as stories, that get us thinking and feeling in ways that we don't think and feel when we unfold the tattered road map in the glove compartment of our car. Yet none of these maps insist upon their radical difference from other maps as maps, only upon the radical difference between the worlds they depict and the real world. On the contrary, they work by insisting upon their own similarity with those maps. And perhaps they teach us that any map, in fact, will do: that we would do well to give that tattered road map a chance.

Notes

1. The Sacred Texts Web site has digitized the portions of Plato's *Critias* and *Timaeus* that gave birth to the myth of Atlantis, as well as numerous texts at the heart of the modern, occultist interest in the lost continent (Internet Sacred Text Archive). For an overview of the vast literature on Atlantis, see Ellis (1998).
2. I follow Pierre Jourde's definition of an imaginary world, which I translate here from the French: "A complex spatial ensemble identified by toponyms

RICARDO PADRÓN

the majority of which are invented, and that constitutes an autonomous structure clearly detached from the space known and explored at the time the author wrote" (1991, 16).

3. Manetti's ideas were recapitulated in an influential 1481 commentary on the *Inferno* written by Cristoforo Landino, and then in a 1506 dialogue written by Girolamo Benivieni.

4. A number of these images are available online, in an exhibition assembled collaboratively by the Newberry Library, the University of Chicago, and the University of Notre Dame. See "Dante's Hell," in Cachey and Jordan (2006).

5. For reproductions of Botticelli's illustrations for the *Divine Comedy*, including the map, see Botticelli (1976, 2000). The latter publication includes several useful interpretative essays. For the details of the commission, see Morello (2000, 318). The drawings may have been intended for a luxurious illustrated manuscript edition of Dante's text, but since the project was left unfinished, we cannot be sure of this, and we do not have any documentation that could shed light on the question (Schulze Altcappenberg 2000, 29).

6. Charles Singleton argues that Dante's "precise measurements are calculated to add realism to the description of Hell, but they in fact show a curious indifference to reality" (see Dante 1980, 2.558). For more on contemporary critical opinion, see Kleiner (1994) and A. Gilbert (1945).

7. As one scholar puts it, "The mapping of social spaces . . . privileges abstraction and the total grid. We need to augment maps with stories . . . While we need not—indeed cannot—get on without maps and clocks, their partial and abstract roles ought always to be kept in mind. Narrative space and narrative time are powerful and meaningful counters to the effects of abstraction created by clocks and maps" (Kort 2004, 165).

8. For one of the classics of the genre, and an important influence on More, see Lucian (1968).

9. So too can the "Utopian alphabet" printed with More's text.

10. Although I will occasionally refer, for convenience's sake, to "Tolkien's maps," I do not mean to suggest that these maps were drawn by J. R. R. Tolkien himself. The maps in *The Hobbit* were, while those in *The Lord of the Rings* and *The Silmarillion* were drawn by his son Christopher Tolkien. For more on the genesis of these maps, see Anderson's commentary in Tolkien (2002, 50), and Hammond and Scull (2005, lv–lxvii).

11. A clipping about Jaro Hess from an unidentified newspaper is available at Rosen-Ducat Imaging (2006). I have been unable to locate other sources about the artist or his piece.

12. Readers new to Benet may want to consult Herzberger (1976).

13. Technically speaking, this is completely untrue. Maps are unlike perspective paintings in that they are drawn as if that ideal onlooker were directly over each and every point. This technicality, however, is beside the point. For more on cartography and perspective, see Edgerton 1987 and Hillis 1994.

CONSUMING MAPS

Diane Dillon

In the mid-eighteenth century, Johann Joachim Kaendler designed a statue of a map peddler to be produced by the porcelain factory at Meissen, Germany (fig. 164). The figurine performs one of the acts essential to the consumption of cartography, selling maps. More subtly, Kaendler's map seller neatly encapsulates many of the issues that have shaped the consumption of maps from the early modern era to the present. First, vendors like the map peddler contributed to the growth of the commercial marketplace that took root in the fifteenth century and has continued through the present day. This expansion provided the context for the development of maps as consumer merchandise. Second, the content of the map in the crier's hand points to the political valence of cartography. It represents an area that loomed large in the news during the War of the Austrian Succession: the duchy of Silesia, with the city of Breslau highlighted in large type. In 1742, the leaders of Prussia and Austria met in Breslau to sign a peace treaty that transferred Silesia to Prussia (*Map Collector* 1988). The diplomats probably used maps, perhaps like the one proffered by the peddler, to clarify the boundaries of the ceded lands. Third, as an ornamental object itself,

FIGURE 164.

Johann Joachim Kaendler, porcelain figurine
of map seller (ca. 1742).

the statue gestures toward the decorative use of maps. Like porcelain figurines, maps have often been sold and used to embellish the interiors of homes, offices, and public buildings, where they could project the social ambitions, local and national pride, and artistic tastes of individuals and communities. Fourth, the map seller hints at the emotional power of cartography. If Kaendler struck a sentimental chord to capture viewers' sympathies, mapmakers have likewise appealed to consumers' heartfelt attachment to place. And finally, by grasping the sheet in his right hand and carrying additional maps in the box in his left and the pack on his back, the peddler demonstrates the bodily bond between a map and its user.

Regarded together, the cluster of meanings conveyed by the map seller's form and content illuminate the complex dynamic between map consumption and personal and group identity. In chapter 1, James Akerman observes that wayfinding maps tell us who we are as well as where we are headed. I extend this line of thought by arguing that by studying map consumption, we can better understand the intertwinement of modes of identity rooted in geography with modes based on the acquisition of goods. The places where we live, the sites we visit, and the goods we acquire in each location exert a powerful influence in shaping the self. As geographic tools and personal possessions, maps guide the construction of identity. Although map consumption has a long history as a global phenomenon, conspicuous in China and Japan as well as the West, in this chapter I focus more narrowly on maps as consumer goods that contributed to the development of capitalist economies in Europe and the United States since the Renaissance. Cartographic products intended for personal use played a social role distinct from those produced to further the ambitions of state leaders or scientists. For consumers, maps served not only to fulfill established needs, but also to create needs and wants and ways to satisfy them (Williams 1983, 79).

To understand map consumption, we need to examine its history in conjunction with the production, marketing, sale, and distribution of cartographic merchandise. Studying this wider network is important not only because it shaped consumption, but also because its history offers clues when direct evidence is lacking. In relatively few cases do we have details about how specific patrons and buyers acquired and used maps. In most situations, we can only extrapolate about patterns of consumption from evidence such as the number of surviving copies of any given map, the quantity of editions, or the attention a map or atlas received by critics and journalists.

The late fifteenth century provides a logical starting point for this narrative for three related reasons. First, the era witnessed the earliest printed map, a small woodcut of a world map that appeared in 1472, some twenty years after the invention of printing with moveable type (Robinson 1975, 1). Before this, maps had to be drawn by hand and were few in number. Printed as a marginal illustration in the *Etymologiae* of Saint Isidore, Bishop of Seville, this map initiated a series of technological innovations that made the ever-widening production and distribution of maps possible. The next benchmark occurred in 1477, when an Italian printer produced the first maps from engraved copperplates, a process permitting larger impressions, finer lines, more variable tones, greater edition quantities, and easier alterations (Verner 1975, 51). To put these European innovations in context, we should note that as early as 1136, mapmakers in China had found a way to distribute maps by taking rubbings of images carved on stone (Yee 2001, 115; see also chapter 1 of the present text, pp. 36–37 and fig. 10).

Second, the early modern era saw the burgeoning of global exploration and trade. As explorers brought new parts of the world to the attention of Europeans, cartographers recorded the latest geographic findings on new maps, which in turn prompted further voyages and more maps. If the advent of the printing press made maps readily multipliable, global expansion provided the necessary catalyst for their production. Motivated by the desire to chart more efficient trade routes, these expeditions stimulated European consumption of not only maps, but spices from the Molucca Islands, silk from China, and carpets from Turkey. These exotic commodities joined an expanding array of local goods to shape a new culture of materialism (Jardine 1996). Maps played a multifaceted role in this material culture, serving as tools of navigation and trade, emblems of learning and status, and objects for display and exchange. Cartographers did not merely contribute to this marketplace. They helped create it, defining its contours visually and metaphorically. In short, maps led the way.

Third, the Renaissance witnessed the economic transformations that led to the development of capitalism. Although the economy of the fifteenth and sixteenth centuries was not yet capitalist, the period generated the heightened level of economic activity needed to support the creation of capitalist enterprises (Braudel 1976). More specifically, during this period maps and other geographic records functioned not only as consumer goods, but as capital goods, facilitating the global travel and commerce that enabled Europe's imperialist nations to accumulate vast wealth. Maps did not need to be printed or find a large audience to play this role; rare manuscript maps and charts worked as capital goods when they transmitted navigational advice and images of the riches of far-away lands to the princes and merchants who sponsored the voyages. When this information was transferred to consumer goods (in the form of widely available printed maps and documents), they spurred a massive expansion of overseas exploration and international trade (Mukerji 1993).

To highlight the larger patterns of map consumption, I have sorted my examples into four broad divisions. The first three categories underscore both the economics of cartography and the role of maps in shaping identities rooted in social class: cartographic luxury goods, competitively priced and mass-produced maps, and maps that are publicly exhibited or distributed without charge. Like class boundaries, the borders between these groups are porous;

like their users, maps often resist easy classification. As a result, some kinds of maps appear in more than one category. The fourth section draws together examples from various segments of the market to analyze the connections between consumers' bodies and the maps they make, modify, and manipulate.

MAPS FOR THE FEW: LUXURIOUS CARTOGRAPHIC OBJECTS

The most magnificent maps that have survived into our own time were made to be saved. They were deluxe objects intended for gifts, household decoration, or public display rather than as working documents. Since early modern times, atlases have perhaps been the cartographic product most appealing to collectors. Their bound format made them convenient to use and store, and they offered broader and deeper geographic information than any single map or globe. Typically combining text and pictorial images along with maps, atlases served as comprehensive reference works.

Enterprising cartographers transformed maps devised for utilitarian purposes into collectibles. For example, beginning in the fifteenth century chart makers restyled portolan charts, created as workaday tools for sailors, into beautiful drawings compiled into sumptuous volumes. Monarchs, popes, diplomats, and petty nobles all owned these portolan atlases, either commissioned for themselves or received as gifts, and most likely used them for their own edification and pleasure (Campbell 1987, 435, 440). At the same time, the volumes served as emblems of prestige, wealth, and cosmopolitanism. Decorative portolan atlases symbolized the expanding global trade and colonization that their more utilitarian counterparts facilitated. This economic development in turn generated the wealth that supported the production and consumption of such luxurious goods.

One of the most spectacular collections of Portuguese portolans, the Miller Atlas (ca. 1519), named for a previous owner, includes charts made for King Manuel by Lopo Homem, Pedro Reinel, and his son Jorge Reinel (fig. 165). Scholars have speculated that the Portuguese ruler commissioned the atlas to be sent to Francis I of France, to demonstrate the scope of Portugal's empire (H. Wallis 1997, 4; Cortesão and Teixeira da Mota 1960–62, 1:55–61). Portuguese flags fly prominently on the charts, identifying King Manuel's territorial claims. The splendid illuminations, probably executed by Gregorio Lopes, accentuate the global reach and richness of the Portuguese lands, picturing places most readers would never see. The artist highlighted economic exchange with bird's-eye views of bustling ports such as Mogadoxo (Mogadishu), on the East African coast north of Madagascar, and Malaqua, the center of the spice trade on the Malay Peninsula. On the oceans, the juxtaposition of Portuguese caravels and Arabic dhows hints at the nascent rivalry between the two empires, while small, unidentified islands glisten like jewels. In these artistic vignettes, the aesthetic

DIANE DILLON

value of the painting metaphorically confirms the economic value of the land. For King Manuel, the atlas could have framed an identity in terms of both imperial geography and princely consumption.

In the later sixteenth century, consumers turned enthusiastically to a new genre to symbolize their sophistication: the deluxe world atlas. The 1570 publication of Abraham Ortelius's *Theatrum orbis terrarum* (Theater of the World) in Antwerp, the most costly book yet published, established this new trend (Broecke 1986; figs. 58 and 75). Ortelius's atlas met the demand for up-to-date maps and geographic descriptions presented in a form that was convenient to study and pleasing to view and handle. Initial sales were so brisk that the printer, Christoffel Plantijn, produced four editions in 1570 (Koeman 1964). Kings, scholars, merchants, and even poets were quick to acquire and praise the *Theatrum*. Originally published in Latin, the text was translated into all the major European vernaculars, appearing by 1608 in Dutch, German, French, English, Italian, and Spanish (Karrow 1993, 9).

Strong demand for the *Theatrum* caused its price to rise quickly. In 1572 one customer, Caesar Orlandius, grumbled to Ortelius that "Franc. Tramezzinus lately sold a copy for ten gold crowns, whereas four months ago it fetched only eight" (quoted in Koeman 1964, 38). Buyers paid higher prices for deluxe editions. Plantijn sold ordinary copies of the first edition for six florins, ten stuivers, while copies on high-quality paper cost seven florins, ten stuivers, and illuminated copies brought sixteen florins. In 1588, Plantijn produced an ultra-luxurious presentation copy of the first Spanish edition for the prince of Spain, demanding the astounding sum of ninety florins. To put these prices in context, the annual earnings for a bricklayer in Antwerp in 1588 were around 140 florins (Koeman 1964, 38–39).

Ortelius and his publishers kept interest in the atlas high by issuing a steady stream of new versions. By his death in 1598, twenty-four editions had hit the market, from large folios to pocket-sized *Epitomes* (Karrow 1993, 9; Koeman 1997, 88). By 1612, at least 7,300 copies had been issued in thirty-one editions (Broecke 1995, 2). For consumers who already owned a copy, Ortelius put out regular supplements that contained new maps and information (Koeman 1997, 76–78). These additions reflected expanding geographic knowledge, comprising both corrected sheets and maps of parts of the world Ortelius had excluded in previous editions (Broecke 1995, 3).

The world atlas reached a new peak as a luxury consumer article with the publication of Joan Blaeu's eleven-volume *Atlas Maior* in 1662. Blaeu's achievement was as much artistic as it was cartographic and scholarly. The atlas owed its fame to the dazzling beauty of its hand-colored maps, the excellence of its lettering, and the luxuriousness of its paper and binding. Blaeu's volumes were the grandest and most expensive books issued in the seventeenth century. Despite a stiff price, the *Atlas Maior* found a ready market, appearing in French editions in 1663 and 1667, in Dutch in 1664, in German around 1667, and in

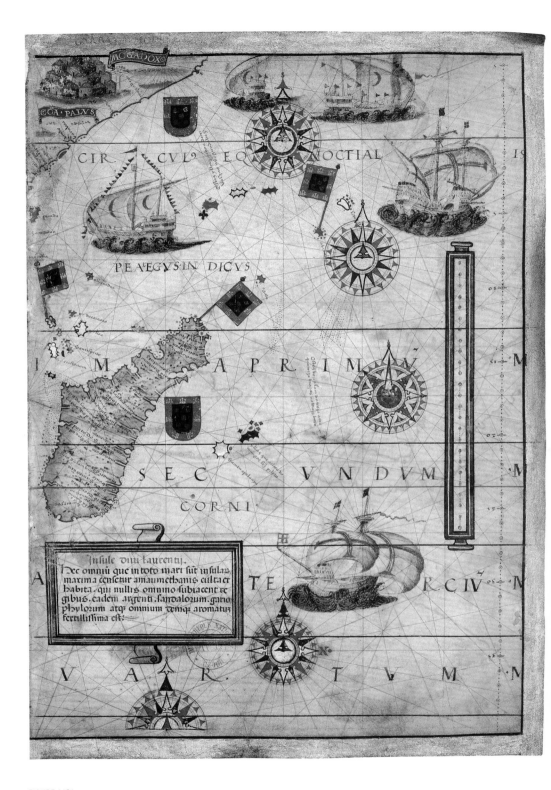

FIGURE 165A.

Chart of the west Indian Ocean with Madagascar.

From the Miller Atlas (ca. 1519).

FIGURE 165B.

Chart of the East Indies. From the Miller Atlas
(ca. 1519).

Spanish in 1672 (Koeman 1997, 90). To personalize the atlas, purchasers could order customized copies. They could replace the standard covers, of limp yellow vellum ornamented with gold, with deluxe bindings in morocco leather or red velvet. Buyers could also have their coat of arms stamped in gold on the front cover (Koeman 1969, 199).

Blaeu's atlas lent itself to still more elaborate forms of personalization. Ambitious collectors used the volumes as portfolios, inserting a wealth of supplemental material between the atlas pages. Amsterdam lawyer Laurens van der Hem enlarged his copy of the *Atlas Maior* with more than 1,800 city views, seascapes, architectural drawings, ethnographic prints, portraits, historical scenes, and maps. Many of his additions were rarer and more original than Blaeu's material. For example, van der Hem inserted a collection of landscape sketches by highly regarded Dutch artists. He also added four volumes of topographic drawings and manuscript maps, which had been part of a secret atlas made for the Dutch East India Company (fig. 166). As the atlas swelled to forty-six volumes, van der Hem established his own order for Blaeu's sheets and his additions. To ensure a harmonious whole, he maintained Blaeu's page dimensions, trimming inserts that were too large and extending small pieces by attaching them to blank leaves, masking the transition with decoration. Van der Hem hired Dirck Jansz. van Santen, a well-known specialist in illumination, to enrich the book further by ornamenting the pages with gold leaf and colored paints (Krogt and Groot 1996, 7; Koeman 1969, 199).

Consumers also pressed the *Atlas Maior* into service as a national symbol. As the culmination of Dutch publishing, Blaeu's book ably represented the cultural achievements of the Republic of the United Netherlands, and thus made an appropriate official gift for heads of state and other high-ranking officials. The republic presented copies to the Sultan of Turkey, to Emperor Leopold II, and to Admiral Michiel de Ruyter following his victory in a naval battle against England (Koeman 1969, 199).

More specialized atlases promoted national and local identities. The beginnings of this trend can be glimpsed in the *isolario*—a specialized atlas composed of maps of islands—pioneered by Christopher Buondelmonti (fig. 167). In the preface to his *Liber insularum Arcipelagi* (Book of the Islands of the Archipelago), Buondelmonti states his goal of describing the past and present of the Greek islands, documenting their rulers and visual appearance on maps (Turner 1987, 13). The undertaking seems to have been prompted by the interests of his Roman patron, Cardinal Giordano Orsini. As Buondelmonti's destinations included prominent libraries and remote monasteries, he may have traveled to Greece in search of early manuscripts for Orsini and produced the *isolario* as a side project (Clutton 1987, 482).

Although the *Liber insularum Arcipelagi* was not printed until the nineteenth century, the book readily found readers well beyond Orsini. The original manuscript, completed by 1420, was frequently copied throughout the

DIANE DILLON

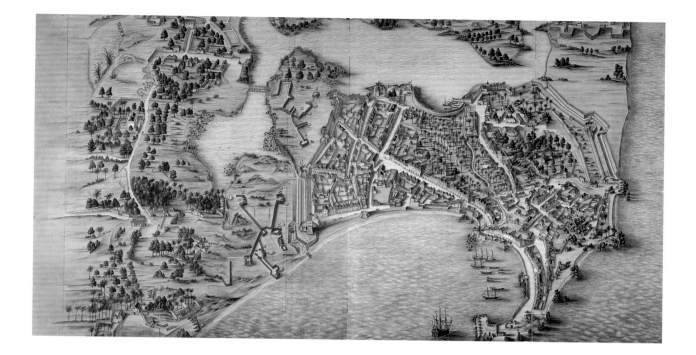

FIGURE 166.

Bird's-eye plan of Colombo, Sri Lanka. From the
Atlas Blaeu–Van der Hem (ca. 1670).

fifteenth century and translated from Latin into Greek, Italian, and English. Most transcribers "improved" the manuscript by altering the text, inserting additional maps, and embellishing the sheets with color and ornamentation. Like van der Hem, these copyists at once expanded and personalized the original atlas. Buondelmonti's island portraits reached a still broader audience when atlas compilers appropriated his work (Turner 1987, 13, 25).

The appeal of the *isolario* stems from both its subject matter and its tone. Interest in the Aegean flowered during the Renaissance as humanists focused attention on classical culture. More specifically, in the imaginations of mainlanders, islands have traditionally loomed as exotic Edens, kept pure and primitive because they were cut off from the corrupting influences of mainland civilization (Gillis 2004; see chapter 6 of the present text, pp. 265–71). The Greek islands drew Homer's Odysseus, Tahiti lured the painter Paul Gauguin, and an imaginary island served as Sir Thomas More's *Utopia*. Like these literary and visual imaginings, Buondelmonti's manuscript was often as fictive as it was compelling. He crafted the text perhaps above all to amuse his patron, incorporating the descriptions of stormy seas and local color that still captivate travel writers and armchair adventurers (Clutton 1987, 482–83).

The grander relevance of the *Liber insularum Arcipelagi* to the history of map consumption may derive from the precedents Buondelmonti established for visualizing political and economic concepts. By showing the islands as small, clearly demarcated landmasses within a larger group of like entities, Buondelmonti unwittingly predicted later representations of territorial states

FIGURE 167.

Christopher Buondelmonti, Lesbos. From *Liber insularum Arcipelagi* (Book of the Islands of the Archipelago) (ca. 1450).

governed by a single sovereign (Steinberg 2005, 254). These conventions help the human community—the nation—residing within the boundaries to identify with both the land and the political entity (Anderson 1991, 170–78; Akerman 1995, 141, 153). Similarly, Buondelmonti's islands foreshadow the modern form of the commodity. The minute scale and distinct outlines of the islands make them seem readily graspable, visually and physically. Especially in the imagi-

DIANE DILLON

nation, it is but a small step from grasping to possessing. In its array of many comparable units, spread before the eye on the map, the archipelago resembles the marketplace, where goods are likewise spread out to be seen and touched. In preparing consumers, however obliquely, to appreciate the kindred forms of the state-in-the-world and the commodity-in-the-market, the *Liber insularum Arcipelagi* hinted at their eventual interpenetration through the commodification of geography and politics.

In addition to atlases, wealthy consumers sought ornate wall maps and globes as emblems of their attainment of status. During the Renaissance, Italian elites frequently set aside rooms in palaces and public buildings for fresco cycles featuring maps and views. In the 1490s, Francesco II Gonzaga, fourth Marquis of Mantua, collected bird's-eye views to serve as the basis for the frescoes in the "Room of the Cities" within his private apartments of the family palace in Gonzaga. One of four map-mural cycles Francesco commissioned for his three princely homes, the eight views pictured prominent Italian cities (Rome, Naples, Venice, and Genoa) along with Constantinople, Cairo, and either Paris or Jerusalem. Francesco's motives for choosing these particular cities are undocumented, but they may have reflected his mercantile interests. More generally, the maps would have impressed visitors with their patron's humanistic learning and commercial ambitions, affirming his place in the larger global picture (Schulz 1987, 116; Bourne 1991, 51–53).

Early modern elites such as Francesco II Gonzaga often acquired cartographic materials to enrich personal collections that also included books, manuscripts, and artworks. The deliberate collection of maps developed during the fifteenth and sixteenth centuries as international exploration and trade stimulated public interest in geography and increased map production (Skelton 1972, 35). Distinctive collections could demonstrate their owners' discriminating tastes more emphatically than individual objects. Whereas some collections served primarily as status symbols, others fulfilled scholarly purposes or provided entertainment. Maps found a place in the private libraries of princes and dukes, but also in the more modest *studioli* of scholars and antiquarians. Learned consumers such as Orsini turned to maps to enhance their knowledge not only of contemporary geography, but also of past events and the culture of the ancient world (Woodward 1996, 88–93). Renaissance map collections signaled the emergence of maps as a genre of consumer goods tied to the construction of personal identity. Collectors established patterns of living with maps that would become widespread in subsequent centuries.

MAPS FOR THE MANY: CARTOGRAPHY ON THE MARKET

By the end of the seventeenth century, the market for the grandest sort of maps, atlases, and globes had waned. The deluxe products that had set previous generations' cartographic benchmarks were already giving way to a new kind of merchandise: competitively priced maps turned out for a broad market. From the Renaissance to the present day, mapmakers have steadily expanded the volume and variety of their goods, producing maps aimed at different classes for a widening range of uses. Publishers issued maps to decorate the walls of homes and workplaces, to aid travelers, and to enlighten the public. As map production and consumption grew, novelty increasingly determined the success of cartographic products. Although newness of geographic knowledge had long determined the value of maps and presenting up-to-date information continued to be important, other, novel characteristics increasingly shaped consumption—new kinds of maps, new manufacturing technologies, new marketing strategies, new audiences.

Maps as decoration

In the fifteenth and sixteenth centuries, when Venice and Rome dominated the European map trade, Italian household inventories reveal that globes and world maps were often displayed in the entrances of private homes. For example, Bartolomeo Falier juxtaposed maps of the four parts of the world with his collection of Turkish jewels in his entry hall, while Vito Vidotto displayed his maps near his silver and a spinet in the portico of his San Simeon home. Visitors to each household would thus be immediately impressed by the owners' wealth, taste, and knowledge. By the later seventeenth century, Italians had spread their cartographic decorations around the house, placing maps and globes in locations ranging from stairwells to living rooms. Geographic representations became increasingly common, though they never attained the ubiquity of religious and classical images, or portraits and landscapes (Woodward 1996, 84).

Although the homes most likely to be adorned with maps were the townhouses and country manors of the landed gentry, the inventories disclose that at least a few workers possessed geographic prints. For example, Andrea Bareta, evidently a wool worker in the late sixteenth century, owned a set of images of the four continents along with some religious pictures. In addition to world maps, Italians displayed maps of individual countries and cities to show their awareness of other centers of civilization. They appear to have avoided local images, however, which may have connoted provincialism or seemed too familiar. Publishers churned out great quantities of maps and views of Venice, but these seem to have been intended for visitors rather than locals (Woodward 1996, 80–83).

Maps proved even more fashionable as decoration in the Low Countries in the seventeenth century. This vogue coincided with the rise of the Netherlandish cartographic industry, which led the world in map production from the late sixteenth century through the end of the seventeenth. The aesthetic achieve-

DIANE DILLON

ment and popularity of decorative mapmaking peaked in this era, blurring the distinction between art and cartography (Alpers 1987, 54). Dutch cartographers and consumers regarded wall maps, atlases, and globes as works of art; paintings and large wall maps sold for similar prices; artists featured maps in almost every sort of interior scene; and shops purveyed maps alongside paintings, books, and prints (Welu 1977, 22, 42; Heijden 1990, 34).

The number and variety of maps and globes that appear in paintings during this period attest to the popularity of cartographic materials as decorative consumer goods. These paintings also show us where and how the Dutch displayed maps, illustrating the wide range of social classes that embraced the trend and the kinds of maps they chose. Numerous artists used maps to adumbrate aspects of the nation's economic life, frequently representing maps in places of business, from cobbler shops to taverns to bordellos. Several recurring genres of maps depicted in paintings also point to The Netherlands' prominent role in international trade. For example, Jacob Ochtervelt included a 1661 map of America by Dancker Danckerts in the aristocratic interior of *A Woman Playing a Virginal*. Quiringh Van Brekelenkan attested that tradesmen shared this interest in the New World by showing Jodocus Hondius's 1630 map of America displayed in a tailor shop, and part of Willem Jansz. Blaeu's four-sheet map of America from 1608 in *The Woolspinner and His Wife* (Welu 1977, 1, 20–22). Dirck Hals alluded to the enterprises of the Dutch East India Company by featuring maps of Asia in several of his paintings, while the world maps that appear in canvases by various artists might also suggest imperial ambitions. In 1636, a German traveler through Holland recounted that tailors and shoemakers learned about the history of the Indies by studying the wall maps hung in their homes (Alpers 1987, 91).

At the same time, pictured maps carried potent civic and national associations. Most frequently, maps in Dutch paintings depict Holland or the Seventeen Provinces (Welu 1977, 13). Moreover, as representations of the contemporary political geography, maps of Holland might connote their owner's pride and identification with the Dutch Republic. Such sentiments were pronounced among burghers, the group most likely to display maps at home (Schama 1988, 7; Welu 1977, 23). Maps of the Seventeen Provinces, in contrast, reflected both nostalgia for the period before Holland and Flanders had separated and hopes that the provinces might be reunited. Hugo Allart expressed this desire in his late seventeenth-century editions of a map of the Seventeen Provinces first printed in 1594. He extensively revised the map's ornamentation to meet contemporary tastes, but left its depiction of the outdated political union intact (Welu 1977, 43–46).

In the late seventeenth and eighteenth centuries in England, map selling and buying flourished in the context of a dramatically expanding consumer marketplace, fueled by the enhanced spending power of all classes (McKendrick, Brewer, and Plumb 1982; Pedley 2005). As in the Low Countries, consumers preferred maps suited to decorative use. In his 1679 catalogue, London map and print dealer John Garrett assured customers that maps were "very pleasant and delightful ornaments for houses, studies or closets." Garrett also provided more specific decorating advice, noting that "a most exact map of the world in four sheets with descriptions at the bottom and pictures on the sides" was "a fit ornament for a chimney piece" (quoted in Barber 1990, 2).

Another map seller, Philip Lea, stressed consumer choice in his catalogue from around 1700. He offered a large wall map of London and Westminster from 1681/82 with or without a "prospect of London" at the top. To embellish the sides, customers could update the map with views of "the new buildings and improvements of these late years" or choose a more general selection of the city's "chief buildings and churches" (quoted in Barber 1990, 2). Lea also pointed out that any of the maps in his inventory could be printed on silk, to be used as curtains for the new sash windows, "or to carry in the pocket . . . as an handkerchiefe" (quoted in Tyacke 1973, 65).

Among the landed gentry, maps encompassing personal landholdings found special favor. Map publishers played to their vanity by offering them the option of having images of their house and garden, along with their coat of arms, reproduced in the borders of county maps (Barber 1990, 3). Estate plans also gained popularity as decoration. As surveyor William Leybourne noted in 1653, a well-drawn plan could be "a neat ornament for the Lord of the Manor to hang in his study or other private place, so that at pleasure he may see his land before him" (quoted in Barber 1990, 3–4). Beginning in the 1740s, Jean Roque, a French Huguenot landscape designer and surveyor living in London, helped extend the fashion beyond actual property owners by producing printed estate maps (Pedley 2005, 42; Andrews 1967). Such plans most likely appealed to the same sort of consumers who acquired the printed views of stately homes and castles also sold by map sellers (Clayton 1997). For the growing middle class, buying views and plans of desirable real estate owned by others could express their economic ambitions and desire to emulate the lifestyles of the wealthy. The maps themselves could have served as a surrogate for an acquisition that was beyond their means.

The commercial production of local maps to adorn the walls of homes reached new heights in nineteenth-century America. Townscapes such as James T. Palmatary's 1857 view of Chicago (fig. 168) may have constituted the most popular genre of printed images of America in that era. The advent of lithography, adopted by commercial printers by the 1820s, facilitated the mass production and dissemination of these views. Printers could make lithographs much more quickly and cheaply than copperplate engravings; the lines were less fine, but tonal effects were more easily achieved. By midcentury, color lithography allowed printers to dispense with the slower, more costly technique of hand coloring (Reps 1984, 3–4).

DIANE DILLON

Printers had been producing bird's-eye views for at least three hundred years by the time urban lithographs became fashionable in the United States, as Jacopo de' Barbari's 1500 *View of Venice* attests (see chapter 3 of the present text, pp. 121 and 124 and fig. 63). Although the American makers imitated the visual form of the earlier views and capitalized on their artistic and cartographic prestige, they democratized the genre in important ways. The itinerant artists who drew the views crossed the country, sketching tiny villages as well as major metropolises. Turning out a staggering volume of prints depicting many different places, they introduced the form to a much-widened audience. The use of bird's-eye views enabled artists to encompass all of a town's buildings and landscape features, ensuring that every local resident would feel represented and thus be more willing to invest in the print (Reps 1984, 65).

In the production and marketing of their work, view makers followed strategies similar to those of earlier mapmakers. After the artist finished a preliminary drawing, agents showed the sketch to potential customers and sold subscriptions to the printed view. As with the English county map and atlas producers, the agents charged additionally for extras, such as giving special prominence to a building either within the townscape or in a border vignette. The producers rarely paid for advertising, relying instead on local newspapers to supply free publicity by reporting on their activities. They also sought endorsements from city officials and business leaders, who recognized the prints' promotional value (Reps 1984, 6–7, 43, 57, 60).

In his view of Chicago, Palmatary ably served the cause of the city's boosters. He emphasized the bustle of trade by filling the harbor with steamers and sailing vessels, showing trains heading north and south along lakefront tracks, and placing the grain elevators and lumberyards at the mouth of the Chicago River near the center of the foreground. The view stresses the city's grand ambitions with its own bigness: the print measures nearly 7 feet (2.1 m). The monumental scale of Palmatary's image made it expensive, however. The view sold to subscribers for $10 and after publication retailed for $12. Its dimensions and price meant that its market would have been composed mostly of organizations rather than individuals, and that it would have been displayed chiefly in public places, such as hotel and bank lobbies, municipal offices, and shops. Subscribers whose buildings appeared in the scene were most likely to hang the print in a prominent place (Reps 1984, 48, 62).

Palmatary's cityscape was the grandest example in a steady stream of Chicago views issued in the second half of the nineteenth century. Whereas view makers typically believed that small-town markets would be exhausted by one scene, they produced revised images of growing metropolises like Chicago on a regular basis. In 1892, Currier and Ives reprinted their 1874 bird's-eye view of Chicago, adding a few recent buildings to the cityscape, most likely in exchange for subscriptions from their owners. The print's smaller dimensions

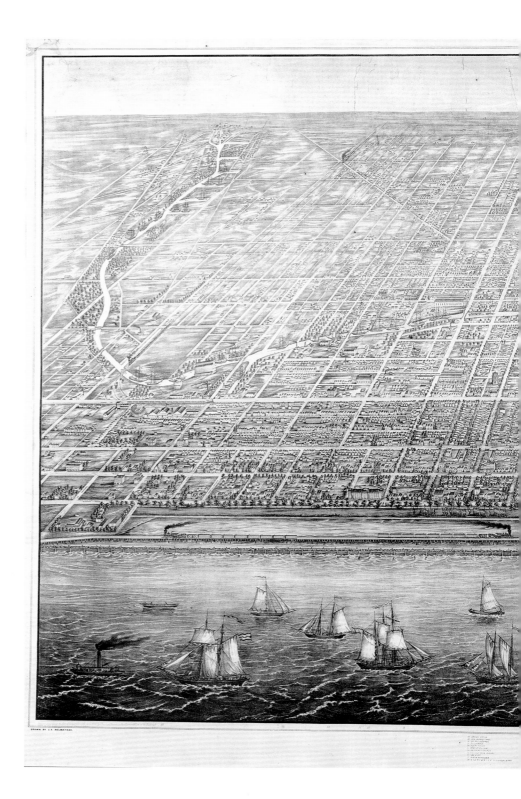

FIGURE 168.
James T. Palmatary, "Chicago" (1857).

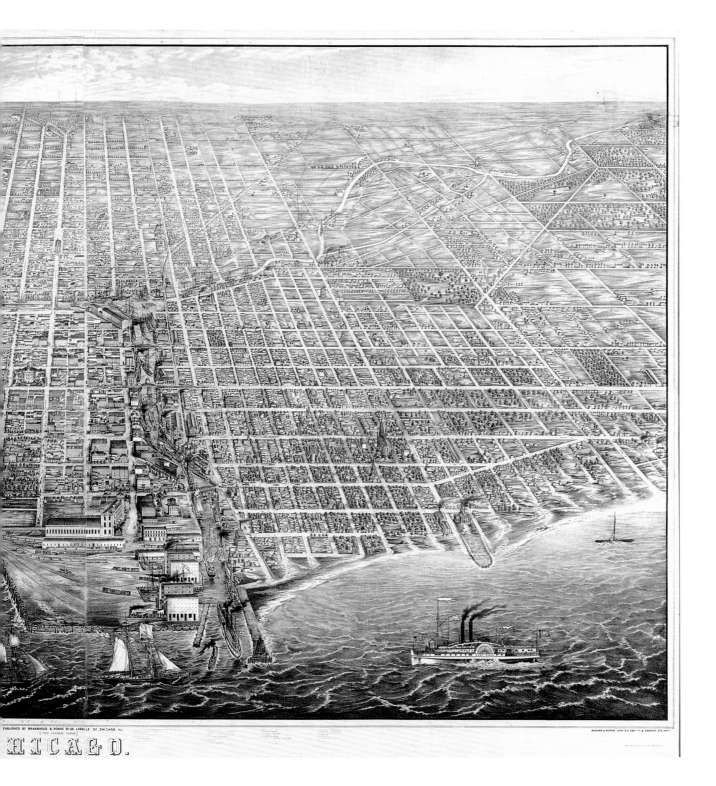

PUBLISHED BY BRAUNHOLD & SONNE 51-55 LASALLE ST. CHICAGO ILL.

MERSINE & HENSEL LITH S.E COR. 7TH & CHESNUT STS. PHIL.ª

HICAGO.

(22 5/8 × 32 1/2 inches [57.5 × 82.5 cm]) made it suitable for home display. In 1893, Rand McNally translated the format into a new guidebook, *Rand, McNally & Co.'s Bird's-Eye Views and Guide to Chicago*, which featured highly detailed engravings and descriptions of parts of the city, block by block (fig. 169). Much as the dimensions of Palmatary's print symbolized Chicago's great ambitions, the elevated vantage point of the views echoed the lofty height of the city's celebrated skyscrapers. Chicago's residents and tourists could gain a bird's-eye perspective by visiting the observation decks atop the roofs of tall buildings. For these consumers, printed views could serve as visual reminders of a breathtaking experience. As souvenirs, the urban views amplified the social function of decorative maps, adding to their established roles as emblems of learning, sophistication, patriotism, and property.

FIGURE 169.
"Region of the 12th Street Railway Station." From *Rand, McNally & Co.'s Bird's-Eye Views and Guide to Chicago* (1893).

Maps for home reference

Consumers who adorned their walls with maps may have occasionally studied them for geographic information, but were more likely to consult atlases

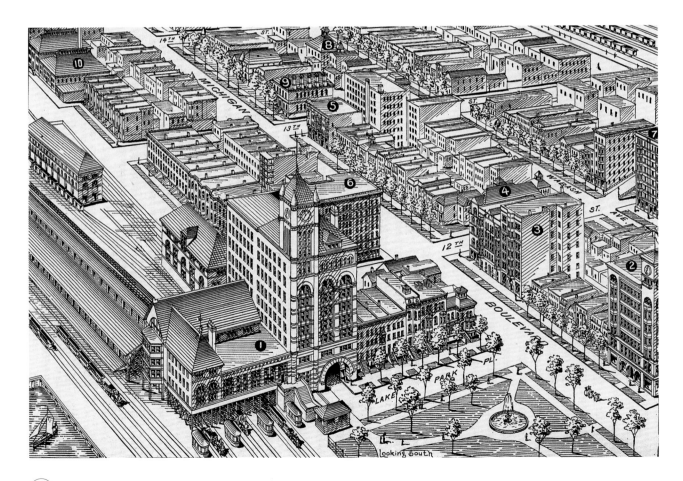

DIANE DILLON

for this purpose. Seeking to widen their market, atlas makers were willing to accommodate customers who wanted to have an atlas in their home as a reference book, but could not afford the deluxe volumes. For example, in 1574 Johannes Castellus wrote Ortelius to acknowledge receipt of the *Theatrum*, explaining that "on account of my poverty, I will pay in three terms. I now send two 'Daalders,' will send more at Christmas, and the rest next Easter" (quoted in Koeman 1964, 38). Many more consumers purchased smaller, less expensive versions of the atlases. In 1577, Ortelius's colleague Philip Galle recast the *Theatrum* as a 5 × 7 inch (12.7 × 17.8 cm) *Epitome* containing about seventy simple maps along with geographic descriptions in rhyme. Its price of one florin, two stuivers (in comparison with the ten-florin price of the *Theatrum* that year) brought the atlas within reach of the lower middle classes. The volume spawned multiple editions, including five in Dutch and nine in French by 1609; it was also translated into English, German, Latin, Spanish, and Italian (Koeman 1964, 44–45).

While world atlases like the *Theatrum* met consumers' demands for comprehensive reference books, markets soon emerged for more specialized works. As territorial concepts of states and nations developed, volumes devoted entirely to individual countries and their internal parts appeared. Land surveyor Christopher Saxton's 1579 *Atlas of England and Wales* was probably the first national atlas (fig. 170). Comprising a general map followed by thirty-four maps, each showing one or several counties, the volume depicted the land in unprecedented detail and precision. By picturing the royal arms prominently on each sheet, the atlas simultaneously fostered local and national pride (Akerman 1995, 146; Evans and Lawrence 1979; Ravenhill 1992).

In the United States, nineteenth-century map publishers represented individual counties in still greater depth in lavishly illustrated atlases. Although the heyday of the volumes lasted only from 1861 to 1910, the American county atlas set an important precedent that continues to shape many consumer-oriented maps: it personalized an industrial product. Whereas earlier consumers ordered customized copies of atlases or altered the volumes themselves, county atlas producers reached out to their highly localized market by incorporating personal information and images directly into the books (Conzen 1984a, 1984b, 1997).

The *Atlas of Peoria County, Illinois*, issued by the innovative Chicago publisher Alfred T. Andreas, exemplified county atlas production in the Midwest. It includes a detailed map of the county, followed by individual maps of each township, likenesses of local men, engraved views of landmarks and scenery, a table of population statistics, and lists of atlas subscribers and area businesses (fig. 171). The atlas contextualizes this portrait of contemporary Peoria with historical data, including an engraving showing the town as it appeared on August 29, 1831, and two extended essays, the general "History of Peoria County, Illinois" and the more biographical "Old Settlers of Peoria County, Illinois" (*Atlas* 1873). The historical selections underscore the county residents'

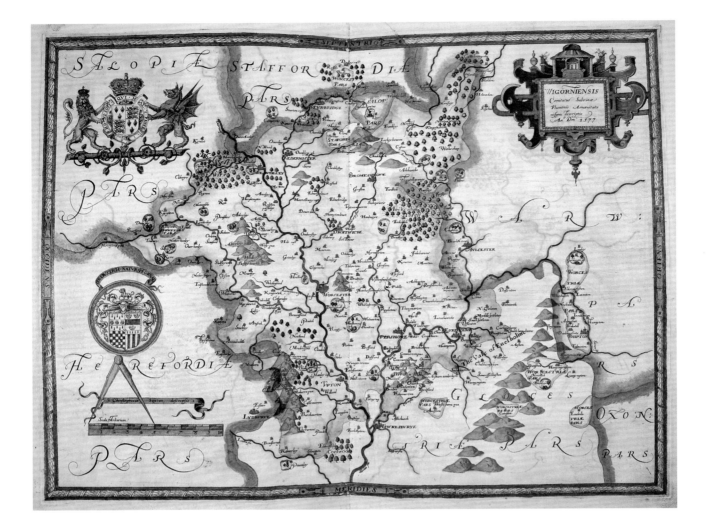

FIGURE 170.
Christopher Saxton, "Wigorniensis"
(Worcestershire) (1579).

achievements in transforming raw prairie into a thriving agricultural and commercial community. Like the makers of bird's-eye lithographs, the atlas producers ascertained their market by hiring agents to secure subscriptions in advance (fig. 172).

The comprehensive scope of the county atlases appealed to the civic as well as the personal pride of consumers, and made them useful to local lawyers, real estate agents, and county officials. The key to their charm and profitability, however, were the paid inserts: the biographies and engraved portraits of individuals, and the views of residences, farms, flour mills, and commercial blocks. These illustrations were elaborations of the vignettes that decorated the borders of county maps. By expanding the sale of the insertions, Andreas thus turned a local commodity with a limited market into a profitable enterprise. Whereas wall maps could only accommodate a limited number of vignettes, the atlases could be expanded to include as many extras as the traveling sales-

DIANE DILLON

men could sell by simply adding pages to the volume. The illustrations were inexpensive to insert, but commanded a pretty price. Subscribers to the Peoria atlas, for example, could receive a copy of the book, featuring their name in the patron's list and on the map that included their land, for $9. But if they wanted the book to include an engraving of their house or farm also, they would have to pay between $28 (for the smallest, 4-inch [10.2-cm] view) and $145 (for a full page). The insertion of portraits ran from $100 to $250, while biographical essays cost 2-1/2¢ per word. The farm views could be further embellished with small likenesses of family members and their livestock around the edges. As Bates Harrington commented, "Anything a man had could be placed on his picture so long as he paid for it" (Conzen 1984a, 50; Harrington 1890, 48, 54).

FIGURE 171.
"Residence of David Kemp." From *Atlas of Peoria County, Illinois* (1873).

FIGURE 172.
"Canvassing a Farmer for an Atlas." From Bates Harrington, *How 'Tis Done* (1890).

The sale of the inserts, like their content, was emphatically personal. The social interaction between the canvasser and the consumer was crucial to the successful transaction. The sale depended on the agent's gift of gab and ability to stroke the customer's vanity "with a delicate touch" (Harrington 1890, 5, 19). To seal the deal, the salesmen often appealed to purchasers' ambitions. A typical pitch assured customers that

any changes or improvements you may contemplate making in the future can be made in the sketch just as you may dictate, and appear in proper form in the years to come, after you have carried out your plans. For instance, you would want that pile of wood near your house left out of the sketch, and the rubbish about the back-yard, which you are about to cart off, should not appear. You had better have a picket fence in front, instead of those rails, as you undoubtedly will have a picket fence there some day. It will only be necessary for you to tell the artist what sort of a picket you prefer, and the thing will be properly done. I would put a pump in that well, and make your barn a little larger. Of course we can make your house look as though it had just been painted, and we can put a grass lawn in front. A few evergreens would look well. (Harrington 1890, 33)

In enumerating the improvements that could be revealed in the engraving, the canvasser effectively advised the farmer about which capital investments he should work toward. The atlases, of course, did not furnish their purchaser with the means to make the improvements, but they offered a plan and a prompt.

This practice took the relationship between atlas makers and buyers in a new direction. Earlier producers like Ortelius and Blaeu encouraged customers to contribute to the volumes' decoration by choosing styles of binding and ornamentation, while van der Hem used his printed atlas as an easel for his own creative project. In persuading customers to show off their accomplishments and ambitions, the canvassers encouraged farmers and businessmen to emulate not the artistry of the atlas makers, but rather their entrepreneurship.

Andreas and other atlas publishers also increased profits by streamlining the data collection and manufacturing processes. Teams of three to five surveyors would descend upon county courthouses to comb landownership records and copy government plats to use as a base map. They then studied the entire county on horseback, filling in details such as the location of schools, churches, mills, and quarries. Although the process was labor intensive, the surveyors soon became efficient, because they repeated the same routine in each county. Andreas expedited production by locating his offices in the Lakeside Building in Chicago, where his fellow tenants could supply all the required services, from engraving, printing, and coloring the maps to binding the volumes and tooling the leather covers (Conzen 1984a, 50, 53).

DIANE DILLON

The mass production of the atlases was offset by the vernacular style of the engravings and the personal connection the canvassers cultivated with customers. The maps and especially the views evidenced more homespun charm than technical skill or aesthetic sophistication. Andreas's county atlases mimicked the ostentatious binding of early modern atlases with leather covers stamped in gold, although the materials and craftsmanship were not as fine. The pages between the covers manifested a different sensibility, however. In comparison with the elegant draftsmanship, delicate coloring, and exquisite printing of the maps and illustrations in the volumes compiled by Ortelius and Blaeu, the images in the county atlases look crude. Where the world atlases exude refinement, the county volumes are provincial in style as well as well as subject.

In form and content, the county atlases expressed the democratic spirit that boosters ascribed to the communities the volumes portrayed. The books enabled ordinary citizens to put themselves literally on the map, inserting their personal achievements into the county's history. Like land in Peoria County and elsewhere in the West, the atlases were readily obtainable. The essays and illustrations in the books were similarly accessible, requiring no specialized knowledge to comprehend. Arcane symbolism and intricate designs were eschewed in favor of directness and legibility. In the farm scenes, for example, the artists rendered all of the buildings, animals, and figures with equal clarity, spreading them neatly across the composition to ensure that all were easily visible. The views, like the landownership maps, read as inventories of possessions rather than realistic landscapes. The layout of the volumes reinforced this sensibility, for the design's hierarchy derived from the size and complexity of the optional images, as dictated by the amount of money subscribers were willing to spend.

Toward the end of the nineteenth century, makers of atlases and wall maps for home reference pioneered another marketing strategy: offering newspaper readers their merchandise at bargain prices through coupons printed in the papers. In 1894, the Chicago *Tribune* made Rand McNally & Co.'s *Indexed Atlas of the World* available to readers in eight installments (fig. 173). Readers could obtain each part for $1 plus three book coupons, to be clipped from the newspaper on different days. The promotion, first published in the February 25, 1894 issue of the *Tribune*, pitched the atlas as a comprehensive reference work: "Not merely a 'collection of maps,' but an encyclopedia, a railway directory, and a gazetteer." Echoing the nationalist bent of earlier atlases, the advertisement boasts that the atlas features 192 maps, 42 statistical diagrams, and "historical and descriptive matter . . . which will assist in giving a thorough understanding of the origin of nations, the causes of their growth, the nature of their institutions, and the extent of their resources."

Like previous purveyors of serial atlases, the *Tribune* offered the publication in installments to make it readily affordable and to cultivate ongoing reader loyalty. The advertisements stressed the deep discount of the offer, not-

ing that the atlas "has never been sold for less than $30.00 or $35.00, according to style of binding." If the atlas offer helped the *Tribune* attract subscribers, the steep markdown hints that the arrangement may also have helped Rand McNally reduce its inventory of slow-selling goods. This was even more likely the case when the Minneapolis *Times* offered *Rand, McNally & Co.'s New General Atlas with Marginal Index* to its readers in ten parts. The paper covers for each installment bear a printed price of 25¢ at the top, but at the bottom, stamped in ink, "Supplied by The Minneapolis Times in 10 Parts at 10¢ Each" appears. Part 1 features the table of contents for the entire set, indicating that the atlas had been completed before the parts were marketed.

The *Indexed Atlas of the World* was the most elaborate—and expensive—of four cartographic products the Chicago *Tribune* offered in 1893 and 1894. The first advertisement, on December 3, 1893, described *The Tribune New General Atlas* as a $4.50 value available for $2 and seven coupons. It introduced the *General Atlas* on December 3, 1893, in tandem with an offer for *Art and Handicraft in the Woman's Building*, a souvenir of the World's Columbian Exposition that was also printed by Rand McNally. The advertisements pitched both volumes as Christmas presents. The next month, on January 10, 1894, the newspaper presented the least expensive of its cartographic bargains, *The Tribune's Handy Census Gazetteer and Atlas of the World*. Available for just 50¢ (and seven coupons), the paper touted the book as "a marvel of condensation and cheapness." At the same time, the advertisements assured potential customers that the book was legible by reproducing a sample page and asserting that it was sturdily "bound in a handsome leatherette flexible cover." The promotion went on to declare, "This little book is an exact copy of the great $35.00 atlas which hitherto has been sold only to those with large purses."

The *Tribune's* steady stream of cartographic offers indicates that they must have met with reasonable success. It seems likely, however, that the lower-priced items were the most sought after. The offer for the most expensive volume, the multipart *Indexed Atlas of the World*, seems to have been terminated prematurely, as the advertisements disappeared after announcing the first two installments. The *Tribune* may have doomed the sales of the *Indexed Atlas* by offering so many competing products at lower prices. In addition to the cheaper atlases and wall map, the *Indexed Atlas* competed for consumers' dollars with numerous other coupon offers for books and portfolios of art prints. These included souvenirs of the World's Columbian Exposition (such as *Neely's Great History of the Parliament of Religions*, *Glimpses of the Fair*, and *Peristyle to Plaisance*) and portfolios of tourist views such as *Picturesque Venice*. Rival newspapers, including the Chicago *Times* and Chicago *Inter-Ocean*, presented comparable coupon offers for souvenir volumes and art portfolios. And finally, the *Tribune's* cartographic goods were in competition with all of the other wares advertised on the same pages. The map offers were interspersed with advertisements for everything from clothing to groceries, brass beds to patent medicines. This stiff competition helps explain why the atlas advertisements put so much emphasis on their bargain prices. At the same time, the *Tribune* differentiated the wall map and atlases from other advertised commodities by casting them as embodiments of learning. The March 12, 1894, announcement for a reversible wall map advised consumers to "Read the domestic and foreign cable news with this map hanging on the wall and you will know more about the world at large than you ever knew before." The *Tribune* billed its cartographic offerings as forms of popular education akin to the newspaper itself. Like the *Epitomes* of the *Theatrum* and the county atlases, maps distributed through newspapers tendered a world of knowledge for a small price.

Maps for travelers and tourists

The culture of tourism pushed the links between cartography and identity in new directions. Broadly defined, a tourist is one who travels for pleasure or personal enrichment. Herodotus, the fifth-century Greek historian, is often described as the first tourist, because he traveled widely just to satisfy his curiosity. Ancient tourists and medieval religious pilgrims, however, relied on verbal directions and written texts rather than maps as guides. For traveling Romans, Pusanias's *Guide to Greece* was a definitive source, while pilgrims of the Middle Ages turned to Sir John Mandeville's *Travels*, which was translated into nine languages after its initial appearance in 1357 (Feifer 1985, 8, 16, 29). For more philosophical wayfarers, Hieronymous Turler offered advice on the "art of travel" in his 1575 *The Traveller* (Delano-Smith 2006, 27). And at least one early modern consumer expressed a desire for a map-filled pocket guide. In 1579, Jacobus Monaw of Breslau complained to Ortelius about the low quality of the maps in the French epitome of his *Theatrum*, noting that the "plan is good, and I wish you would yourself have such a work executed by a good engraver, which we could use as a guide on our travels" (Koeman 1964, 45).

Still, as we have seen in chapter 1, with only a few exceptions European travelers had to wait until the end of the seventeenth century for maps designed specifically for use on the road. The consumption of maps by travelers expanded exponentially in the mid-nineteenth century, when tourism and cartography both developed into mass-market phenomena. Their rise was part of the larger growth of capitalism, which enabled the industrial production and distribution of maps and transformed the pursuit of leisure into a booming business.

Leisure travel enabled tourists to solidify their identities through exposure to the exotic. Sightseers could better appreciate who they were after encountering those whom they clearly were *not*. Maps and guidebooks mediated the encounter with new places and people by presenting foreign environments in familiar terms, using language and symbols consumers could readily understand. Guidebooks reminded tourists of their own identity through their

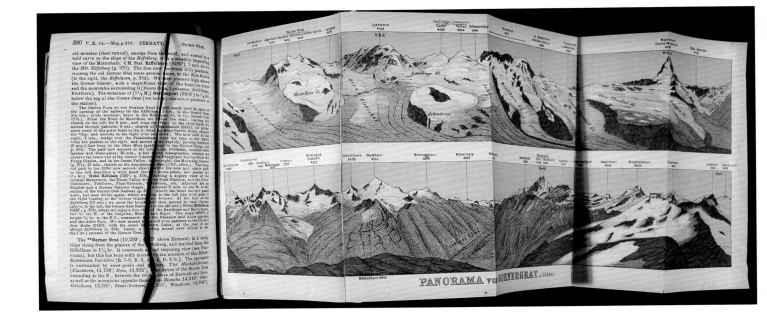

FIGURE 174.

"Panorama vom Gornergrat" (Panorama of Gorner-grat). From Baedeker's *Switzerland* (1905).

nationalist slant. For example, Murray's 1838 *Handbook to Switzerland* featured exuberant quotations from Tennyson and Byron to appeal to British readers, while the Baedeker guidebook reflected its German sensibility through a more somber tone and pithy quotes from Goethe. Although the two publishers competed for much of the century, by 1900 Baedeker dominated the market with guides that were cheaper, more compact, and celebrated for their high-quality maps (Palmowski 2002, 107–8, 120) (fig. 174).

Tourists could get a taste of the world without traveling the globe by visiting the international expositions staged in Europe and North America during the nineteenth and twentieth centuries. The fairs drew together displays of art and architecture, educational exhibits, and entertainment, but above all were designed as showcases for the latest technological innovations and consumer goods. In Walter Benjamin's words, "World exhibitions are places of pilgrimage to the commodity fetish" (1999, 7). The expositions cast consumption in nationalist terms, grouping the exhibits by country and according pride of place to those of the host country.

To guide tourists to and through the exposition grounds, mapmakers issued specialized items, which often rivaled the inventiveness and visual allure of the displays themselves. For the first modern world's fair, the Great Exhibition held in London in 1851, George Shove configured a map of London on a lady's glove (fig. 175). The glove highlighted the location of the exposition with a sketch of its distinctive building, the iron-and-glass Crystal Palace designed by James Paxton, near the base of the palm (Hyde 1986, 47; H. Wal-

DIANE DILLON

lis and Robinson 1987, 265–66). Recognizing that most out-of-town visitors would want to tour the city as well as the exhibition, Shove illustrated the relative positions of other prominent destinations, picturing St. Paul's Cathedral across two fingers, the Colosseum on the thumb, and Kensington Gardens near the wrist. This wider territory meant that the glove could continue to serve tourists after the exhibition closed. If the glove advertised London's major attractions, they in turn enhanced the glove's desirability as a fashionable accessory.

Shove's choice of a glove in particular was essential to its function as a status symbol and a commodity. In the Victorian era, gloves gained importance as indicators of social boundaries. The upper and middle classes had the luxury of preserving their hands, while those who performed manual labor did not. Moreover, gloves tended to be expensive, and thus beyond the normal reach of working people. If high prices typically restricted glove wearing to people of means, those consumers bought them frequently. As ladies and gentlemen covered their hands whenever they stepped outside, gloves quickly wore out. The need for regular replacement urged glove makers toward faddish designs (Levitt 1986, 131, 136). Because consumers expected gloves to be novel, they would have been prepared to accept Shove's unusual design. If Shove's glove met the market for stylish novelties, its value as an exhibition memento made it more likely to be carefully preserved rather than worn out and discarded.

If the glove map slid most readily onto the hands of middle- and upper-class consumers, it may also have appealed to less wealthy consumers as a special purchase (Judd and Fainstein 1999, 16). The promoters of the Great

FIGURE 175.

George Shove, map of London created on a glove for the Great Exhibition (1851).

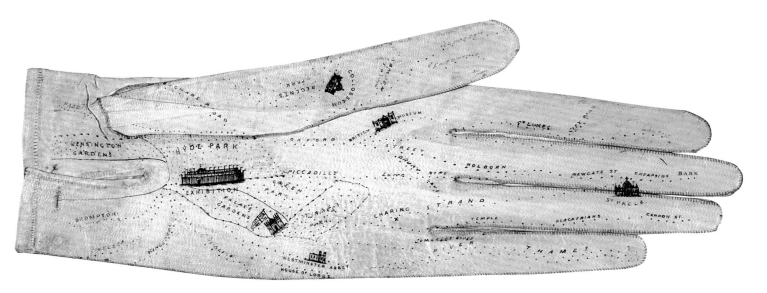

Exhibition urged the attendance of the working classes by initiating savings clubs, offering excursion travel packages, and reducing the admission price to a shilling after the first few weeks (Auerbach 1999, 129, 137). For these tourists, Shove's map might have been valued as a souvenir, but also as a way to try on a new class identity. The glove, like tourist excursions to the Great Exhibition and other destinations, could have taken consumers beyond the confines of their daily lives, encouraging their fantasies and ambitions.

Maps in wartime

As Kaendler's map seller implies, wars have long stimulated public interest in the geography of conflict-torn regions, prompting surges in cartographic production and consumption. For example, during the siege of Malta in 1565, Niccolò Nelli published a broadsheet map of the island and its immediate surroundings. As the conflict progressed, Nelli updated the sheet continually with handwritten annotations indicating the latest developments (Woodward 1996, 94–99). Altogether, the siege of Malta engendered nearly 150 printed maps within a few months of the event. A few years later, the battle of Lepanto resulted in another flurry of published maps and views to help Europeans celebrate the decisive defeat of the Ottoman navy. In the seventeenth century, mapmakers such as Pierre Duval issued inexpensive atlases chronicling France's territorial expansion, offering details about military campaigns and battles (Akerman 1995, 152), while Claes Jansz. Visscher published siege plans accompanied by news reports during the war between The Netherlands and Spain (T. Campbell 1968, 5). The desire of Englishmen to have current maps of the progress of the revolt in the American colonies was met with dozens of special news maps from the presses of William Faden, some drawn, engraved, and published in London shortly after the events they depict (Harley, Petchenik, and Towner 1978, 44, 85, 86, 92, 94).

As recently as 1898, a novel war-related product appeared: the mass-market war atlas. Both Rand McNally and Cram introduced special atlases to take advantage of rabid interest in the Spanish-American War. In addition to familiarizing readers with newly relevant parts of the world, these atlases reinforced the government's economic rationale for the war by stressing the natural resources and commercial products of Cuba, Puerto Rico, and the Philippines (Schulten 2001, 39).

The two world wars prompted even more staggering cartographic booms. During the First World War, Americans extended established map consumption patterns, seeking international maps that stressed the same sorts of practical information they found most useful on maps of the United States. As A. F. Henning of Dallas, Texas, wrote to the National Geographic Society, "I should like to have an atlas showing every TOWN in EUROPE and ASIA big enough to have a POST OFFICE and every STREAM long enough to have a NAME" (quoted in Schulten 2001, 184). After the Nazis invaded Poland on September 1, 1939, Rand McNally sold more European maps and atlases in two weeks than it had in the years since 1918. A weekend of map-buying frenzy erupted on Friday, February 20, 1942, when President Franklin Roosevelt encouraged Americans to buy world maps on which they could trace the strategies he would explain over the radio on Monday. To follow both world wars, consumers modified their interior decor by tacking maps to the walls of offices, kitchens, and living rooms. Most of the maps and atlases snapped up during the wars did not illustrate the shifting political relationships and national borders wrought by the conflicts. Instead of producing new maps, commercial firms repackaged renderings of the relevant parts of the world that they already had on hand (Schulten 2001, 180, 204, 207–9).

War maps and atlases multiplied the connections between the state-in-the-world and the commodity-in-the-market envisioned in Buondelmonti's *Liber insularum Arcipelagi*. Most broadly, cartographers exploited the heightened nationalism of wartime to sell maps. During the Spanish-American War and two world wars, mapmakers entwined politics and economics even more tightly. In relying on existing stock to satisfy consumers' demands for maps of war zones, cartographers often provided maps that highlighted the area's commercial resources. In so doing, the maps helped recast the private economic motivations underlying the wars—the quest for foreign sources of raw materials and markets for goods—as patriotic goals. More broadly, war maps, like reference atlases, tourist itineraries, and decorative maps, helped their users move between the exotic and the familiar. Whether studying images of faraway lands amid the comforts of home or consulting travel guides while adventuring abroad, people have turned to maps to define the contours of their everyday lives as citizens and consumers.

MAPS FOR ALL: PUBLIC AND FREE CARTOGRAPHY

Since ancient times, maps have been displayed in public places. The eighteenth century brought a related novelty, the mass-distributed free map. Both types of maps reached broad audiences without money changing hands. Although free of charge, such maps were rarely free of vested interest (Wood and Fels, 1992, 4–27). They have almost always promoted something, be it civic spirit, religious beliefs, or consumption.

Maps for public display

One of the oldest surviving public maps is the *Forma urbis Romae* (Plan of the City of Rome), a gigantic marble map of Rome created between 203 CE and 211 CE (fig. 176). The map, approximately 10%–15% of which survives in fragments,

probably decorated the wall of a grand room in the Temple of Peace, which
contained the city's property ownership records. Tracing the footprint of every
building and monument in Rome at a scale of 1:240, the map replicated the official cadastral plans on papyrus that were stored in the temple. The papyrus
scrolls depicted the city on the same scale as the stone map, adding details of
measurement and landownership (Reynolds 1996).

    The marble plan thus transformed mundane administrative records into a
magnificent showpiece. When mounted on the wall, the map enabled Romans
to see their city as a whole, all at once. Like the islands pictured by Buondelmonti, the *Forma urbis Romae* rendered its subject as an entity that could be

DIANE DILLON

readily grasped, intellectually if not physically. Whereas the papyrus scrolls documented the private ownership of individual parcels of land, the stone plan presented the entire city for public consumption. By depicting Rome for the Romans on a monumental scale, the plan may have bolstered civic pride and helped solidify the locals' identification with the city and the empire.

During the Middle Ages, the most prominent public maps were probably monumental *mappae mundi* hung on the walls of churches and palaces. Like the *Forma urbis Romae*, the symbolic importance of these world maps outweighed their value as geographic information. In contrast with the Roman plan's civic and imperial messages, *mappae mundi* served a religious purpose, offering moralizing instruction about biblical events and places. For example, the largest known *mappa mundi*, the thirteenth-century Ebsdorf map, shows the burial sites of the apostles. To fill a secondary purpose, the map included travel directions for pilgrims to the Holy Land (Harley and Woodward 1987, 335, 286, 309; on *mappae mundi*, see chapter 2 of the present text).

By the later sixteenth century, public maps frequently served as emblems of political power and religious authority. In 1562, when Duke Alfonso II d'Este visited Venice, his palace hosts hung tapestries decorated with perspective views of the cities in his dominion. For the Galleria delle Carte Geographiche in the Vatican (1580–81), Egnazio Danti designed a suite of maps of Italian cities and provinces, along with the papal domain in France, casting them all as religious grounds under the benevolent rule of the pope (Schulz 1978, 107, 120–21). To decorate his palace at Marly, the French king Louis XIV commissioned Vincenzo Maria Coronelli to fabricate a spectacular pair of globes in the early 1680s. The handmade terrestrial and celestial spheres, each nearly 13 feet (3.9 m) in diameter, intimated the king's ambition to extend his dominion far beyond the borders of France (Dekker and van der Krogt 1993, 63).

More recently, public map exhibits have promoted the prestige and geographic identity of corporations. In the twentieth century, numerous businesses commissioned Rand McNally to create maps and globes for display in the lobbies and windows of their corporate headquarters, in railroad and airline terminals, and at expositions. According to a note on "The Chicago Tribune Automobile Road Map of Alabama, Georgia" (1930), the newspaper maintained a giant wall map of Greater Chicago in its public service office, updated daily to reflect the latest information about roads in the metropolitan area. The map at once attested to the paper's cartographic authority and promoted the goods and services offered by its Touring Bureau, which ranged from sightseeing tips to maps of the region and of every state in the country. For companies such as National Cash Register, International Harvester, and American Airways, public map displays burnished their reputation by stressing the global reach of their business (Schulten 2001, 201–2). If maps displayed and consulted at home shaped personal selves, maps exhibited in public fostered the construction of collective identities.

Maps in the media

In the 1730s, English publisher Edward Cave included maps in the *Gentleman's Magazine*, partly to add original content to his publication, which he had begun as a digest (fig. 177). Competitors quickly followed suit. For these magazine publishers, maps proved particularly desirable as accompaniments to news of foreign wars, whose locations may have been unfamiliar to readers. In 1747, John Hinton, Cave's rival at the *Universal Magazine*, initiated a novel program by publishing a map and description of Berkshire County as the first in a series. By purchasing every issue of the magazine containing the maps, readers could assemble a complete county atlas. Although the series was protracted across twenty years, the marketing tool proved effective and was imitated by two other rivals, the *London Magazine* and *Political Magazine* (Klein 1989, viii–xi).

With the rise of graphic journalism in the nineteenth century, maps gained prominence in newspapers as well as magazines. American publications such as *Harper's Weekly* and *Frank Leslie's Weekly* broke new ground by featuring profuse illustrations that served as information in their own right. Along with engravings of current events, portraits, and scenic views, maps fed readers' sharpening appetite for visual imagery (Bosse 1993, 3–4). During the American Civil War, detailed maps became a standard feature in daily newspapers. Editors pressed field correspondents into service as cartographers, whose maps enabled readers to follow campaign routes and comprehend defense strategies. The New York *Herald* instructed reporters to "send maps and plans with all descriptions of battles, though they may be but hastily and rudely drawn, they will be of great value. The *Herald* is designed to be a complete history of the war and nothing should be omitted by you" (quoted in Bosse 1993, 7). The newspapers' fierce competition to be the first to print the latest news, however, often drove correspondents to dash off maps compromised by errors (Bosse 1993, 14).

In the years since the Civil War, journalistic maps have proliferated, becoming standard fare for newspapers, magazines, television broadcasts, and Web sites (Monmonier 1989). Maps in the media continue to deliver war news, but also summarize weather forecasts, chart the course of floods, fires, and traffic delays, schematize voting patterns, and target crime scenes and tourist destinations. Although all of these maps serve the journalistic mandate to inform and explain, they remain, like their eighteenth-century predecessors, fundamental to marketing. It is no coincidence that the paper with the largest national circulation, *USA Today*, devotes its entire back page to its eye-catching daily weather map (Monmonier 1999, 171–74).

Maps promoting transportation and tourism

Since the second half of the nineteenth century, an abundance of free maps have vied for the attention of travelers. Distributed to promote transporta-

DIANE DILLON

tion, hotels, restaurants, attractions, and anything else that might command a tourist's purse, complimentary maps have become major advertising vehicles. Readily available, they offer stiff competition to maps produced for sale.

In the United States, the railroads took the lead in this endeavor, producing tens of thousands of free maps from about 1830 through the beginning of the twentieth century. Rail companies issued maps to attract corporate investors, to promote the sale and settlement of land along their routes, and to facilitate shipping. The most widely consumed railroad maps, by far, were those created for passengers to promote leisure travel (fig. 8). As we learned in chapter 1, the maps themselves were typically small and schematic, as tourists only needed to know the general trajectory of the route and the location or the stops. They often were inserted into a promotional leaflet or booklet, whose pictures and descriptive text captured consumers' attention (Musich 2006, 98, 115, 141).

In the early 1890s, the Michigan Central Railway issued a map for tourists planning to visit the World's Columbian Exposition in Chicago in 1893. Much as Shove exploited the London exhibition as a gimmick to sell gloves, the Michigan Central looked to the Chicago fair to stimulate ticket sales. Printed on both sides of a 32 × 16 7/8 inch (81.3 × 42.9 cm) sheet, the flyer incorporated a prodigious amount of information and images, approximating the function of a guidebook. Although the upcoming exposition served as a timely hook to capture the attention of potential travelers, the map's main purpose was to promote the Michigan Central's "Niagara Falls Route" from Boston and New York to Chicago and St. Louis. The largest illustrations are a detailed map of the route and an indexed bird's-eye view of Niagara Falls and the Niagara River. Most of the descriptive text is likewise devoted to Niagara Falls and other points of interest along the way. For the Michigan Central, the exposition added glamour to its western route, offering a man-made wonder to complement the natural splendor of Niagara Falls. The alternation of views of the fair buildings and the falls on the map's folded panels emphasized this correspondence.

Of course, the abundance of world's fair imagery and Chicago information on the map meant that it also promoted the exposition itself. The event's managers recognized that they could reap free publicity from railroads, newspapers, and other firms for *their* product—the exposition—and so were quick to supply free illustrations and descriptive matter. Like the glove map, the Michigan Central flyer encouraged tourists to visit other sites in the city beyond the exposition. The sheet included inset maps of downtown Chicago and the fairgrounds to help tourists plan both portions of their trip.

If railroad brochures, like guidebooks, nudged tourists toward a standardized vacation experience, consumers were often quick to put the maps to their own purposes. A writer in *Housekeeper's Weekly* wondered, "How many of our readers have ever made any use of the United States maps issued by the various railroad companies?" and recounted her own adaptation:

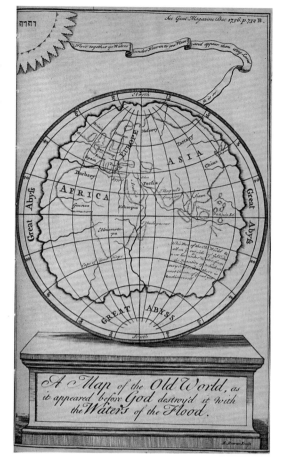

FIGURE 177.
"A Map of the Old World, as It Appeared before God Destroy'd It with the Waters of the Flood."
From *Gentleman's Magazine* (1736).

Some of them have the states very clearly designed, and I have tacked such a one to the kitchen wall very near the work-table. While busy there my little boys have their lesson in United States geography. The bright colors first attracted them, then I said a few words explaining what it was, pointing out the locations of their home, the river in which they watched the fish last Summer, the bay on which they went boating, and the map had a special interest for them. Then we would hunt out the State and city where some friend lived, and finally the elder (aged seven) would spell out the places for himself. Only a few questions were answered at a time, and the next day we had a review. Almost unconsciously he had learned every State in the Union, nearly all the capitals, the lakes, rivers, mountains, capes and bays.

This anecdote underscores the appeal to map consumers of places infused with personal associations. Although the writer has transformed the railroad advertisement into an educational tool, the rest of the story makes it clear that economic imperatives were not excluded from the geography lessons:

Twenty-five cents procured a globe, and we are now going "all around the world" as the children say. Nearly every article used in the kitchen has been utilized to make our "little talks" attractive to them. The tea, coffee, spices, fruits, etc., are first talked about, then their native lands hunted out, and I think such a lesson has never been forgotten. The baby of four years is as much interested as his elder brother, and I think I have laid a foundation of real love for geography that even the driest text book and most indifferent teacher will not eradicate. (Quoted in Anonymous 1893)

In this account, the commodification of cartography has come full circle. Instead of using maps to promote commercial products, the mother uses commodities such as tea and spices to deepen the geographic knowledge conveyed by the map. Because the groceries are at once exotic, at hand in the kitchen, and literally consumable, they symbolize the power of maps to deliver the world to the consumer's household and make it palatable.

By the late 1920s, as automobiles superseded trains for leisure travel, road maps replaced railroad timetables as the most ubiquitous source of free tourist advice. Consumers found these complimentary maps at gasoline stations or obtained them through motor clubs, highway associations, and state offices. Unlike their railroad precursors, the road maps devoted the largest share of their space to detailed, practical cartography, designed for travelers who were now navigating their own vehicle (fig. 1). At the same time, the maps followed railroad examples in relying on eye-catching graphics. Maps illustrated with distinctive artwork and corporate logos worked along with standardized service station architecture to establish oil company brands with consumers. This strategy was crucial, because the petroleum products they sold varied so little from brand to brand (Akerman 2002, 175, 179, 181).

The artful covers of the folded maps enhanced the brands by glamorizing automobile travel. They pictured fashionably clad motorists being served by smiling gas station attendants or visiting high-toned attractions. The front of a 1958 *Washington D.C., and Vicinity Map and Visitor's Guide*, produced by General Drafting for the Esso Standard Oil Company, showed a vacationing family on the National Mall in front of the Smithsonian Institution (Akerman 2002 and 2006b; Yorke, Margolies, and Baker 1996). The illustration implied that by buying the right brand of gasoline and touring the right spots, consumers could become as happy and charming as the sightseers in the picture. Like the glove map, the road maps tantalized status-conscious consumers with the fantasy of improving their image and upgrading their social standing.

Road maps enhanced this appeal to consumers' personal identity with a more generalized promotion of national identity. The proponents of automobile tourism built on the idea advanced by the railroads that by traveling the country, Americans could become better citizens. Visiting national parks could help tourists appreciate the nation's natural resources, while taking in heritage sites could improve their understanding of American history. The maps contributed to the formation of a national tourist industry, which framed seeing the country as a patriotic duty (Akerman 2006b; Shaffer 2001).

Road maps also fostered consumers' sense of self through their function as capital goods. Because the maps were free, they were not commodities on the market. But like the early modern charts that facilitated global trade, they generated capital by promoting the consumption of numerous other goods and services by motoring tourists. For example, Rand McNally's popular *Auto Trails* maps from the 1920s highlighted the names of hotels, garages, filling stations, restaurants, car dealers, and automotive supply stores by printing them in red ink, with arrows pointing to their exact location (Rand McNally 1925; Akerman 2002, 182). By studying the maps, consumers would learn that they needed to shop at all of these places in order to *be* a tourist. After staking their identity to these acquisitions, tourists could drive on to Yellowstone or Mount Rushmore and fortify their patriotic leanings. In this context, improved citizenship is contingent on membership in consumer society.

Regarded together, the various kinds of consumption promoted by railroad and automobile maps make up the tourist experience. Tourism itself became a new kind of commodity, whose value is less a reflection of the labor that goes into its making than of the quality of the experience it promises. In turn, the value of the traditional material commodities involved becomes determined, at least in part, by their ability to deliver a satisfying experience. Hotels need to provide comfortable beds and hot showers, restaurants need to serve tasty meals, shops need to offer novel souvenirs—and maps need to guide tourists smoothly through their sequence of destinations. In the tourist economy, the circulation of people replaces the circulation of goods. This economy prospers not only by exploiting the labor of those who produce essential goods and services, but also by exploiting the leisure of the tourists; they need time as well as money to savor the experience (MacCannell 1976, 23, 28; Debord 1994, 120).

Maps and advertising

The use of maps in advertising, common since the nineteenth century, amplifies the connections between cartography and capitalism illustrated by tourist maps. Advertisers have used maps as graphics to plug a limitless array of goods and services, urging consumers to convert their interest in geography or their identification with particular places to product loyalties. We can sort these uses into three main categories: maps featuring advertisements; advertisements and product packaging incorporating maps; and maps distributed as bonuses to promote other goods.

Many of the earliest advertisements on maps focused on commodities directly related to the geography depicted. For example, real estate agents used maps to promote property sales. New York lithographer Peter A. Mesier printed a map of Chicago showing the lots to be auctioned on October 23, 1835, by James A. Bleecker & Sons, whose name appears prominently in the upper right-hand corner. In the early 1890s, Chicago property and mortgage brokers such B. F. Jacobs and H. C. Van Schaack bought standard city maps in bulk from Rufus Blanchard and other mapmakers to distribute to potential clients (fig. 178). The firms routinely added their company name and bits of promotional text to the sheets. Purchasing mass-produced maps was not only less costly than commissioning specially made ones, but enabled the real estate agents to use them to promote their business more generally. Customers were more likely to keep a general city map, because it could be used for purposes beyond scouting out a specific property; but they would nonetheless be reminded of the agent's name every time they consulted the sheet.

Property agents such as Jacobs and Van Schaack packaged the city as a commodity, depicting Chicago as a landscape that existed to be bought, developed, and sold. They stressed the commercial character of the land by marking their subdivisions in red; Van Schaack added a summary of the firm's "choice mortgages for sale." Eschewing the decorative flourishes of earlier consumer-oriented maps, the real estate sheets define the city as a grid of streets, which mirrors the geometric configuration of the individual blocks and lots that would be visible at a smaller scale. The form of the grid is readily expandable, making the city seem capable of growing indefinitely to the west, north, and south. The Chicago maps also highlight features that influence property values, such as the location of the city's parks and landscaped boulevards, and

DIANE DILLON

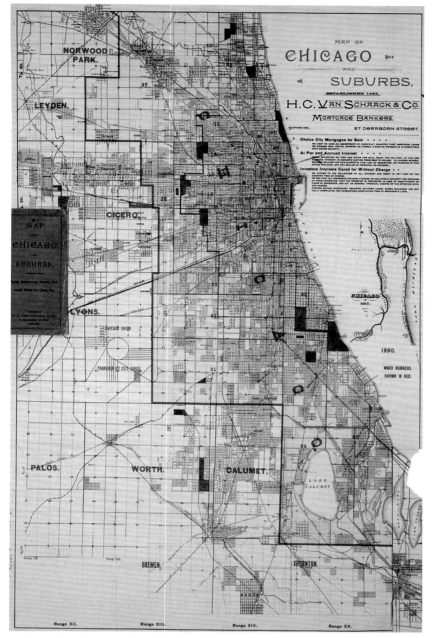

FIGURE 178.

H. C. Van Schaack, "Map of Chicago and Suburbs" (1890). Used by permission of the Chicago History Museum (ICHi-31126).

the routes of rail and horse-car lines. Jacobs added an inset map of the World's Columbian Exposition, acknowledging the belief shared by many brokers that the fair would stimulate the local real estate market (Dillon 2003).

The second genre, advertisements incorporating maps, began to proliferate in the twentieth century as advertising itself became more sophisticated and ubiquitous. Promoters have continued to seize the power of maps to sell

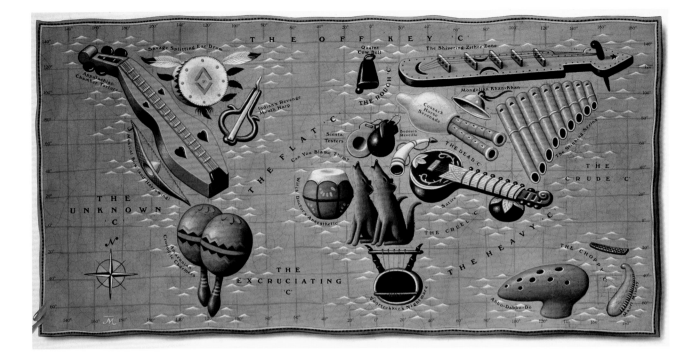

FIGURE 179.

James Marsh, map of the world made for Yamaha
Corporation (1987).

an endless stream of goods, services, and events. Some advertising maps are simple and utilitarian, designed to direct consumers to store locations or diagram the network of airline routes. Others are more dazzling, conceived to capture attention in locations ranging from billboards to magazines to the Internet. This genre of advertisements contains some of the most imaginative and visually striking maps ever produced.

For its centennial in 1987, the musical instrument maker Yamaha commissioned James Marsh to create a pictorial map featuring traditional instruments arranged across an oceanic backdrop in the configuration of the world's landmasses. Amazonian castanets represent South America, cowbells indicate the United Kingdom, and a dulcimer marks the Appalachian Mountains (fig. 179). Marsh added to the map's humor with his punning place-names. The oceans all roar in the key of "C," Siberia appears as "The Shivering Zither Zone," and the sitar denoting India is dubbed "Satire." In using musical commodities to represent the topography of his map, Marsh wittily illustrates the interpenetration of consumption and cartography. Despite the picture's visual harmony, the cacophony that would reach our ears ("The Excruciating 'C'") if this mismatched orchestra played aloud hints that the marriage of maps and commerce is not without discord.

In addition to using cartographic imagery in media advertisements, manufacturers have incorporated maps into the packaging of their products. The maker of Kleenex brand tissue featured a world map on its travel-size box

DIANE DILLON

(dubbed "Little Travelers") to stress the practicality of the item to tourists. Pete's brand of "Wicked Lager" pictures a nineteenth-century German map of the Bay of Naples on its six-pack container, assuring consumers that the ale is genuinely Old World, steeped in venerable brewing traditions (Smits 2006). Although visually varied, most advertisements use maps for limited purposes, chiefly to illustrate the geographic range of their products. World maps proclaim global reach, while national, state, and city maps suggest local flavor (Holmes 1991, 98–99, 51, 78).

The third category, maps distributed with goods as premiums, echoes the promotional strategy of newspaper coupon incentives in soliciting customer loyalty. The practice grew out of the nineteenth-century tradition of shopkeepers handing out trade cards to customers. In France, retailers moved on to giving away pictorial cards, urging customers to return each week to collect all the images in a set (Murray 1989, 10–13). By the end of the century, manufacturers in the United States as well as Europe commonly inserted colorful cards into their product wrappers.

In 1889, the Arbuckle Bros. Coffee Co. copyrighted two cartographic trade card sets: a series of fifty chromolithographs representing the world's prominent nations, and another showing the American states and territories. The 3 × 5 inch (7.6 × 12.7 cm) cards juxtapose topographic maps of the selected areas with images of local scenery, natural resources, and industries. Set off in decorative borders, the maps indicate principal mountain ranges, rivers, and cities. Some of the pictures illustrate the translation of nature's bounty into commodities, such as Newfoundland fishermen cleaning their catch and Minnesotans packing flour into barrels. The entry for Italy implies the process of manufacture by picturing grapevines and olive branches next to bottles of olive oil and wine. Brazil's card focuses on the company's own product, coffee (fig. 180), Egypt's highlights historic attractions such as the Sphinx, and Ireland's represents landscape tourists in Killarney. The explanation on the reverse side of the cards lauds their "instructive and artistic" value, noting that the maps are "correct" and the pictures "made by the best modern artists." The text goes on to claim, "Teachers in the public schools are unanimous in their praise of our object lesson cards and pronounce them one of the happiest, and most impressive mediums for improving instruction of all classes of students." The artists added to the card's didactic value by incorporating population statistics, geographic dimensions, and forms of government into the design.

The educational and artistic appeal of the cards partly disguised their role as advertisements. The company included its name only on the reverse, where a description of Arbuckle's coffee appears alongside the explanation of the picture on the front. The advertising purpose, however, informs all of the imagery, styling each of the pictured subjects—from maps and natural scenery to historic sites and agricultural produce—as commodities akin to the

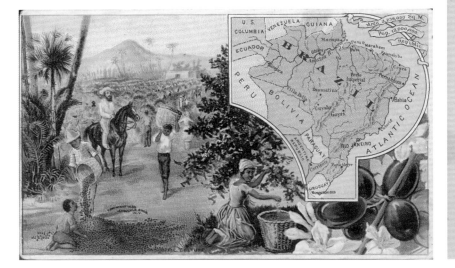

company's coffee. The company recruited the glamour of geographic imagery to promote a mundane household staple. By encouraging customers to collect the cards, Arbuckle's prompted them to not only buy the coffee, but identify with the brand.

Cartographic advertisements like the road maps and Arbuckle cards bid so aggressively for the map user's interest because securing customer loyalty is their primary goal. Their undisguised pitch alerts us to the more subtle forms of persuasion inherent in other publicly displayed and freely distributed maps. These maps exploit cartography's aura of usefulness, learning, and authority for ends that are most often economic. Even when they primarily promote social ambition, civic pride, or patriotism, the maps share the rhetoric of consumption.

BODIES AND MAPS

Turning once again to Kaendler's statue, if the intimacy it depicts between the peddler's body and his wares is often pivotal to map selling, it is even more essential to map consumption. Since ancient times, cartographers have visu-

DIANE DILLON

alized this bond by including human figures on maps and representing land-masses in the form of people and animals. By humanizing geography, such images invite viewers to envision themselves on the map, as part of the territory. The map-body connection becomes all the more central when we recognize that users experience cartographic materials haptically (through the sense of touch) as well as optically. Sightseers wave tourist maps as they make their way around landmarks, collectors finger the embossed leather bindings of luxury atlases, game-players clutch the pieces of cartographic board games and jigsaw puzzle maps. Whereas cartographers project the earth's three-dimensional surface onto two-dimensional images, consumers return to the third dimension through their physical manipulation of maps. The haptic experience of maps involves movement through space—like the travel maps often facilitate. The experience becomes all the more personal when the map is self-made, whether by drawing, sewing, or tapping a computer keyboard. When consumers become producers, they gain a feeling of control, a sense of power over both maps and the landscapes they represent.

Figures on maps

Many of the earliest known geographic representations incorporate figural imagery. A diagram of the Fayum district west of the Nile, drawn on papyrus around 330–32 BC, features stylized portraits of Egyptian elites and animal-shaped deities (Walters Art Gallery 1952, entry 3). In the sixteenth century, figures began to animate the decorative borders of maps made in the Southern Netherlands. Ortelius produced one of the first examples in his 1586 map illustrating the life and travels of the biblical figure Abraham. For the twenty-two rondels framing the map, Maerten de Vos represented Abraham at various points in his journey. Ortelius published the map in his *Parergon*, an atlas of ancient geography that was attached to the later editions of his *Theatrum* (Schilder 2000, 55–56).

The decorative borders of subsequent maps gave readers of all classes figures they could relate to. Portraits of kings and queens on national maps fostered consumers' identification with their land through their rulers, while depictions of nobles, bourgeois, and peasants allowed people of different classes to find themselves in the picture. Cartographers frequently animated maps with types native to the geography represented. Blaeu surrounded his 1605 world map with vignettes of peoples from around the globe, while the images compiled by Georg Braun and Frans Hogenberg for their six-volume atlas of city views, the *Civitates orbis terrarum* (Cities of the World; 1572–1617), usually included foreground figures wearing local garb (fig. 181). Joris Hoefnagel and the other engravers employed by Braun and Hogenberg often drew analogies between the contours of the costumes and those of the topography and architecture within the scene (Wilson 1999, 167). By making the types seem to belong

CONSUMING MAPS

to the landscape, the artists reinforced the larger idea of the identification of people with place. If we interpret the figures as stand-ins for consumers, they urge map readers to likewise identify with the mapped territory.

Despite the clever integration of the bodies into the landscapes in the *Civitates*, most of the figures depicted on maps remain ancillary to the cartographic content. Mapmakers included them as decorative embellishments pointedly to make their wares more appealing to consumers. By serving at once as surrogates for the user and as promoters of consumption, the figures might have prodded map readers to see themselves as similarly aligned with consumption. In other words, by urging users to identify specifically with the consumer-oriented parts of maps, the figures encouraged them to identify with maps *as* consumer goods. In this way, the figures promoted identification in terms of both geography and commodities.

Metamorphic maps

The bodily bond between map and user takes another form in the tradition of anthropomorphic and zoomorphic cartography. The medieval Ebsdorf map configured the world as the body of Christ, picturing his head, hands, and feet at the four cardinal directions (Harley and Woodward 1987, 290). In 1537, Joannes Bucius shaped a map of Europe to resemble a female figure. Such metamorphic maps probably grew out of the age-old tendency to interpret landscape forms, from rocky outcroppings to the moon's surface, as representations of humans or animals. Although the motives prompting metamorphic maps vary, many were intended as political or social allegories (Hill 1978, 39). The Yamaha map pushed this tradition in a new direction by transforming continents into musical instruments and promoting goods rather than ideas.

Among the most enduring metamorphic conventions is a map of the Low Countries in the shape of a lion, known as "Leo Belgicus." Austrian cartographer Michael von Aitzing inaugurated the genre as an illustration for his historical and topographical account of The Netherlands, *De Leone Belgico*, published in Cologne in 1583. In the preface, Aitzing explains his patriotic choice of the lion, comparing "the strongest of animals" with Julius Caesar's characterization of the "Belgae" as "the strongest of all tribes." He also points out that "virtually all provinces carry a lion in their coat of arms." The vaguely leonine shape of the Low Countries also inspired his design (Heijden 1990, 16–18). Published after the bond between the Northern and Southern Netherlands had begun to fray, "Leo Belgicus," like the wall maps of the Seventeen Provinces, expressed a desire for a united country.

Subsequent variations on Aitzing's design reflect the evolving political landscape. Johan van Doetecum added an explanatory text to his 1598 version, which recognized the disintegrating union. Between 1608 and 1611, Hessel Gerritsz cast the lion roaring toward the southwest to confront the threats repre-

DIANE DILLON

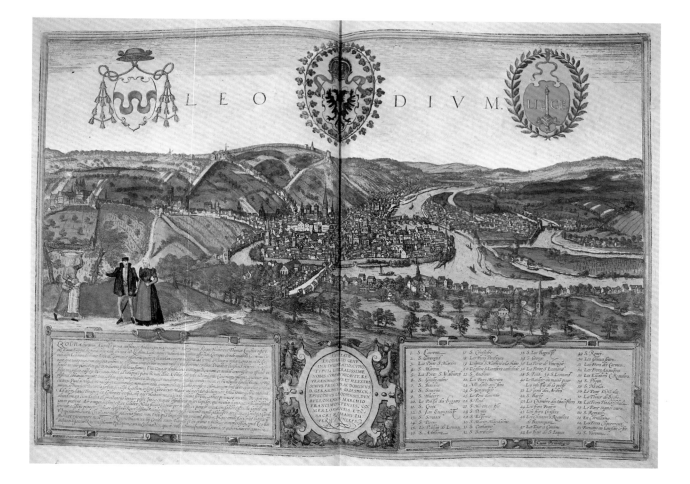

FIGURE 181.

"Leodium" (Liege). From Georg Braun and Frans Hogenberg, *Civitates orbis terrarum* (Cities of the World) (1572).

sented first by the Spanish in the Southern Netherlands and then by the French. During the "Twelve Years Truce" (1609–21), Claes Jansz. Visscher showed the lion sitting peacefully, with the image of war asleep in the lower right-hand corner. In the fourth version, "Leo Hollandicus," Visscher expressed the rise of patriotism after Holland captured the "lion's share" in the war with Spain (Heijden 1990, 18–19). Cartographers in the Low Countries continued to tap the symbolic power of "Leo Belgicus" into the nineteenth century (Tooley 1963, 16). The longevity of the design underscores its popularity with consumers. As an emblem of the storied strength of their ancestors and their country, "Leo Belgicus" allowed Netherlandish people to frame their identity in terms of a glorious past and confront the future with confidence.

A more playful, yet no less political, series of metamorphic maps appeared in the 1869 book *Geographical Fun: Being Humourous Outlines of Various Countries* (fig. 182). The volume features twelve vividly colored lithographs showing a variety of humans and animals twisted into the shapes of European countries. For example, the cover illustration contorts a bear and a figure of the czar into

ITALY.

FIGURE 182.
Aleph [William Harvey], "Italy" (1869).

a representation of Russia. Publishers Hodder and Stoughton reportedly found the inspiration for *Geographical Fun* in drawings made by a fifteen-year-old girl to entertain her sick brother. She in turn seems to have adopted the idea from an allegorical sketch of England as Punch atop a dolphin (Slowther 1991, 49). William Harvey, a London journalist who wrote the accompanying text under the pseudonym Aleph, states in the introduction that the book (like its inspiration) was intended to amuse and educate children. As he puts it, "It is believed that illustrations of Geography may be rendered educational, and prove of service to young scholars who commonly think Globes and Maps but wearisome aids to knowledge" (Aleph 1869).

The maps' sharp satirical edges, however, contradict Harvey's claim that the book was solely for children. England appears as an elegant Queen Victoria,

DIANE DILLON

seated serenely atop a shield bearing the British royal coat of arms with ships representing her empire at her knees, while Ireland and Scotland are considerably less dignified. The caption below the figure of Ireland, a sturdy peasant woman with a baby strapped to her back, notes she lacks fortune, but is "rich in native grace, Herrings, potatoes, and a joyous face." Scotland appears as a doddering piper in a windswept kilt, struggling through a bog with a broken wind bag. Other illustrations poke fun at the weaknesses plaguing the rest of Europe. In caricaturing Italy as a "model chieftain" (probably Garibaldi or Cavour) hovering over a tiny figure of the pope, positioned as Corsica, the map makes light of the belief that Italy's unification was predicated on the overthrow of the papal state (Slowther 1991, 50). Maps such as *Leo Belgicus* and those in *Geographical Fun* are doubly metamorphic. They not only shape geography into bodies, but also transform politics into art and humor.

Self-made maps

If cartographic consumption generally depends on the bond forged between the user and the map, the link becomes all the more vital when the map is created or altered by the consumer. Amateur cartographers have made maps for a variety of practical uses, as well as for their own edification and pleasure, from time immemorial (Akerman 2000, 27–28). Although rarely preserved, such self-made maps narrate an important chapter in the history of map consumption, offering insight into ways maps helped define the lives of ordinary people.

In the eighteenth century, specialized vernacular traditions of mapmaking proliferated. During the Seven Years' War and the Revolutionary War in North America, soldiers used their jackknives to engrave maps and other decorations on their powder horns. Such scratched designs are the oldest form of graphic art, and the soldiers' handiwork was a counterpart to sailors' scrimshaw artistry. Professional engravers, gunsmiths, and comb makers occasionally decorated horns, but most often the individual owners executed the work (Grancsay 1945, 1, 3).

Eighteenth-century soldiers carved maps for practical reasons. Much of the territory contested in the Seven Years' War was uncharted, and printed maps of the area were rare. It thus seems likely that powder horns recorded some geographic information for the first time. The horn in figure 183 features the two most frequently represented paths from New York to Canada, one through the Hudson River and Lake Champlain and the other by way of the Hudson and Mohawk rivers and Lake Ontario. Both were strategic routes during the conflict and essential to the trade with Native Americans. The horn enumerates many of the forts and towns along these paths, while also representing New York City, the British royal coat of arms, and an assortment of boats, houses, and windmills (Grancsay 1945, 4, 12; Thompson 1902, 1009; Walters Art Gallery 1952).

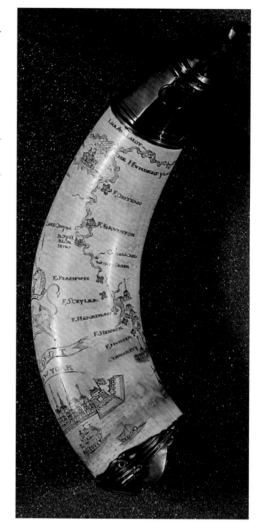

FIGURE 183.
Powder horn incised with map of Hudson and Mohawk river valleys (ca. 1757).

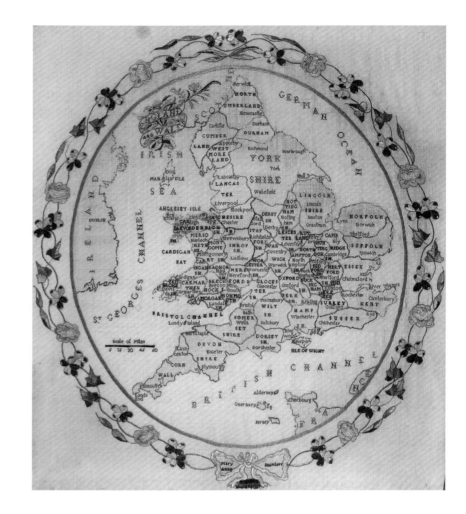

For soldiers and hunters alike, powder horns were intimate possessions, kept close to the body. Owners held them in their hands as they incised decorations or filled them with powder, and carried them in the pockets of breeches or slung over their shoulder by a leather thong. Along with the musket, the powder horn became a romantic extension of the self, that affirmed the owner's masculine identity. When not in use, the horns were often hung with the guns over kitchen fireplaces, where they remained in daily view (Grancsay 1945, 2, 6).

During the same period in which engraved powder horns became popular, another distinctly gendered craft tradition flourished: the production of embroidered maps and globes by girls and young women. Like the horns, needlework maps demonstrated artistry, dexterity, and technical skill, but were made to be useful. Map embroidery must have been established in Europe by the 1620s, when Johann Friedrich Greuter made a mythological print after a design by Giovanni Lanfranco, showing women embroidering a world map, a

DIANE DILLON

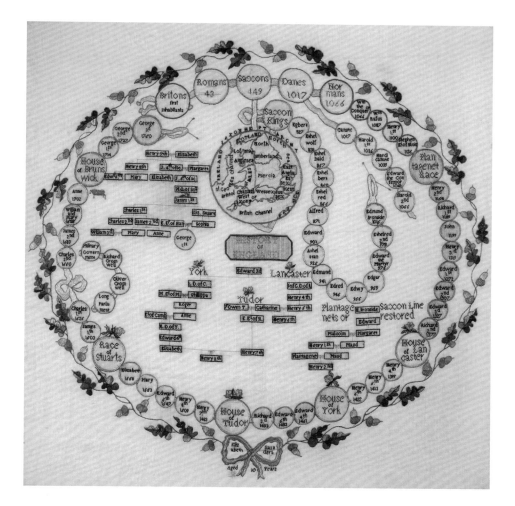

celestial globe, and a nautical chart. The net of geographic coordinates visible on the globe and world map implies a parallel between the weaver's warp and woof threads and the geographer's latitude and longitude lines. The image suggests that cartography, like needlework, is at once aesthetic, practical, and productive (Woodward 1996, 87).

Embroidered maps emerged in greater numbers in Britain during the 1770s and soon spread to North America. At home and in school, girls made needlework samplers to learn embroidery stitches, the alphabet, and geography. Most surviving examples from Britain, like the pair made by the sisters Mary Anne and Elizabeth Saunders, depicted the British Isles or parts thereof (figs. 184A and 184B). Students or teachers drew some of the base maps (often inaccurately), traced them from printed maps, or relied on patterns published in ladies' magazines or printed directly on the fabric to be embroidered. The completed images often served as decorative accessories in the form of fire screens or framed samplers hung by the proud parents of their young makers. In at least one instance, an embroidered map was used to record land use on an English farm (Tyner 1994, 2, 5–6).

We can find additional evidence of the cartographic forays of amateurs in inscriptions on printed maps. Early twentieth-century automobile tourists frequently personalized road maps by tracing out their routes, adding notes about where they camped or how much time and money they spent for lunch. Map companies soon capitalized on this impulse by offering professionally customized maps and itineraries. In 1936, the Continental Oil Co. collaborated with mapmaker H. M. Gousha to produce "Touraides," spiral-bound booklets containing specially selected maps highlighting suggested roads, lists of hotels, and tourist attractions, as well as charts for travelers to record stops and expenses. The front covers feature the customer's name, lettered by hand as a personal touch. After World War II, the American Automobile Association introduced its more famous variation, the TripTik. Mapmakers produced both Touraides and TripTiks in a highly mechanical way, but the result seemed hand-tailored to consumers—a profitable combination pioneered by the American county atlas makers of the previous century (Akerman 2000).

These individualized guides set the precedent for the customized electronic maps that proliferated at the beginning of the twenty-first century. Like TripTiks, Web-based cartographic tools such as Google Maps and MapQuest add a layer of personal travel advice to standardized maps. Quick, convenient, and free, these services have prompted more people to use more maps more often. The ease of generating route maps and travel directions encourages Internet-connected consumers to create maps for even simple journeys. To enhance the user-friendliness of electronic maps, their designers have incorporated features that mimic conventional patterns of map use. For example, with Google Maps one can drag the map across the screen, much as one would move a paper map spread out on a table. The Zoom tool emulates a magnifying glass, albeit with greater power and the ability to preserve the enlarged view.

For many consumers, however, the appeal of using maps on the Web turns on functions that go beyond those of conventional maps. Electronic cartography not only offers instant access to more details about more places, but also permits users to combine maps readily with other kinds of images and information. One click on Google Maps brings a satellite photograph of the mapped area to your screen; another click combines the map and the photograph, adding street names to the aerial view. Moreover, the linking of digital maps to search engines has made it easy to create a dizzying variety of thematic maps. Using maps from Google or Yahoo, consumers can add layers to a basic city map to identify the closest restaurants or theaters, while reviews, hours/showtimes, and phone numbers pop up in windows. The map then becomes a transit guide when the user adds bus and train lines. Weather maps can be

DIANE DILLON

customized to show the temperature, precipitation, and pollen forecasts for any given location.

The expanding consumption of digital cartography has led to important shifts in the way we acquire, use, and think about maps. The possibility of synthesizing and representing so many different kinds of information has made maps a popular tool for individuals, groups, and businesses to plan work and leisure activities, organize research, and communicate findings. Maps can be generated in conjunction with any project that either has a street address or involves some other sort of spatial information. These maps can readily supplement—or replace—scribbled to-do lists, typed memos, and printed reports. In turn, the lure of electronically customized maps has prompted consumers to prioritize geographic information, considering location ahead of many other issues in any given context.

The appeal of digital cartography may derive as much from the power it grants consumers to visually transform maps—and the pleasure this yields—as from the usefulness of the final product. Like earlier anthropomorphic and zoomorphic images, in which cartographers transformed maps into bodies, electronic cartography invites consumers to play with maps, to exercise their creativity and experience the thrill of visualizing geography in a new way. The use of digital maps on handheld electronic devices recreates the intimate, haptic experience of maps on powder horns and gloves. In making geography fun, digital maps extend the long tradition of cartographic amusements.

Cartographic games and toys

In permitting amateurs to draw, assemble, or manipulate geographic images, cartographic toys and games emphasize the haptic nature of self-made maps. Historically, most of these amusements have been quintessential consumer goods, invented and produced for profit. Although conceived for entertainment rather than practical use, map games have followed many of the larger trends in cartographic consumption: they have been sold in different versions at a range of prices; they have imitated specialized products, such as travel guides and war maps; they have capitalized on public interest in imperialism and exotic places; they often have been decorative; and they have been adapted to educational purposes.

Most board games are cartographic in that they involve moving pieces across two-dimensional representations of real or imaginary spaces. The earliest known geographic table amusement was the Game of the Goose, a gambling game popular in western Europe from around the sixteenth century. Players pursued one another by moving tokens along a spiral track, their advance determined by a roll of dice. The first player to reach the final square (after overcoming all the hazards along the route marked by the goose) collected all the stake money. In 1645, Pierre du Val, geographer to the youthful French king Louis XIV, created a more cartographic version, filling the sixty-three circles along the track with maps of different parts of the world. Du Val added maps of the four known continents in each corner of the board and used a map of France as the winner's circle. The game inspired countless variations over the years, surviving into the present as Snakes and Ladders (Hill 1978, 7; H. Wallis and Robinson 1987, 40; Shefrin 1999, 13).

In 1759, English geographer and writing master John Jeffreys introduced the first game with a map as a playing surface: A Journey through Europe, or The Play of Geography. The rules still followed the Game of the Goose, but the players moved through the countries of Europe in geographic sequence. In a nationalistic gesture, Jeffreys made London the winner's destination. A Journey through Europe established the formula travel games would follow for the next eighty years. Many traced the popular routes of the grand tour, incorporating references to Thomas Nugent's 1749 guidebook, *The Grand Tour; or, A Journey through the Netherlands, Germany, Italy, and France.* Around 1845, Edward Wallis produced an exotic variation, Wanderers in the Wilderness, where players explored the jungles of South America. The Great Exhibition of 1851 inspired Smith Evans to create The Crystal Palace Game, a Voyage Round the World, an Entertaining Excursion in Search of Knowledge, whereby Geography Is Made Easy. The game does not lead players through the Crystal Palace, but to places around the globe that had been represented in the exhibits. Like the world's fair itself, the game offered a virtual tour of the globe. Its playing board was a colorful map of the world, surrounded by scenes of thrilling adventures such as a Tiger Hunt in India. By replacing gambling with geographic education, English makers of games cast their products as wholesome family amusements. Yet they retained economic overtones in their promotion of tourism. The rule booklet for The Crystal Palace Game even itemized trip expenses, including the cost of provisions aboard ship. The tokens players pushed around the board, sometimes called travellers and shaped like chess pawns, mediated the link between player and map and served as surrogates for the participant's body (Hill 1978, 10, 12–14; Shefrin 1999, 15).

In the twentieth century, new map games focusing on entrepreneurship or military imperialism gave players a sense of power and control as they reshaped the world by reconfiguring the map. In British Rails (1984), players construct railroads, using a grease pencil to draw rail lines on a map of Great Britain as they amass fortunes by hauling freight and passengers. Players vie for control of a map of the world loosely based on Mercator's projection in *La conquête du monde*, introduced by filmmaker Albert Lamorisse in France in 1957 and subsequently sold in the United States as Risk (fig. 185). More complicated strategic games are often based on historic wars, such as Kingmaker (1976), set during the fifteenth-century War of the Roses, and Russian Front (1985), set during the Second World War.

DIANE DILLON

FIGURE 185.

Game board and pieces for Risk (1968).

In contrast with these fictionalized war games introduced long after the conflicts had been resolved, in 1914 the Chicago *Tribune* published a game designed to be played as the First World War progressed. The newspaper included a 19-1/2 × 33 inch (49.5 × 83.8 cm) War Game World Map in its color section on August 23, 1914, accompanied by playing rules. The instructions directed readers to mount the map on cloth or cardboard and assemble pins, color coded to match the countries on the map, to represent armies and battleships. The directions, published under the map in the Chicago *Tribune* on August 23, 1914, continued:

Now, with the pins prepared and the map pasted up you are ready to begin the war game. Turn to the news section of the paper and from the dispatches of the day you will readily learn where the opposing troops and ships are located as accurately as the censored cables can tell.

With pins you can blockade ports, follow the fleet cruisers out in the steamship lanes and mark the battle formations, wherever they may be.

With little tags of paper or pasteboard impaled on the pins, the numerical strength of the armies and navies may be designated with figures.

Take THE TRIBUNE each day and move your pin armies and navies. When the game is over your pins may mark new national boundaries. And you may know more about history, geography, and war than ever you learned at school.

The rather shocking callousness of using the word *game* in connection with the carnage of the First World War is explained by the game's early publication date. At the end of August 1914, when the *Tribune* published this feature, most combatants were convinced the war would be over in a few months. Although intended to promote sales of the newspaper, the "game" encouraged readers to digest the war news systematically. In tracing "new national boundaries" with the pins, players effectively redrew the world map. Mapmakers including Rand McNally and Hammond revived the idea during the Second World War by including flags, pins, and tacks with their maps (Schulten 2001, 207).

Cartographic jigsaw puzzles—originally called dissected maps—were probably the most popular didactic map toys in the eighteenth and nineteenth centuries. John Spilsbury produced the first example in 1759, mounting a printed map on thin sheets of mahogany and then cutting it into pieces along national borders. Spilsbury parlayed his innovation into a successful business by specializing in a variety of related products he could sell at a range of prices. The year after he made the original dissection, he set up his own London business as an engraver, mapmaker, and retailer. In addition to puzzles he sold prints, stationery, books, and maps printed on paper and handkerchief silk. Spilsbury realized he could maximize the value of his copperplates by using them repeatedly to print maps, puzzle sheets, and handkerchiefs (Shefrin 1999, 8, 36 note 3).

By 1763, Spilsbury gave shoppers a choice of twenty-eight dissected maps, including maps of the world, most European countries, the West Indies, and ancient geography. He sold the cheapest versions for seven shillings, six pence, "in a chip box with[ou]t the sea"—and, it seems, without the cartouche. For ten shillings, six pence, a buyer could obtain the map with the sea, and for twelve shillings he could get the puzzle in a fancier square box. At the top of his product line in 1766–67, Spilsbury sold a customized set of five dissected maps in a handsomely crafted mahogany cabinet. Each puzzle had its own locking drawer, inset with an ivory label indicating the map's subject (Europe, Asia, Africa, the Americas, Ireland). The inclusion of Ireland rather than England and Wales reflected the customer's preference (Shefrin 1999, 8–9, 25).

Dissected maps, like the later caricatures in *Geographical Fun* and the railroad map used by the mother quoted in *Housekeeper's Weekly*, enlivened the study of a subject long dulled by an emphasis on rote memorization. Progressive educators quickly caught on to their potential. On July 3, 1765, Thomas Grimstone, a student at William Gilpin's school at Cheam, wrote his father John to report that he had "lost one of the countis [*sic*] of my wooden map" and asked for a replacement (quoted in Hannas 1972, 60). In Robert Davidson's *Elements of Geography, Short and Plain* (1787), the introductory letter advises young readers, "Persuade your parents to reward your diligence with a dissected map of England, which you may put together in five minutes, and learn its principal manufacturing towns and counties in one week" (quoted in Shefrin 1999, 21).

DIANE DILLON

For the status conscious, mastering the puzzles or strategic board games could symbolize civilized attainment. In Jane Austen's *Mansfield Park*, Fanny's cousins consider her stupid because she "cannot put the map of Europe together" (Austen 1966, 54).

Cartifacts

Map toys and games form part of an expansive class of consumer goods that resemble maps, yet whose function is not primarily cartographic. Often dubbed "cartifacts," such objects have proliferated as mass-market novelties in the twentieth and twenty-first centuries (Post 2001). Despite their variety, most cartifacts retain the connotations that have inspired map consumption since the Renaissance. Because few, if any, are necessities, consumers acquire these goods for their symbolic or entertainment value. Their primary purpose is often to express the buyer's sense of self.

Beyond games and advertisements, cartifacts fall into several overlapping categories. Souvenirs, such as coffee mugs bearing the Tour de France route, are commemorative. They descend from more rarefied objects, such as the silver medal produced in 1581 depicting the route of Sir Francis Drake's circumnavigation of the world between 1577 and 1580. A second group, picking up on the toys' learned overtones, includes desk accessories such as pencil boxes, paperweights, stationery, and bookends. Home furnishings form another set, encompassing clocks, shower curtains, pillows, refrigerator magnets, and posters. Like the desk accessories, these decorations connote worldliness and sophistication; they might picture a place the owner has visited, or hopes to visit one day.

Images of maps on clothing and fashion accessories convey similar meanings in a more intimate, body-centered way. Some of the earliest examples, dating back to sixteenth-century Japan, were maps painted on fans. An important part of the ceremonial dress of aristocrats and courtiers, collapsible fans typically featured brightly painted decorations; most were pictorial, but a few featured maps. By the nineteenth century, printed maps appeared on fans in Japan and China (H. Wallis and Robinson 1987, 25) (fig. 186). Souvenir manufacturers in the United States expanded on the trend, producing maps on fans as world's fair mementos (Harris 1993, 15). Clothes decorated with historical maps preserve a sense of the Asian fans' dignity. Visitors to Hawaii can buy shirts featuring maps of Captain James Cook's eighteenth-century voyage to the islands, while more formal dressers might favor silk ties or scarves printed with antique maps. Most wearable maps, however, are more frivolous, such as the tube socks decorated with maps of the London Underground.

One of the largest groups of cartifacts amplifies the map-body connection by integrating the commercial consumption of maps and the physical consumption of food. Roadside restaurants often include maps on their menus,

place mats, and coasters, which patrons are encouraged to take with them as reminders to come back. Home cooks can buy muffin tins and ice cube trays in the shape of Texas, plates depicting continents, and glasses etched with world maps. Winemakers often feature maps on their product's label to assure consumers that the contents of the bottle are indeed from Italy or California. All of these artifacts invite diners to consume maps with their eyes while they ingest food and drink.

As maps on powder horns, jigsaw puzzles, scarves, and placemats attest, the intimacy between consumers and cartography is configured both optically and haptically. Some maps visualize the connection through embellishments that remain secondary to their geographic message, such as human figures within border decorations. In others, such as "Leo Belgicus," body and map become one. Self-made maps, games, and cartifacts often rely on the sense of touch to create the bond. As leisure pursuits and tourist souvenirs, these ob-

DIANE DILLON

jects appeal to consumers at least partly because they do not take their subject too seriously. By making geography amusing, they transform grave themes like war and nation-making into palatable fare.

CONCLUSIONS

English caricaturist James Gillray wittily satirized map consumption in his 1805 cartoon "The Plumb-Pudding in Danger;—or—State Epicures Taking un Petit Souper." Gillray depicted British prime minister William Pitt and French emperor Napoleon Bonaparte facing each other at dinner, lustily carving up a juicy, globe-shaped plum pudding, whose raisin marbling has morphed into the outlines of continents. Pitt slices his trident neatly through the North Atlantic, just to the west of the prominently featured British Isles. Napoleon carves out a chunk encompassing Holland, France, Switzerland, Italy, and Spain with a sword, piercing Hanover with a fork. Gillray patriotically gives the British leader the upper hand, literally and figuratively. As he stabs the top of the globe with his fork, Pitt appears to say to Napoleon—as a contemporary commentator in a German journal put it—"You take the continent of Europe, and I'll take world trade, and all the coasts washed by the ocean will quake with fear at the roar of the English lion of the seas." Gillray reinforced the imperial iconography on the diners' chair backs. On Napoleon's chair, a French legionary eagle builds his nest atop a Jacobean red bonnet, while on Pitt's the British lion waves an admiral's flag. Gold dinner plates hint at the enormous wealth at stake on the table, reiterating the diners' national identities in coats of arms akin to those often represented on maps (Banerji and Donald 1999, 215–21).

Like Kaendler's map seller, Gillray's cartoon adumbrates the complex nature of map consumption. Within the image, the globe is the main object of consumption, rendering the countries of the world as delectable commodities. Gillray stresses the connections between maps and bodies, portraying Pitt and Napoleon as embodiments of their respective nations who prepare to ingest additional parts of the globe. The two leaders reshape political geography with their utensils, acting as amateur cartographers as well as map users. Like the tourist maps, the print casts empire-building and patriotism as consumer activities. And like the cartographic games, it makes light of sober international affairs. Moreover, the etching itself was a mass-produced consumer item, sold by subscriptions through print dealers as well as disseminated through periodicals (Banerji and Donald 1999, 4–5). As with individual maps, subscribers could hang the framed print on the wall of their home or workplace to project their politics as well their artistic tastes, intellectual acumen, and sense of humor. In this context, we might see Napoleon and Pitt as surrogates for the print buyers who, like the world leaders, identify themselves with geography by consuming it.

REFERENCES AND SELECTED BIBLIOGRAPHY

Ahmed, S. Maqbul. 1992. "Cartography of al-Sharif al-Idrisi." In Harley and Woodward 1992, 156–74.

Aiken, Charles S. 1977. "Faulkner's Yoknapatawpha County: Geographical Fact into Fiction." *Geographical Review* 67 (1): 1–21.

———. 1979. "Faulkner's Yoknapatawpha County: A Place in the American South." *Geographical Review* 69 (3): 331–48.

Akerman, James R. 1995. "The Structuring of Political Territory in Early Printed Atlases." *Imago Mundi* 47:138–54.

———. 2000. "Private Journeys on Public Maps: A Look at Inscribed Road Maps." *Cartographic Perspectives* 35:27–47.

———. 2002. "American Promotional Road Mapping in the Twentieth Century." *Cartography and Geographic Information Science* 29:175–91.

———, ed. 2006a. *Cartographies of Travel and Navigation.* Chicago: University of Chicago Press.

———. 2006b. "Twentieth-Century American Road Maps and the Making of a National Motorized Space." In Akerman 2006a, 151–206.

Albu, Emily. 2005. "Imperial Geography and the Medieval Peutinger Map." *Imago Mundi* 57 (2): 136–48.

Aleph [William Harvey]. 1869. Introduction to *Geographical Fun: Being Humorous Outlines of Various Countries.* London: Hodder and Stoughton.

Alpers, Svetlana. 1983. *The Art of Describing: Dutch Art in the Seventeenth Century.* Chicago: University of Chicago Press.

———. 1987. "The Mapping Impulse in Dutch Art." In Woodward 1987.

Anderson, Benedict. 1991. *Imagined Communities: Reflections on the Origin and Spread of Nationalism.* Rev. ed. London: Verso.

Anderson, Margo. 1980. *The American Census: A Social History.* New Haven, CT: Yale University Press.

Andrews, J. H. 1967. "The French School

of Dublin Land Surveyors." *Irish Geography* 5:275–92.

———. 1975. *A Paper Landscape: The Ordnance Survey in Nineteenth-Century Ireland*. Oxford: Clarendon Press. Reprint, Dublin: Four Courts Press, 2002.

Angeville, Adolphe d'. 1836. *Essai sur la Statistique de la Population française*. Bourg-en-Bresse, France: F. Doufour.

Anonymous. 1893. "Geography Learned without Tears." *Western Rural* 31, no. 9 (March 4): 137.

Anonymous. 1944. "The Society's New Map of Southeast Asia." *The National Geographic Magazine* 86 (4): 449–50.

Arbuckle's. 2006. "Arbuckle's Ariosa Trade Cards." http://home.att.net/ ~arbycards/arbmain.htm (accessed Aug. 21, 2006).

Atkin, Tony, and Joseph Rykwert, eds. 2005. *Structure and Meaning in Human Settlement*. Philadelphia: University of Pennsylvania Museum of Archaeology and Anthropology.

Atlas of Peoria County, Illinois. 1873. Chicago: A. T. Andreas.

Auerbach, Jeffrey A. 1999. *The Great Exhibition of 1851: A Nation on Display*. New Haven, CT: Yale University Press.

Aujac, Germain, J. B. Harley, and David Woodward. 1987. "The Foundations of Theoretical Cartography in Archaic and Classical Greece." In Harley and Woodward 1987, 130–47.

Austen, Jane. 1966. *Mansfield Park*. London: Penguin Books.

Bagrow, Leo. 1964. *History of Cartography*. Rev. and enlarged by R. A. Skelton. Cambridge, MA: Harvard University Press.

Baker, Robert. 1833. *Report of the Leeds Board of Health, MDCCCXXXIII*. Leeds, UK: Hernaman and Perring.

Baker-Smith, Dominic. 1991. "The Location of Utopia: Narrative Devices in Renaissance Fiction." In *Addressing Frank Kermode: Essays in Criticism and Interpretation*, ed. Margaret Tudeau-Clayton and Martin Warner, 109–23. Houndmills, UK: Macmillan Academic and Professional.

Balbi, Adriano, and André-Michel Guerry. 1829. *Statistique comparée de l'état de l'instruction et du nombre des crimes dans les divers arrondissements des Académies et des Cours Royales de France*. Paris.

Baldwin, Marc D. 1991–92. "Faulkner's Cartographic Method: Producing the Land through Cognitive Mapping." *The Faulkner Journal* 7 (1–2): 193–214.

Banerji, Christiane, and Diana Donald, trans. and eds. 1999. *Gillray Observed: The Earliest Account of His Caricatures in London and Paris*. Cambridge: Cambridge University Press.

Barber, Peter. 1990. "Necessary and Ornamental: Map Use in England under the Later Stuarts, 1660–1714." *Eighteenth Century Life* 14:1–28.

———, ed. 2005. *The Map Book*. London: Weidenfeld & Nicolson.

Barber, Peter, and Christopher Board, eds. 1993. *Tales from the Map Room: Fact and Fiction about Maps and Their Makers*. London: BBC Books.

Barral, Jean-Augustin. 1861. *Atlas du Cosmos contenant les cartes astronomiques, physiques, thermiques, magnétiques, géologiques relatives aux œuvres de A. de Humboldt et F. Arago*. Paris: Gide.

Barron, Roderick M. 2006. "Map of Matrimony c1840–50." http://www. barron.co.uk/?pgid=176&artid=730 (accessed Apr. 17, 2006).

Bartle, Richard A. 2004. *Designing Virtual Worlds*. Indianapolis: New Riders Pub.

Bartlett, Richard A. 1962. *Great Surveys of the American West*. Norman: University of Oklahoma Press.

Bartolovich, Crystal. 1994. "Spatial Sto-ries: *The Surveyor* and the Politics of Transition." In *Place and Displacement in the Renaissance*, ed. Alvin Vos, 255–83. Medieval & Renaissance Texts & Studies 132. Binghamton, NY: Center for Medieval and Early Modern Studies, State University of New York.

Bassett, Thomas J. 1998. "Indigenous Mapmaking in Intertropical Africa." In Woodward and Lewis 1998, 24–50.

Bassett, Thomas J., and Philip W. Porter. 1991. "'From the Best Authorities': The Mountains of Kong in the Cartography of West Africa." *Journal of African History* 32 (3): 367–413.

Bellin, Jacques-Nicolas. 1765. "Carte des variations de la boussole et des vents généraux que l'on trouve dans les mers les plus fréquentées." Paris: Dépôt des Cartes de la Marine.

Benet, Juan. 1983. *Herrumbrosas lanzas*. Literatura Alfaguara, vol. 122. Madrid: Alfaguara.

———. 1985. *Return to Región*. New York: Columbia University Press.

———. 1996. *Volverás a Región*. Ed. Victor G. de la Concha. Clásicos contemporáneos comentados, vol. 16. Barcelona: Destino.

Benjamin, Walter. 1999. "Paris, the Capital of the Nineteenth Century." In *Arcades Project*, trans. Howard Eiland and Kevin McLaughlin. Cambridge, MA: Harvard University Press.

Beresford, M. W. 1986. "Inclesmoor, West Riding of Yorkshire, *circa* 1407." In *Local Maps and Plans from Medieval England*, ed. R. A. Skelton and P. D. A. Harvey, 147–61. Oxford: Clarendon Press.

Berger, Harry. 1965. "The Renaissance Imagination: Second World and Green World." *Centennial Review* 9:40.

Berggren, J. Lennart, and Alexander Jones. 2000. *Ptolemy's "Geography": An Annotated Translation of the Theoretical*

Chapters. Princeton, NJ: Princeton University Press.

Berghaus, Heinrich. 1838–48. *Physikalischer Atlas.* Gotha, Germany: Justus Perthes.

Bertin, Jacques. 1967. *Sémiologie Graphique: Les diagrammes, les réseaux, les cartes.* Paris: Gauthier-Villars; La Haye, France: Mouton.

Blake, John. 2004. *The Sea Chart: The Illustrated History of Nautical Maps and Navigational Charts.* London: Conway Maritime.

Blansett, Lisa. 2005. "John Smith Maps Virginia: Knowledge, Rhetoric, and Politics." In *Envisioning an English Empire: Jamestown and the Making of the North Atlantic World*, ed. Robert Appelbaum and John Wood Sweet. Philadelphia: University of Pennsylvania Press.

Bloch, Marc. 1966. *French Rural History: An Essay on Its Basic Characteristics.* Trans. Janet Sondheimer. Berkeley and Los Angeles: University of California Press.

Bony, Alain. 1977. "Fabula, Tabula: L'Utopie de More et l'image du monde." *Etudes Anglaises* 30:1–19.

Boone, Elizabeth Hill. 1998. "Maps of Territory, History, and Community in Aztec Mexico." In *Cartographic Encounters: Perspectives on Native American Mapmaking and Map Use*, ed. G. Malcolm Lewis, 111–33. Chicago: University of Chicago Press.

Booth, Charles. 1889. *Charles Booth's Descriptive Map of London Poverty 1889 by Charles Booth; Introduction by David A. Reeder.* Reprint, London: London Topographical Society, 1984.

———. 1889–91. *Labour and Life of the People.* 2nd ed. 2 vols. London: Macmillan & Co.

Borges, Jorge Luis. 1964. *Dreamtigers.* Trans. Mildred Boyer and Harold

Morland. Austin: University of Texas Press.

Bosse, David. 1993. *Civil War Newspaper Maps.* Baltimore: Johns Hopkins University Press.

Botticelli, Sandro. 1976. *The Drawings by Sandro Botticelli for Dante's "Divine Comedy": After the Originals in the Berlin Museums and the Vatican.* 1st U.S. ed. New York: Harper & Row.

———. 2000. *Sandro Botticelli: The Drawings for Dante's "Divine Comedy."* London: Royal Academy of Arts.

Bourne, Molly. 1999. "Francesco II Gonzaga and Maps as Palace Decoration in Renaissance Mantua." *Imago Mundi* 51:51–82.

Bracher, Frederick. 1944. "The Maps in *Gulliver's Travels.*" *Huntington Library Quarterly* 8 (1): 59–74.

Branco, Rui Miguel Carvalhinho. 2005. "The Cornerstones of Modern Government: Maps, Weights and Measures and Census in Liberal Portugal (19th Century)." Ph.D. diss., European University Institute, Florence.

Braudel, Fernand. 1976. *Afterthoughts on Material Civilization and Capitalism.* Baltimore: Johns Hopkins University Press.

Brod, Raymond M. 1995. "The Art of Persuasion: John Smith's *New England* and *Virginia* Maps." *Historical Geography* 24 (1–2): 91–106.

Broecke, Marcel P. R. van der. 1986. "How Rare Is a Map and the Atlas It Comes From? Facts and Speculations on Production and Survival of Ortelius' *Theatrum Orbis Terrarum* and Its Maps." *Map Collector* 36:2–12.

———. 1995. "Unstable Editions of Ortelius' Atlas." *Map Collector* 70:2–8.

Brooks, Cleanth. 1990. *William Faulkner: The Yoknapatawpha Country.* Louisiana pbk. ed. Baton Rouge: Louisiana State University Press.

Brotton, Jerry. 1997. *Trading Territories: Mapping the Early Modern World.* London: Reaktion Books.

Brown, Calvin S. 1962. "Faulkner's Geography and Topography." *PMLA* 77:652–59.

Brown, Lloyd A. 1949. *The Story of Maps.* Boston: Little, Brown.

Brückner, Martin. 2006. *The Geographic Revolution in Early America: Maps, Literacy, and National Identity.* Chapel Hill: University of North Carolina Press for the Omohundro Institute of Early American History and Culture.

Buache, Philippe. 1752. "Carte minéralogique où l'on voit la nature des terreins du Canada et de la Louisiane." Paris: Dresbrulins fils sculps.

———. ca. 1770. "Profils représentants la crüe et la diminution des eaux de la Seine et des rivières qu'elle reçoit dans le païs haut au dessus de Paris." In *Cartes et tables de la géographie physique ou naturelle*, ed. G. d. L'Isle and P. Buache. Paris.

Buch, Leopold von. 1826. "Geognostische Karte von Deutschland und den umliegenden Staaten." Berlin: S. Schropp & Co.

Buisseret, David, ed. 1990. *From Sea Charts to Satellite Images: Interpreting North American History through Maps.* Chicago: University of Chicago Press.

———, ed. 1992. *Monarchs, Ministers, and Maps: The Emergence of Cartography as a Tool of Government in Early Modern Europe.* Chicago: University of Chicago Press.

———. 2003. *The Mapmakers' Quest: Depicting New Worlds in Renaissance Europe.* New York: Oxford University Press.

Cachey, Theodore, and Louis Jordan. 2006. "Renaissance Dante in Print (1472–1629)." http://www.nd.edu/

~italnet/Dante/index.html (accessed Apr. 6, 2006).

Campbell, Brian. 2000. *The Writings of the Roman Land Surveyors: Introduction, Text, Translation and Commentary.* *Journal of Roman Studies*, Monograph 9. London: Society for the Promotion of Roman Studies.

Campbell, Tony. 1968. *Claesz Jansz. Visscher: A Hundred Maps Described.* Map Collector's Series, no. 46. London: The Map Collector's Circle.

———. 1987. "Portolan Charts from the Late Thirteenth Century to 1500," In Harley and Woodward 1987, 371–463.

Carlucci, April, and Peter Barber, eds. 2001. *Lie of the Land: The Secret Life of Maps.* London: British Library.

Carroll, Lewis. 1936. *The Complete Works of Lewis Carroll.* New York: Modern Library.

Castronova, Edward. 2005. *Synthetic Worlds: The Business and Culture of Online Games.* Chicago: University of Chicago Press.

Certeau, Michel de. 1982. *La fable mystique.* Vol. 1, *XVIe–XVIIe siècle.* Paris: Gallimard.

Charpentier, Johann F. W. 1778. *Mineralogische Geographie der Chursächsischen Lande.* Leipzig: Crusius.

Clarke, Susanna. 2004. *Jonathan Strange & Mr. Norrell.* London: Bloomsbury.

Clayton, Timothy. 1997. *The English Print.* New Haven, CT: Yale University Press.

Clutton, Elizabeth. 1983. "On the Nature of Thematic Maps and Their History." *Map Collector* 22:42–43.

———. 1987. "The *Isolarii*: Buondelmonti's *Liber Insularum Arcipelagi*," in "Local and Regional Cartography in Medieval Europe," by P. D. A. Harvey. In Harley and Woodward 1987: 464–501.

Cohen, Paul E. 1998. "Michel Capitaine du

Chesnoy, the Marquis de Lafayette's Cartographer." *The Magazine Antiques* 153 (1): 170–78.

Conley, Tom. 1996. *The Self-Made Map: Cartographic Writing in Early Modern France.* Minneapolis: University of Minnesota Press.

Connolly, Daniel K. 1999. "Imagined Pilgrimage in the Itinerary Maps of Matthew Paris." *The Art Bulletin* 81 (4): 598–622.

Conzen, Michael P. 1984a. "Maps for the Masses: Alfred T. Andreas and the Midwestern County Atlas Trade." In *Chicago Mapmakers: Essays on the Rise of the City's Map Trade*, ed. Michael P. Conzen, 47–63. Chicago: Chicago Historical Society, for the Chicago Map Society.

———. 1984b. "The County Landownership Map in America: Its Commercial Development and Social Transformation 1814–1939." *Imago Mundi* 36: 9–31.

———. 1997. "The All-American County Atlas: Styles of Commercial Landownership Mapping and American Culture." In Wolter and Grim 1997, 331–65.

Cook, Andrew S. 2006. "Surveying the Seas: Establishing the Sea Route to the East Indies." In Akerman 2006a, 69–96.

Cortesão, Armando, and Avelino Teixeira da Mota. 1960–62. *Portugaliae monumenta cartographica.* Lisbon: Comisão Executiva das Comemorações do V Centenário da Morte do Infante D. Henrique. Reprinted at reduced size, Lisbon: Imprensa Nacional-Casa da Moeda, 1987.

Cosgrove, Denis, ed. 1999a. *Mappings.* London: Reaktion Books.

———. 1999b. "Introduction: Mapping and Meaning." In Cosgrove 1999a, 1–23.

———. 1999c. "Global Illumination and Enlightenment in the Geographies of Vincenzo Coronelli and Athanasius Kircher." In *Geography and Enlightenment*, ed. David N. Livingstone and Charles W. J. Withers, 33–66. Chicago: University of Chicago Press.

———. 2001. *Apollo's Eye: A Cartographic Genealogy of the Earth in the Western Imagination.* Baltimore: Johns Hopkins University Press.

———. 2003. "Globalism and Tolerance in Early Modern Geography." *Annals, Association of American Geographers* 93:852–70.

———. 2005. "Maps, Mapping, Modernity: Art and Cartography in the Twentieth Century." *Imago Mundi* 57 (1): 35–54.

Craib, Raymond B. 2004. *Cartographic Mexico: A History of State Fixations and Fugitive Landscapes.* Durham, NC: Duke University Press.

Crampton, Jeremy W. 2003. *The Political Mapping of Cyberspace.* Chicago: University of Chicago Press.

———. 2006. "The Cartographic Calculation of Space: Race Mapping and the Balkans at the Paris Peace Conference of 1919." *Social & Cultural Geography* 7 (5): 731–52.

Cresswell, Tim. 2004. *Place: A Short Introduction.* Oxford: Blackwell.

Crome, August F. W. 1785. *Über die Grösse und Bevölkerung der Sämtlichen Europäischen Staaten.* Leipzig: Wegand.

———. 1819. "A Map of the Statistical Relations of Europe Serving as a View and Comparison of the Extent of Surface, Population, and Other Public Revenue of All States of Europe." London: Engraved under the direction of A. Arrowsmith.

Cumming, William P. 1998. *The Southeast in Early Maps.* Rev. by Louis de Vorsey

Jr. Chapel Hill: University of North Carolina Press.

Curnow, Wystan. 1999. "Mapping and the Expanded Field of Contemporary Art." In Cosgrove 1999a, 253–68.

Cuvier, Georges, and Alexandre Brongniart. 1808. "Essai sur la géographie minéralogique des environs de Paris, avec une carte géognostique et des coupes de terrain." *Annales du Muséum d'Histoire Naturelle de Paris* 11:293–326.

Dahl, Edward H., and Jean-Francois Gauvin. 2000. *Sphaerae Mundi*. Montreal: Septentrion, McGill-Queen's University Press.

Dainville, François de. 1970. "Cartes et contestations au XVe siècle." *Imago Mundi* 24:99–121.

———. 1972. "Les bases d'une cartographie industrielle de l'Europe au dix-neuvième siècle." In *L'industrialisation en Europe au 19ème siècle: Cartographie et typologie; Lyon 7–10 octobre 1970*, ed. P. Léon, F. Crouzet, and R. Gascon. Paris: Bibliothèque nationale.

Dangeau, Louis de. 1697. *Nouvelle méthode de géographie historique*. Paris: A. Lambin.

Dante. 1980. *Inferno*. Trans. Charles Southward Singleton. Princeton, NJ: Princeton University Press.

D'Aulaire, Ingri, and Edgar Parin D'Aulaire. 2005. *D'Aulaires' Book of Norse Myths*. New York: New York Review Books.

Debord, Guy. 1994. *The Society of the Spectacle*. Trans. Donald Nicholson-Smith. New York: Zone Books.

Dechen, Heinrich von. 1838. "Geognostische Übersichts-Karte von Deutschland, Frankreich, England und den angrenzenden Laendern." Berlin: S. Schropp & Co.

Dee, John. 1947. *The Hieroglyphic Monad*. Trans. J. W. Hamilton-Jones. London: J. M. Watkins.

Dekker, Elly, and Peter van der Krogt. 1993. *Globes from the Western World*. London: Zwemmer.

Delano-Smith, Catherine. 1995. "Map Ownership in Sixteenth-Century Cambridge: The Evidence of Probate Inventories." *Imago Mundi* 47:67–93.

———. 2001. "The Map as a Commodity." In *Approaches and Challenges in a World-Wide History of Cartography*, ed. David Woodward, Catherine Delano Smith, and Cordell D. K. Yee. Barcelona: Institut Cartographic de Catalunya.

———. 2006. "Milieus of Mobility: Itineraries, Route Maps and Road Maps." In Akerman 2006a, 16–68.

Delano-Smith, Catherine, and Roger J. P. Kain. 1999. *English Maps: A History*. Toronto: University of Toronto Press for the British Library.

Descartes, René. 1996. *Règles pour la direction de l'esprit*. Paris: Gallimard.

Dewdney, Selwyn. 1975. *The Sacred Scrolls of the Southern Ojibway*. Toronto: University of Toronto Press.

Dilke, O. A. W. 1985. *Greek and Roman Maps*. Ithaca, NY: Cornell University Press.

———. 1987a. "Roman Large-Scale Mapping in the Early Empire." In Harley and Woodward 1987, 212–33.

———. 1987b. "Itineraries and Geographical Maps in the Early and Late Roman Empire." In Harley and Woodward 1987, 234–57.

Dillon, Diane. 1994. "'The Fair as a Spectacle': American Art and Culture at the 1893 World's Fair." Ph.D. diss., Yale University.

———. 2003. "Mapping Enterprise: Cartography and Commodification at the 1893 World's Columbian Exposition." In *Nineteenth-Century Geographies: The Transformation of Space from the Victorian Age to the American Century*, ed. Helena Michie and Ronald R. Thomas, 75–98. New Brunswick, NJ: Rutgers University Press.

Dirks, Nicholas. 1994. "Guiltless Spoliations: Picturesque Beauty, Colonial Knowledge, and Colin Mackenzie's Survey of India." In *Perceptions of South Asia's Visual Past*, ed. Catherine B. Asher and Thomas R. Metcalf, 211–32. New Delhi: Oxford and IBH Publishing for the American Institute of Indian Studies.

Doel, Ronald E., Tanya J. Levin, and Mason K. Marker. 2006. "Extending Modern Cartography to the Ocean Depths: Military Patronage, Cold War Priorities, and the Heezen-Tharp Mapping Project, 1952–1959." *Journal of Historical Geography* 32:605–26.

Dufrénoy, Ours Pierre, and Jean-Baptiste Elie de Beaumont. 1840. "Carte géologique de la France." Paris: Ministère des Travaux Publics.

Dupain-Triel (Père). 1781. "Carte générale des fleuves, des rivières et des principaux ruisseaux de France avec les canaux existants ou même projettés." Paris: Dupain-Triel.

Dupin, Charles. 1826. *Effets de l'enseignement populaire de la lecture, de l'écriture et de l'arithmétique, de la géométrie et de la mécanique appliquées aux arts, sur les prospérités de la France*. Paris: Bachelier.

———. 1827. *Forces productives et commerciales de la France*. Paris: Bachelier.

Duvert, Elizabeth. 1986. "Faulkner's Map of Time." *Faulkner Journal* 2 (1): 14–28.

Dykes, Jason, Alan M. MacEachren, and Menno-Jan Kraak. 2005. *Exploring Geovisualisation*. New York: Elsevier.

Eco, Umberto. 1972. *La structure absente: Introduction à la recherche sémiotique*. Paris: Mercure de France.

———. 1985. "Map of the Empire." Trans. S. Eugene Scalia. *Literary Review* 28 (2): 233–38.

———. 1994. "On the Impossibility of Drawing a Map of the Empire on a Scale of 1 to 1." In *How to Travel with a Salmon & Other Essays*, 95–110. New York: Harvest.

Edgerton, Samuel Y. 1987. "From Mental Matrix to Mappamundi to Christian Empire: The Heritage of Ptolemaic Cartography in the Renaissance." In *Art and Cartography: Six Historical Essays*, ed. David Woodward, 10–50. Chicago: University of Chicago Press.

Edney, Matthew H. 1997. *Mapping an Empire: The Geographical Construction of British India, 1765–1843*. Chicago: University of Chicago Press. Reprint, New Delhi: Oxford University Press, 1999.

———. 2003. "Bringing India to Hand: Mapping Empires, Denying Space." In *The Global Eighteenth Century*, ed. Felicity Nussbaum, 65–78, 334–36. Baltimore: Johns Hopkins University Press.

———. 2005. "The Origins and Development of J. B. Harley's Cartographic Theories." Special issue, *Cartographica* 40 (1, 2): monograph 54. Toronto: University of Toronto Press.

Edson, Evelyn, and Emilie Savage-Smith. 2004. *Medieval Views of the Cosmos: Picturing the Universe in the Christian and Islamic Middle Ages*. Oxford: Bodleian Library.

Ehrenberg, Ralph. 2003. "Rand McNally and Aviation Cartography: Changing Markets and Marketing Strategies." Paper presented at the 20th International Conference on the History of Cartography, Cambridge, MA, and Portland, ME, June 15–20.

———. 2005. "The Expeditions of John Charles Fremont." Paper presented at the 24th International Symposium of the International Map Collectors' Society, Denver CO, Sept. 20.

———. 2006a. *Mapping the World: An Illustrated History of Cartography*. Washington, DC: National Geographic.

———. 2006b. "'Up in the air in more ways than one': The Emergence of Aeronautical Charts in the United States." In Akerman 2006a, 207–59.

Ellis, Richard. 1998. *Imagining Atlantis*. 1st ed. New York: Alfred A. Knopf.

Emmerson, Donald K. 1984. "'Southeast Asia': What's in a Name?" *Journal of Southeast Asian Studies* 15:1–21.

Englisch, Brigitte. 1996. "Erhard Etzlaub's Projection and Methods of Mapping." *Imago Mundi* 48:103–23.

EQ Atlas. 2006. http://www.eqatlas.com/ (accessed July 17, 2006).

Evans, Ivor M., and Heather Lawrence. 1979. *Christopher Saxton, Elizabethan Map-maker*. Wakefield, UK: Wakefield Historical Publications.

Fabricant, Carole. 1995. "History, Narrativity, and Swift's Project to 'Mend the World.'" In Jonathan Swift, *Gulliver's Travels: Complete, Authoritative Text with Biographical and Historical Contexts, Critical History, and Essays from Five Contemporary Critical Perspectives*, ed. Christopher Fox, 348–65. Boston: St. Martin's Press, Bedford Books.

Feifer, Maxine. 1985. *Tourism in History*. New York: Stein and Day.

Finkelstein, J. J. 1962. "Mesopotamia." *Journal of Near Eastern Studies* 21 (2): 73–92.

Finney, Ben. 1998. "Nautical Cartography and Traditional Navigation in Oceania." In Woodward and Lewis 1998, 443–92.

Fite, Emerson D., and Archibald Freeman. 1926. *A Book of Old Maps Delineating American History*. Cambridge, MA: Harvard University Press. Reprint, New York: Dover Publications, 1969.

Fletcher, Angus. 2005. Introduction to *Treasure Island*, by Robert Louis Stevenson. New York: Barnes & Noble Classics.

Fletcher, David H. 1995. *The Emergence of Estate Maps: Christ Church, Oxford, 1600 to 1840*. Christ Church Papers, 4. Oxford: Clarendon Press.

Fletcher, Joseph. 1847–49. "Moral and Educational Statistics of England and Wales." *Journal of the Statistical Society of London* 10:193–21 and 12:151–76, 189–335.

Fonstad, Karen Wynn. 1991. *The Atlas of Middle-Earth*. Boston: Houghton Mifflin.

Fourcroy, Charles de. 1782. *Essai d'une table poléométrique, ou amusement d'un amateur de plans sur la grandeur de quelques villes*. Paris: Dupain-Triel.

Francaviglia, Richard V. 1996. *The Shape of Texas: Maps as Metaphors*. College Station: Texas A&M University Press.

François, Jean. 1652. *La science de la géographie, divisée en trois parties, qui expliquent les divisions, les universalitez, et les particularitez du globe terrestre*. Renne, France: J. Hardy.

Franklin, Benjamin. 1768. "A Chart of the Gulf Stream." London: J. Mount and T. Page.

Franzwa, Gregory M. 1982. *Maps of the Oregon Trail*. Gerald, MO: Patrice Press.

Frère de Montizon, Armand Joseph. 1830. "Carte philosophique figurant la population de la France." Paris.

Friedman, John Block. 2000. *The Monstrous Races in Medieval Art and Thought*. Syracuse, NY: Syracuse University Press.

Friendly, Michael. 1994. "Mosaic Displays for Multi-Way Contingency Tables." *Journal of the American Statistical Association* 89:190–200.

———. 2002. "A Brief History of the Mosaic Display." *Journal of Computational and Graphical Statistics* 11 (1): 89–107.

———. 2007. "A Brief History of Data Visualization." In *Handbook of Computational Statistics: Data Visualization*, ed. C. Chen, W. Härdle, and A. Unwin. Heidelberg: Springer-Verlag.

Friendly, Michael, and Dan Denis. 2000. "The Roots and Branches of Statistical Graphics." *Journal de la Société Française de Statistique* 141 (4): 51–60.

———. 2006. *Milestones in the History of Thematic Cartography, Statistical Graphics, and Data Visualization: An Illustrated Chronology of Innovations.* http://www.math.yorku.ca/SCS/Gallery/milestone/.

Friendly, Michael, and Ernest Kwan. 2003. "Effect Ordering for Data Displays." *Computational Statistics and Data Analysis* 43 (4): 509–39.

Funkhouser, Howard Gray. 1937. "Historical Development of the Graphical Representation of Statistical Data." *Osiris* 3 (1): 269–405.

Galton, Francis. 1863. *Meteorographica; or, Methods of Mapping the Weather.* London: Macmillan.

———. 1886. "Regression towards Mediocrity in Hereditary Stature." *Journal of the Anthropological Institute* 15:246–63.

Garland, Ken. 1994. *Mr. Beck's Underground Map.* Harrow Weald, UK: Capital Transport Publishing.

Gartner, William Gustav. 1998. "Mapmaking in the Central Andes." In Woodward and Lewis 1998, 257–300.

Gastner, Michael, Cosma Shalizi, and Mark Newman. 2004. "Maps and Cartograms of the 2004 US Presidential Election Results." University of Michigan. http://www-personal.umich.edu/~mejn/election/

Gilbert, Allan H. 1945. "Can Dante's Inferno Be Exactly Charted?" *PMLA* 60 (2): 287–306.

Gilbert, Pamela K. 2004. *Mapping the Victorian Social Body.* Albany: State University of New York Press.

Gillis, John. 2004. *Islands of the Mind: How the Human Imagination Created the Atlantic World.* New York: Palgrave Macmillan.

Gläser, Friedrich G. 1775. *Versuch einer mineralogischen Beschreibung der gefürsteten Grafschaft Henneberg.* Leipzig: Crusius.

Godlewska, Anne. 1988. *The Napoleonic Survey of Egypt: A Masterpiece of Cartographic Compilation and Early Nineteenth-Century Fieldwork. Cartographica* 25 (1–2): Monographs 38–39. Toronto: University of Toronto Press.

Goetzmann, William. 1959. *Army Exploration in the American West, 1803-1863.* New Haven, CT: Yale University Press.

Gole, Susan. 1976. *Early Maps of India.* New York: Humanities Press.

———. 1989. *Indian Maps and Plans: From Earliest Times to the Advent of European Surveys.* New Delhi: Manohar.

Goodey, Brian R. 1970. "Mapping 'Utopia': A Comment on the Geography of Sir Thomas More." *Geographical Review* 60 (1): 15–30.

Grancsay, Stephen V. 1945. *American Engraved Powder Horns.* New York: Metropolitan Museum of Art.

Greenberg, John L. 1994. *The Problem of the Earth's Shape from Newton to Clairaut: The Rise of Mathematical Science in Eighteenth-Century Paris and the Fall of "Normal" Science.* Cambridge: Cambridge University Press.

Greenblatt, Stephen Jay. 1980. *Renaissance Self-Fashioning: From More to Shakespeare.* Chicago: University of Chicago Press.

Greenhood, David. 1944. *Down to Earth: Mapping for Everybody.* New York: Holiday House. Reprint, Chicago, as *Mapping:* University of Chicago Press, 1964.

Greenough, George B. 1820. "A Geological Map of England and Wales." London: Longman, Hurst, Rees, Orme & Browne.

Griaule, Marcel. 1949. "L'image du monde au Soudan." *Journal de la Société des Africanistes* 19:81–87.

Guerry, André-Michel. 1833. *Essai sur la statistique morale de la France.* Paris: Crochard.

———. 1864. *Statistique morale de l'Angleterre comparée avec la statistique morale de la France, d'après les comptes de l'administration de la justice criminelle en Angleterre et en France, etc.* Paris: J.-B. Baillière et fils.

Guettard, Jean-Étienne. 1746. "Mémoire et carte minéralogique sur la nature et la situation des terreins qui traversent la France et l'Angleterre." *Mémoire de l'Académie des Sciences [de Paris] pour* 1751:363–92.

Hacking, Ian. 1990. *The Taming of Chance.* Cambridge: Cambridge University Press.

Halley, Edmund. 1686. "An Historical Account of the Trade Winds, and Monsoons, Observable in the Seas between and near the Tropicks, with an Attempt to Assign the Physical Cause of the Said Winds." *Philosophical Transactions* 16:153–68.

———. 1701. "A New and Correct Chart Shewing the Variations of the Compass in the Western and Southern Ocean." London.

Hammond, Wayne G., and Christina Scull. 2005. *"The Lord of the Rings": A Reader's Companion.* Boston: Houghton Mifflin Co.

Hannas, Linda. 1972. *The English Jigsaw Puzzle, 1760-1890.* London: Wayland.

Harley, J. B. 1987. "The Map as Biography: Thoughts on Ordnance Survey Map, Six'Inch Sheet Devonshire CIX, SE, Newton Abbot." *Map Collector* 41:18–20.

———. 1988. "Silences and Secrecy: The Hidden Agenda of Cartography in Early Modern Europe." *Imago Mundi* 40:57–76. Reprinted as "Silences and Secrets: The Hidden Agenda of Cartography in Early Modern Europe" in Harley 2001, 83–107, 241–51.

———. 1990. *Maps and the Columbian Encounter.* Milwaukee: The Golda Meir Library, University of Wisconsin-Milwaukee.

———. 1997. "Power and Legitimation in the English Geographical Atlases of the Eighteenth Century." In Wolter and Grim 1997, 161–204.

———. 2001. *The New Nature of Maps: Essays in the History of Cartography.* Ed. Paul Laxton. Baltimore: Johns Hopkins University Press.

Harley, J. B., Barbara Bartz Petchenik, and Lawrence W. Towner. 1978. *Mapping the American Revolutionary War.* Chicago: University of Chicago Press.

Harley, J. B., and David Woodward, eds. 1987. *Cartography in Prehistoric, Ancient, and Medieval Europe and the Mediterranean.* Vol. 1 of *The History of Cartography.* Chicago: University of Chicago Press.

———, eds. 1992. *Cartography in the Traditional Islamic and South Asian Societies.* Vol. 2, book 1 of *The History of Cartography.* Chicago: University of Chicago Press.

———, eds. 1994. *Cartography in the Traditional East and Southeast Asian Societies.* Vol. 2, book 2 of *The History of Cartography.* Chicago: University of Chicago Press.

Harness, Henry D. 1838. *Atlas to Accompany the Second Report of the Railway Commissioners, Ireland.* Dublin: H.M.S.O.

Harrell, J. A., and V. M. Brown 1992. "The World's Oldest Surviving Geological Map—The 1150 BC Turin Papyrus from Egypt." *Journal of Geology* 100:3–18.

Harrington, Bates. 1890. *How 'Tis Done: A Thorough Ventilation of the Numerous Schemes Conducted by Wandering Canvassers, Together with the Various Advertising Dodges for the Swindling of the Public.* Syracuse, NY: W. I. Pattison.

Harris, Neil. 1993. "Memory and the White City." In *Grand Illusions: Chicago's World's Fair of 1893,* by Neil Harris et al., 1–40. Chicago: Chicago Historical Society.

Hart, Albert Bushnell. 1891. *Epoch Maps Illustrating American History.* New York: Longmans, Green, and Co.

Harvey, P. D. A. 1980. *The History of Topographical Maps: Symbols, Pictures and Surveys.* London: Thames and Hudson.

———. 1987. "Local and Regional Cartography in Medieval Europe." In Harley and Woodward 1987, 482–83.

Heijden, H. A. M. van der. 1990. *Leo Belgicus: An Illustrated and Annotated Cartobibliography.* Alphen aan den Rijn, The Netherlands: Canaletto.

Helgerson, Richard. 1992. *Forms of Nationhood: The Elizabethan Writing of England.* Chicago: University of Chicago Press.

Henderson, John B. 1994. "Chinese Cosmographical Thought: The High Intellectual Tradition." In Harley and Woodward 1994, 203–27.

Herzberger, David K. 1976. *The Novelistic World of Juan Benet.* Clear Creek, IN: The American Hispanist.

Hill, Gillian. 1978. *Cartographical Curiosities.* London: British Museum Publications.

Hillis, Ken. 1994. "The Power of Disembodied Imagination: Perspective's Role in Cartography." *Cartographica* 31 (3): 1–17.

Hoefer, Ferdinand. 1858–78. *Nouvelle biographie générale depuis les temps les plus reculés jusqu'à nos jours.* 46 vols. Paris: Firmin-Didot.

Hofmann, Catherine. 2001–2. "The Globe as Symbol in Emblem Books in the West, Sixteenth and Seventeenth Centuries." *Globe Studies* 49–50:81–120.

Holmes, Nigel. 1991. *Pictorial Maps.* New York: Watson-Guptill Publications.

Hopkins, Martha E., Michael Buscher, and Library of Congress. 1999. *Language of the Land: The Library of Congress Book of Literary Maps.* Washington, DC: Library of Congress.

Hostetler, Laura. 2001. *Qing Colonial Enterprise: Ethnography and Cartography in Early Modern China.* Chicago: University of Chicago Press.

Hsu, Mei-Ling. 1993. "The Qin Maps: A Clue to Later Chinese Cartographic Development." *Imago Mundi* 45:90–100.

Humboldt, Alexander von. 1817. "Sur les lignes isothermes." *Annales de chimie et de physique,* 2nd ser., 5:102–11.

Hunter, J. Paul. 2003. "Gulliver's Travels and the Later Writings." In *The Cambridge Companion to Jonathan Swift,* ed. Christopher Fox, 216–40. New York: Cambridge University Press.

Hyde, Ralph. 1986. "A 'Handy' Map." *Map Collector* 35:47.

———. 1988. *Panoramania! The Art and Entertainment of the "All-Embracing" View.* London: Trefoil Publications in association with Barbican Art Gallery.

Ingram, Elizabeth M. 1993. "Maps as Readers' Aids: Maps and Plans in Geneva Bibles." *Imago Mundi* 45:29–44.

Internet Sacred Text Archive. n.d. "Atlantis." http://www.sacred-texts.com/atl/index.htm (accessed Apr. 4, 2006).

Jacob, Christian. 1996. "Towards a Cultural History of Cartography." *Imago Mundi* 48:191–98.

———. 1999. "Mapping in the Mind: The Earth from Ancient Alexandria." In Cosgrove 1999a, 24–49.

Jardine, Lisa. 1996. *Worldly Goods: A New History of the Renaissance*. New York: Doubleday.

Jauffret, Louis-François. 1800. *Zoographie des diverses régions, tant de l'ancien que du nouveau continent*. Paris: de Crapelet.

Jevons, William Stanley. 1863. *A Serious Fall in the Value of Gold Ascertained, and Its Social Effects Set Forth*. London.

Johns, Jeremy, and Emilie Savage-Smith. 2003. "*The Book of Curiosities*: A Newly Discovered Series of Islamic Maps." *Imago Mundi* 55:7–24.

Johnston, Alexander Keith. 1848. *Physical Atlas*. London: William Blackwood.

Joint Committee on Standards for Graphic Presentation. 1914. "Preliminary Report Published for the Purpose of Inviting Suggestions for the Benefit of the Committee." *Publications of the American Statistical Association* 14 (112): 790–97.

Jourde, Pierre. 1991. *Géographies imaginaires de quelques inventeurs de mondes au XXe siècle: Gracq, Borges, Michaux, Tolkien*. Paris: J. Corti.

Judd, Dennis R., and Susan S. Fainstein, eds. 1999. *The Tourist City*. New Haven, CT: Yale University Press.

Jung, Carl G. 1964. *Man and His Symbols*. Garden City, NY: Doubleday.

Kagan, Richard L. 1998. "*Urbs* and *Civitas* in Sixteenth- and Seventeenth-Century Spain." In *Envisioning the City: Six Studies in Urban Cartography*, ed. David Buisseret, 75–108. Chicago: University of Chicago Press.

Kain, Roger J. P., and Elizabeth Baigent. 1992. *The Cadastral Map in the Service of the State: A History of Property Mapping*. Chicago: University of Chicago Press.

Karrow, Robert W., Jr. 1993. *Mapmakers of the Sixteenth Century and Their Maps*. Chicago: Speculum Orbis Press.

Kasson, John F. 1990. *Rudeness & Civility: Manners in Nineteenth-Century Urban America*. New York: Hill and Wang.

Kidron, Michael, and Ronald Segal. 1981. *The State of the World Atlas*. New York: Simon and Schuster.

King, Russell. 1990. *Visions of the World and the Language of Maps*. Trinity Papers in Geography 1. Dublin: Trinity College.

Klein, Bernhard. 2001. *Maps and the Writing of Space in Early Modern England and Ireland*. Basingstoke, UK: Palgrave.

Klein, Christopher M. 1989. *Maps in Eighteenth-Century British Magazines: A Checklist*. Chicago: The Newberry Library.

Kleiner, John. 1994. *Mismapping the Underworld: Daring and Error in Dante's Comedy*. Stanford, CA: Stanford University Press.

Koch, Tom. 2005. *Cartographies of Disease: Maps, Mapping, and Medicine*. Redlands, CA: ESRI Press.

Koeman, Cornelis. 1964. *The History of Abraham Ortelius and His "Theatrum Orbis Terrarum."* New York: American Elsevier Publishing Co.

———. 1967–85. *Atlantes Neerlandici*. Amsterdam: Theatrum Orbis Terrarum. New ed.: see Krogt (1997–).

———. 1997. "Atlas Cartography in the Low Countries in the Sixteenth, Seventeenth, and Eighteenth Centuries." In Wolter and Grim 1997.

Konvitz, Josef. 1980. "Remplir la carte." In *Cartes et Figures de la Terre*. Exhibition catalogue, Centre G. Pompidou. L'édition artistique. Paris: Centre G. Pompidou.

———. 1987. *Cartography in France, 1660–1848: Science, Engineering, and Statecraft*. Chicago: University of Chicago Press.

Kornblith, Gary J. 2003. "Rethinking the Coming of the Civil War: A Counterfactual Exercise." *Journal of American History* 90 (1): 76–105.

Kort, Wesley A. 2004. *Place and Space in Modern Fiction*. Gainesville: University Press of Florida.

Korzybski, Alfred. 1933. *Science and Sanity: An Introduction to Non-Aristotelian Systems and General Semantics*. Princeton, NJ: Princeton University Press.

Krogt, Peter van der. 1997–. *Koeman's "Atlantes Neerlandici."* 't Goy-Houten, The Netherlands: HES Publishers.

———. 2000. Introduction to *Sphaerae Mundi*, ed. Edward H. Dahl and Jean-François Gauvin, 11–23. Montreal: Septentrion, McGill-Queen's University Press.

Krogt, Peter van der, and Erlend de Groot, eds. 1996–. *The "Atlas Blaeu—Van der Hem" of the Austrian National Library*. 't Goy-Houten, The Netherlands: HES Publishers.

Kula, Witold. 1986. *Measures and Men*. Trans. R. Szreter. Princeton, NJ: Princeton University Press.

Lagarde, Lucie. 1989. "Le Passage du Nord-Ouest et la Mer de l'Ouest dans la cartographie française du 18e siècle: Contribution l'étude de l'oeuvre des Delisle." *Imago Mundi* 41:19–43.

Lalanne, Léon. 1846. "Mémoire sur les tables graphiques et sur la géométrie anamorphique appliquées à diverses questions qui se rattachent a l'art de l'Ingénieur." *Annales des Ponts et Chausées*, 2nd ser., 11:1–69.

Lallemand, Charles. 1885. *Les abaques hexagonaux: Nouvelle méthode générale de calcul graphique, avec de nombreux exemples d'application*. Paris: Ministère des travaux publics, Comité du nivellement général de la France.

Lane, Belden C. 2002. *Landscapes of the Sacred: Geography and Narrative in American Spirituality*. Baltimore: Johns Hopkins University Press.

Langren, Michael Florent van. 1644. *La Verdadera Longitud por Mar y Tierra*. Antwerp.

Lanman, Jonathan T. 1987. *On the Origin of Portolan Charts*. The Hermon Dunlap Smith Center for the History of Cartography, Occasional Publication, 2. Chicago: The Newberry Library.

Lavie, Smadar, and Ted Swedenburg, eds. 1996. *Displacement, Diaspora, and Geographies of Identity*. Durham, NC: Duke University Press.

Ledyard, Gari. 1994. "Cartography in Korea." In Harley and Woodward 1994, 235–345.

León-Portilla, Miguel. 1991. Entry 356. In Levenson 1991.

Lepore, Jill. 1998. *The Name of War: King Philip's War and the Origins of American Identity*. New York: Alfred A. Knopf.

Leskiw, Adrian. 2006. The Map Realm. http://www-personal.umich.edu/~aleskiw/maps/home.htm (accessed Apr. 20, 2006).

Lestringant, Frank. 1980. "Insulaires." In *Cartes et figures de la terre*, xv, 479. Exhibition catalogue, Centre Georges Pompidou, Paris, May 24–Nov. 17, 1980. Paris: Centre Georges Pompidou et Centre de création industrielle.

Levallois, Jean Jacques, ed. 1988. *Mesurer la terre: 300 ans de géodesie française de la toise du Châtelet au satellite*. Paris: Presses de l'école nationale des Ponts et Chaussées.

Levasseur, Émile. 1885. "La statistique graphique." *Journal of the Royal Statistical Society* Jubilee volume: 218–50.

Levenson, Jay A., ed. 1991. *Circa 1492: Art in the Age of Exploration*. Washington, DC: National Gallery of Art; New Haven, CT: Yale University Press.

Levitt, Sarah. 1986. *Victorians Unbuttoned: Registered Designs for Clothing, Their Makers and Wearers*. London: George Allen & Unwin.

Lewis, G. Malcolm. 1998. "Maps, Mapmaking, and Map Use by Native North Americans." In Woodward and Lewis 1998, 51–182.

Lewis, Martin W., and Karen Weigen. 1997. *The Myth of Continents*. Berkeley and Los Angeles: University of California Press.

Lewis, Suzanne. 1987. *The Art of Matthew Paris in the Chronica Majora*. Berkeley and Los Angeles: University of California Press, in collaboration with Corpus Christi College, Cambridge.

Lommer, Christian. 1768. *Mineralogische Bemerckungen bey einer Reisse von Freiberg bis an das Riesen-Gebirge*. [Dresden?]

Lozovsky, Natalia. 2000. *"The Earth Is Our Book": Geographical Knowledge in the Latin West ca. 400–1000*. Ann Arbor: University of Michigan Press.

Lubin, Antoine. 1678. *Mercure géographique, ou le guide des curieux des cartes géographiques*. Paris: C. Rémy.

Lucian. 1968. "A True Story." In *Selected Satires of Lucian*. New York: W. W. Norton.

MacCannell, Dean. *The Tourist*. 1976. New York: Schocken Books.

Mangani, Giogio. 1998. "Abraham Ortelius and the Hermetic Meaning of the Cordiform Projection." *Imago Mundi* 50:59–83.

Map Collector. 1988. Cover illustration. *Map Collector* 45:1.

Margenot, John B. 1988. "Cartography in the Fiction of Juan Benet." *Letras Peninsulares* 1 (3): 331–44.

———. 1991. *Zonas y sombras: Aproximación a Región de Juan Benet*. Madrid: Pliegos.

Marin, Louis. 1984. *Utopics: The Semiological Play of Textual Spaces*. Trans. Robert A. Vollrath. Atlantic Highlands, NJ: Humanities Press.

Marshall, Alfred. 1885. "On the Graphic Method of Statistics." *Journal of the Royal Statistical Society* Jubilee volume: 251–60.

Martin, Lawrence W. 1972. "John Mitchell's Map of the British and French Dominions in North America." In Ristow 1972, 102–13.

Mathews, Richard. 1997. *Fantasy: The Liberation of Imagination*. Studies in Literary Themes and Genres, vol. 16. New York: Twayne Publishers, Prentice Hall International, 1997.

Maunder, Edward Walter. 1904. "Note on the Distribution of Sun-Spots in Heliographic Latitude, 1874 to 1902." *Royal Astronomical Society Monthly Notices* 64:747–61.

Mayhew, Henry. 1851–62. *London Labour and the London Poor*. 4 vols. London.

McClung, William A. 1994. "Designing Utopia." *Moreana* 31 (118–19): 9–28.

McKendrick, Neil, John Brewer, and J. H. Plumb. 1982. *The Birth of a Consumer Society: The Commercialization of Eighteenth-Century England*. Bloomington: Indiana University Press.

Meinig, Donald. 1993. *Continental America, 1800–1867*. Vol. 2 of *The Shaping of America*. New Haven, CT: Yale University Press.

Mendeleev, Dmitri. 1889. "The Periodic Law of the Chemical Elements." *Journal of the Chemical Society* 55:634–56.

Miller, J. Hillis. 1995. "Ideology and Topography in Faulkner's *Absalom, Absalom!*" In *Faulkner and Ideology*, ed. Donald M. Kartiganer and Ann J.

Abadie, 253–77. Jackson: University Press of Mississippi.

Minard, Charles Joseph. 1845a. "Carte de la circulation des voyageurs par voitures publiques sur les routes de la contrée où sera placé le chemin de fer de Dijon à Mulhouse." Paris: Regnier et Dourdet.

———. 1845b. "Tableau figuratif du mouvement commercial du canal du Centre en 1844." Paris.

———. 1861. *Des tableaux graphiques et des cartes figuratives.* Paris: E. Thunot et Cie.

———. 1863. "Carte figurative et approximative des quantités de coton en laine importées en Europe en 1858 et en 1862." Paris: Regnier et Dourdet.

———. 1869. "Carte figurative des pertes successives en hommes de l'armée française dans la campagne de Russie, 1812–1813." Paris: Regnier et Dourdet.

Modelski, Andrew M. 1984. *Railroad Maps of North America: The First Hundred Years.* Washington, DC: Library of Congress.

Mollat Du Jourdin, Michel, and Monique de La Roncière. 1984. *Sea Charts of the Early Explorers: 13th to 17th Century.* Trans. L. Le R. Dethan. New York: Thames and Hudson.

Monmonier, Mark. 1989. *Maps with the News: The Development of Journalistic Cartography.* Chicago: University of Chicago Press.

———. 1997. "The Rise of the National Atlas." In Wolter and Grim 1997, 369–99.

———. 1999. *Air Apparent.* Chicago: University of Chicago Press.

———. 2005. *Rhumb Lines and Map Wars: A Social History of the Mercator Projection.* Chicago: University of Chicago Press.

More, Thomas. 1964. *Utopia.* Trans. and ed. Edward L. Surtz. New Haven, CT: Yale University Press.

———. 1992. *Utopia: A New Translation, Backgrounds, Criticism.* Trans. and ed. Robert Martin Adams. 2nd ed. New York: Norton.

Morello, Giovanni. 2000. "The Chart of Hell." In Botticelli 2000, 318–25.

Moseley, Henry. 1913. "The High Frequency Spectra of the Elements." *Philosophical Magazine* 26:1024–34.

Mouat, Frederic J. 1885. "History of the Statistical Society of London." *Journal of the Royal Statistical Society* Jubilee volume: 14–62.

Mukerji, Chandra. 1993. *From Graven Images: Patterns of Modern Materialism.* New York: Columbia University Press.

Mundy, Barbara E. 1996. *The Mapping of New Spain: Indigenous Cartography and the Maps of the Relaciones geograficas.* Chicago: University of Chicago Press.

———. 1998. "Mesoamerican Cartography." In Woodward and Lewis 1998, 182–256.

Murray, Martin. 1989. "Maps on Cigarette Cards." *Map Collector* 49:10–13.

Musich, Gerald. 2006. "Mapping and Transcontinental Nation: Nineteenth- and Early Twentieth-Century American Rail Travel Cartography." In Akerman 2006a, 97–150.

Nassar, Eugene Paul. 1994. *Illustrations to Dante's "Inferno."* Rutherford: Fairleigh Dickinson University Press; London: Associated University Presses.

Nebenzahl, Kenneth. 1986. *Maps of the Holy Lands: Images of Terra Sancta through Two Millennia.* New York: Abbeville Press.

———. 1990. *Atlas of Columbus and the Great Discoveries.* Chicago: Rand McNally.

———. 2004. *Mapping the Silk Road and Beyond: 2,000 Years of Exploring the East.* London: Phaidon.

Nebenzahl, Kenneth, and Don Higginbotham. 1974. *Atlas of the American Revolution.* Chicago: Rand McNally.

Nicolet, Hercule. 1855. *Atlas de physique et de météorologie agricoles.* Paris: Bachelier.

Nightingale, Florence. 1857. *Mortality of the British Army.* London: Harrison and Sons.

Nischer-Falkenhof, Ernst von. 1937. "The Survey by the Austrian General Staff under the Empress Maria Theresa and the Emperor Joseph II, and the Subsequent Initial Surveys of Neighbouring Territories during the Years 1749–1854." *Imago Mundi* 2:83–88.

Norden, John. 1607. *The Surveyors Dialogue.* London: H. Astley.

Nuti, Lucia. 1999. "Mapping Places: Chorography and Vision in the Renaissance." In Cosgrove 1999a, 90–108.

Ocagne, Maurice d'. 1885. *Coordonnées Parallèles et Axiales: Méthode de transformation géométrique et procédé nouveau de calcul graphique déduits de la considération des coordonnées parallèles.* Paris: Gauthier-Villars.

———. 1899. *Traité de nomographie: Théorie des abaques, applications pratiques.* Paris: Gauthier-Villars.

Orbigny, Alcide d'. 1842. "Carte générale de la République de Bolivia dressée par Alcide d'Orbigny, d'après des itinéraires relevés au cours des années 1830, 1831, 1832 et 1833: Carte géologique." Paris: Pithois Levrault et Cie.

Packe, Christopher. 1743. "A New Philosophico Chorographical Chart of East-Kent." [Canterbury]: C. Packe.

Padrón, Ricardo. 2004. *The Spacious Word: Cartography, Literature, and Empire in Early Modern Spain.* Chicago: University of Chicago Press.

Palazzo reale di Milano. 2001. *Segni e sogni della terra: Il disegno del mondo dal mito di Atlante alla geografia delle reti.* Exhibition catalogue. Novara, Italy: De Agostini.

Palmowski, Jan. 2002. "Travels with Baedeker—The Guidebook and the Middle Classes in Victorian and Edwardian Britain." In *Histories of Leisure*, ed. Rudy Koshar. Oxford: Berg.

Palsky, Gilles. 1995. "La cartographie médicale et anthropologique. In *Le XIXe siècle: Science, politique et tradition*, ed. I. Poutrin. Paris: Berger-Levrault.

———. 1996. *Des chiffres et des cartes: Naissance et développement de la cartographie quantitative français au XIXe siècle.* Paris: Comité des Travaux Historiques et Scientifiques.

———. 2003. "Cartes topographiques et cartes thématiques au XIXe siècle." In *La cartografia europea tra primo Rinascimento e fine dell'Illuminismo*, ed. D. R. Curto, A. Cattaneo, and A. F. d'Almeida, 275–89. Florence: Leo S. Olschki.

Parent-Duchatelet, Alexandre Jean Baptiste. 1836. *De la prostitution dans la ville de Paris.* 3rd ed. Paris: J. B. Baillière.

Pastoureau, Mireille. 1997. "French School Atlases: Sixteenth to Eighteenth Centuries." In Wolter and Grim 1997, 109–34.

Paullin, Charles O. 1932. *Atlas of the Historical Geography of the United States.* New York: American Geographical Society and the Carnegie Institution.

Pavel, Thomas G. 1986. *Fictional Worlds.* Cambridge, MA: Harvard University Press.

Pearce, Margaret Wickens. 1998. "Native Mapping in Southern New England Indian Deeds." In *Cartographic Encounters: Perspectives on Native American Mapmaking and Map Use*, ed.

G. Malcolm Lewis, 157–86. Chicago: University of Chicago Press.

Pearson, Karl. 1901. "On Lines and Planes of Closest Fit to Systems of Points in Space." *Philosophical Magazine* 6 (2): 559–72.

———. 1914. *The Life, Letters and Labours of Francis Galton.* Cambridge: Cambridge University Press.

Pedley, Mary. 1998. "Map Wars: The Role of Maps in the Nova Scotia/Acadia Boundary Disputes of 1750." *Imago Mundi* 50:96–104.

———. 2005. *The Commerce of Cartography: Making and Marketing Maps in Eighteenth-Century France and England.* Chicago: University of Chicago Press.

Pelletier, Monique. 1982. "Les globes du Louis XIV: Les sources françaises de l'oeuvre de Coronelli." *Imago Mundi* 34:72–89.

———. 1990. *La carte de Cassini: L'extraordinaire aventure de la Carte de France.* Paris: Presses de l'École nationale des ponts et chausées.

Petchenik, Barbara B. 1979. "From Place to Space: The Psychological Achievement of Thematic Mapping." *The American Cartographer* 6 (1): 5–12.

Petermann, Augustus. 1850. *The Atlas of Physical Geography.* London: Wm. S. Orr & Co.

Peters, Jeffrey N. 2004. *Mapping Discord: Allegorical Cartography in Early Modern French Writing.* Newark: University of Delaware Press.

Peuchet, Jacques. 1805. *Statistique élémentaire de la France.* Paris: Gilbert.

Playfair, William. 1786. *Commercial and Political Atlas.* London: Corry.

———. 1801. *Statistical Breviary; Shewing, on a Principle Entirely New, the Resources of Every State and Kingdom in Europe.* London: Wallis.

Portinaro, Pierluigi, and Franco Knirsch.

1987. *The Cartography of North America 1500–1800.* New York: Facts on File and Bison Books.

Post, Jeremiah Benjamin. 1973. *An Atlas of Fantasy.* Baltimore: Mirage Press.

———. 1979. *An Atlas of Fantasy.* Rev. ed. New York: Ballantine Books.

———. 2001. "A Map by Any Other Name." *Mercator's World* 6:36–37.

Postnikov, Aleksei V. 1996. *Russia in Maps: A History of the Geographical Study and Cartography of the Country.* Moscow: Nash Dom-l'Age d'Homme.

Priestley, Joseph. 1765. "A Chart of Biography." London.

———. 1769. "A New Chart of History." London: Thomas Jeffreys.

Pritchard, Margaret Beck, and Henry G. Taliaferro. 2002. *Degrees of Latitude: Mapping Colonial America.* Williamsburg: Colonial Williamsburg Foundation.

Quetelet, Adolphe. 1831. *Recherches sur le penchant au crime aux différens ages.* Brussels: Hayez.

———. 1835. *Sur l'homme et le développement de ses facultés, ou Essai d'une physique sociale.* Paris: Bachelier.

Rageh, Rawya. 2006. "Map of Gold a Symbol of National Unity." *Chicago Sun-Times*, Sept. 3.

Raisz, Erwin. 1938. *General Cartography.* New York: McGraw Hill.

———. 1944. *Atlas of Global Geography.* New York: Harper & Brothers, Global Press.

Ramaswamy, Sumathi. 2001. "Maps and Mother Goddesses in Modern India." *Imago Mundi* 53:97–114.

———. 2004. *The Lost Land of Lemuria: Fabulous Geographies, Catastrophic Histories.* Berkeley and Los Angeles: University of California Press.

Rand McNally. 1925. *Rand McNally Official 1925 Auto Trails Map: Illinois.* Chicago: Rand McNally & Company.

Ravenhill, William. 1992. *Christopher Saxton's 16th Century Maps: The Counties of England and Wales.* Shrewsbury, UK: Chatsworth Library.

Reid, Ira De Augustine, 1929. *The Negro Population of Denver, Colorado: A Survey of Its Economic and Social Status.* Denver: Lincoln Press.

Reinhartz, Dennis. 1997. *The Cartographer and the Literati: Herman Moll and His Intellectual Circle.* Lewiston, NY: Edwin Mellen.

Reps, John W. 1984. *Views and Viewmakers of Urban America.* Columbia: University of Missouri Press.

Residents of Hull-House. 1895. *Hull-House Maps and Papers.* New York: Thomas Y. Crowell & Co.

Reynolds, David West. 1996. "Forma Urbis Romae: The Severan Marble Plan and the Urban Form of Ancient Rome." Ph.D. diss., University of Michigan.

Ristow, Walter, ed. 1972. *A La Carte: Selected Papers on Maps and Atlases.* Washington, D.C.: Library of Congress.

———. 1985. *American Maps and Mapmakers: Commercial Cartography in the Nineteenth Century.* Detroit: Wayne State University Press.

Ritter, Carl. 1806. *Sechs Karten von Europa, mit erklaerenden Texte.* Schnepfenthal, Germany: Buchhandlung der Erziehungsanstalt.

Roberts, Adam. 2006. *The History of Science Fiction.* New York: Palgrave Macmillan.

Robinson, Arthur H. 1952. *The Look of Maps: An Examination of Cartographic Design.* Madison: University of Wisconsin Press.

———. 1975. "Mapmaking and Map Printing: The Evolution of a Working Relationship." In Woodward 1975, 1–23.

———. 1982. *Early Thematic Mapping in the History of Cartography.* Chicago: University of Chicago Press.

Roche, Daniel. 2000. *A History of Everyday Things: The Birth of Consumption in France, 1600-1800.* Trans. Brian Pearce. Cambridge: Cambridge University Press.

Rogers, J. M. 1992. "Itineraries and Town Views in Ottoman Histories." In Harley and Woodward 1992, 228–55.

Romanelli, Giandomenico, Susanna Biadene, and Camillo Tonini, eds. 1999. *A volo d'uccello: Jacopo de' Barbari e le rappresentazioni di città nell'Europa del Rinascimento.* Venice: Arsenale Editrice.

Romm, John S. 1992. *The Edges of the Earth in Ancient Thought: Geography, Exploration and Fiction.* Princeton, NJ: Princeton University Press.

Rosen-Ducat Imaging. 2006. "Rosen-Ducat Imaging Presents the Land of Make Believe by Jaro Hess." http://www.jarohesslomb.com/aboutjaro.htm (accessed Apr. 11, 2006).

Saltonstall, Wye. 1635. "The Preface to the Courteous Reader." In *Historia Mundi; or, Mercator's Atlas,* by Jodocus Hondius, x–xiii. London: T. Cotes for Michael Sparkes and Samuell Cartwright.

Salway, Benet. 2005. "The Nature and Genesis of the Peutinger Map." *Imago Mundi* 57 (2): 119–35.

Sanson, Nicolas. 1632. "Carte géographique des postes qui traversent la France." Paris: Tavernier.

———. 1634. "Carte des rivières de la France curieusement recherchée." Paris: Tavernier.

Sawday, Jonathan. 1995. *The Body Emblazoned: Dissection and the Human Body in the Renaissance.* London: Routledge.

Scafi, Alessandro. 2006. *Mapping Paradise: A History of Heaven on Earth.* London: British Library.

Scarborough, Charles. 2006. "RE: World of Warcraft." E-mail to Ricardo Padrón.

Schaer, Roland, Gregory Claeys, and Lyman Tower Sargent. 2000. *Utopia: The Search for the Ideal Society in the Western World.* New York: New York Public Library and Oxford University Press.

Schama, Simon. 1988. *The Embarrassment of Riches.* Berkeley and Los Angeles: University of California Press.

Scheiner, Christophe. 1612. *Tres epistolae de maculis solaribus scriptae ad Marcum Welserum.* Augsburg: Marc Welser.

———. 1626. *Rosa ursina sive sol ex admirando facularum & macularum suarum phoenomeno varius.* Bracciano, Italy: Andream Phaeum.

Schilder, Günter. 1986–. *Monumenta Cartographica Neerlandica.* 7 vols. Alphen aan den Rijn, The Netherlands: Canaletto.

Schilder, Günter, and Klaus Stopp. 2000. *Dutch Folio-Sized Single Sheet Maps with Decorative Borders.* Monumenta Cartographical Neerlandica, 6. Alphen aan den Rijn, The Netherlands: Uitgeverij Canaletto and Repro-Holland.

Schulten, Susan. 2001. *The Geographical Imagination in America, 1880-1950.* Chicago: University of Chicago Press.

Schulz, Juergen. 1978. "Jacopo de' Barbari's View of Venice: Map Making, City Views, and Moralized Cartography before the Year 1500." *Art Bulletin* 60: 425–74.

———. 1987. "Maps as Metaphors: Mural Map Cycles of the Italian Renaissance." In Woodward 1987, 97–122.

Schulze Altcappenberg, Hein-Th. 2000. Introduction to Botticelli 2000, 13–36.

Schwartz, Seymour I., and Ralph E. Ehrenberg. 1980. *The Mapping of America*. New York: Harry N. Abrams, Inc.

Schwartzberg, Joseph E. 1992a. "Introduction to South Asian Cartography." In Harley and Woodward 1992, 295–331.

———. 1992b. "Geographical Mapping." In Harley and Woodward 1992, 388–493.

———. 1994a. "Maps of Greater Tibet." In Harley and Woodward 1994, 607–81.

———. 1994b. "Introduction to Southeast Asian Cartography." In Harley and Woodward 1994, 689–700.

Scrope, George P. 1833. *Principles of Political Economy, Deduced from the Natural Laws of Social Welfare, and Applied to the Present State of Britain*. Longmans.

Seaman, Valentine. 1798. "An Inquiry into the Cause of the Prevalence of the Yellow Fever in New York." *Medical Repository* 1 (3): 303–23.

Shaffer, Marguerite S. 2001. *See America First: Tourism and National Identity, 1880–1940*. Washington, DC: Smithsonian Institution Press.

Shefrin, Jill. 1999. *Neatly Dissected for the Instruction of Young Ladies and Gentlemen in the Knowledge of Geography: John Spilsbury and Early Dissected Puzzles*. Los Angeles: Cotsen Occasional Press.

Shirley, Rodney W. 1980. *Early Printed Maps of the British Isles: A Bibliography, 1477–1650*. King of Prussia, PA: W. Graham Arader; [London]: Holland Press. Rev. ed., East Grinstead, UK: Antique Atlas, 1991.

———. 1983. *The Mapping of the World: Early Printed World Maps, 1472–1700*. London: New Holland. Rev. eds., various publishers, 1984, 1987, 1993, and 2001.

Shore, A. F. 1987. "Egyptian Cartography." In Harley and Woodward 1987, 117–29.

Short, John Rennie. 2001. *Representing the Republic: Mapping the United States, 1600–1900*. London: Reaktion Books.

Simmel, Georg. 1959. "The Adventure." In *Georg Simmel, 1858–1918*, ed. Kurt H. Wolff, 243–58. Columbus: Ohio State University Press.

Skelton, R. A. 1964–70. *County Atlases of the British Isles, 1579–1850: A Bibliography*. Map Collectors' Series, 9, 14, 41, 49, 63. London: Map Collectors' Circle.

———. 1966. Introduction to Georg Braun and Frans Hogenberg, *Civitates Orbis Terrarum: "The Towns of the World," 1572–1618*, ed. R. A. Skelton and A. O. Vietor, 1:vii–xlvi. 3 vols. Mirror of the World: A Series of Books on the History of Urbanization, 1st ser., 1–3. Cleveland: World Publishing Company.

———. 1972. *Maps: A Historical Survey of Their Study and Collecting*. Chicago: University of Chicago Press.

Slowther, Catherine. 1991. "Compass Point." *Map Collector* 16:48–50.

Smith, William. 1815. "A Delineation of the Strata of England and Wales, with Part of Scotland; Exhibiting the Collieries and Mines, the Marshes and Fenlands Originally Overflowed by the Sea, and the Varieties of Soil according to the Substrata, Illustrated by the Most Descriptive Names." London: John Cary. See also http://www.unh.edu/esci/greatmap.html.

———. 1816. "Geological Table of British Organized Fossils." London: John Cary. See also http://www.unh.edu/esci/table.html.

Smits, Jan. 2006. "Cartifacts Gallery of Jan Smits." www.kb.nl/persons/jan-smits/cartifcats/list.html (accessed Sept. 26, 2006).

Snow, John. 1855. *On the Mode of Communication of Cholera*. 2nd ed. London.

Sobel, Dava. 1996. *Longitude: The True Story of a Lone Genius Who Solved the Greatest Scientific Problem of His Time*. New York: Penguin.

Società Geografica Italiana. 2002. *Carte di riso: Genti, paesaggi, colori dell'estremo oriente nelle collezioni della Società Geografica Italiana*. Rome: Società Geografica Italiana.

Somerhausen, H. 1827. "Carte figurative de l'instruction populaire des Pays-bas." [Brussels]: Jobard frères.

Southworth, Michael, and Susan Southworth. 1982. *Maps: A Visual Survey and Design Guide*. Boston: Little, Brown.

Standard Oil of Company of New Jersey. 1932. *Map of the Principal Events in the Life of George Washington*. Convent Station, NJ: General Drafting Company for Standard Oil Company of New Jersey.

Steinberg, Philip E. 2005. "Insularity, Sovereignty and Statehood: The Representation of Islands on Portolan Charts and the Construction of the Territorial State." *Geografiska Annalar* 87 B (4): 253–65.

Stevenson, Edward Luther. 1914. *William Janszoon Blaeu 1571–1638*. New York: Hispanic Society of America.

Strachey, Barbara, and J. R. R. Tolkien. 1981. *Journeys of Frodo: An Atlas of J. R. R. Tolkien's "The Lord of the Rings."* New York: Ballantine Books.

Sullivan, Garrett A., Jr. 1998. *The Drama of Landscape: Land, Property, and Social Relations on the Early Modern Stage*. Stanford, CA: Stanford University Press.

Sutton, Peter. 1998. "Icons of Country: Topographic Representations in Classical Aboriginal Traditions." In Woodward and Lewis 1998, 352–86.

Swaaij, Louise Van, and Jean Klare. 2000. *The Atlas of Experience*. Trans. David

Winner and Isabel Verdurme. New York: Bloomsbury.

Swift, Jonathan. 2004. *Gulliver's Travels and Other Writings*. Ed. Clement Hawes. Boston: Houghton Mifflin.

Talbert, Richard. 2004. "Cartography and Taste in Peutinger's Roman Map." In *Space in the Roman World: Its Perception and Representation*, ed. Richard Talbert and Kai Brodersen, 113–31. Münster: LIT Verlag.

———. 2006. "The Roman World View: Beyond Recovery?" Paper presented at *Geography, Ethnography, and Perceptions of the World in Antiquity, the Middle Ages, and the Renaissance*, Brown University, Mar. 17–19.

Taylor, Eva Germaine Rivington. 1956. *The Haven-Finding Art: A History of Navigation from Odysseus to Captain Cook*. London: Hollis & Carter.

Thompson, Gilbert. 1902. "Historical Military Powder Horns." *American Monthly Magazine* 20 (6): 1005–28.

Thrower, Norman J. W. 1999. *Maps and Civilization: Cartography in Culture and Society*. 2nd ed. Chicago: University of Chicago Press.

Tibbetts, Gerald R. 1992a. "The Balkhi School of Geographers." In Harley and Woodward 1992, 108–36.

———. 1992b. "Later Cartographic Developments." In Harley and Woodward 1992, 137–55.

Togan, Ahmed Zeki Velidi, ed. 1941. *Bīrūnī's Picture of the World*. Memoirs of the Archaeological Survey of India, no. 53. Delhi.

Tolkien, J. R. R. 1996a. *The Hobbit; or, There and Back Again*. Boston: Houghton Mifflin.

———. 1996b. *The Two Towers: Being the Second Part of "The Lord of the Rings."* Boston: Houghton Mifflin.

———. 2002. *The Annotated Hobbit*. Ed. Douglas A. Anderson. Rev. and expanded ed., Boston: Houghton Mifflin.

Tooley, R. V. 1963. *Leo Belgicus: An Illustrated List*. Map Collectors' Series, 7. London: Map Collectors' Circle.

Tuan, Yi-Fu. 1977. *Space and Place: The Perspective of Experience*. Minneapolis: University of Minnesota Press.

———. 1994. *Cosmos and Hearth: The Perspective of Experience*. Minneapolis: University of Minnesota Press.

Tufte, Edward R. 1983. *The Visual Display of Quantitative Information*. Cheshire, CT: Graphics Press.

———. 1997. *Visual Explanations*. Cheshire, CT: Graphics Press.

Tukey, John Wilder. 1962. "The Future of Data Analysis." *Annals of Mathematical Statistics* 33:1–67, 81.

———. 1972. "Some Graphic and Semigraphic Displays." In *Statistical Papers in Honor of George W. Snedecor*, ed. T. A. Bancroft, 292–316. Ames: Iowa State University Press.

Turnbull, David, ed. 1993. *Maps Are Territories, Science Is an Atlas: A Portfolio of Exhibits*. Chicago: University of Chicago Press.

Turner, Hilary Louise. 1987. "Christopher Buondelmonti and the Isolario." *Terrae Incognitae* 19:11–28.

Turow, Joseph. 1997. *Breaking Up America: Advertising and the New Media World*. Chicago: University of Chicago Press.

Tyacke, Sarah. 1973. "Map-Sellers and the London Trade c 1650–1710." In *My Head Is a Map: Essays and Memoirs in Honor of R. V. Tooley*, ed. Helen Wallis and Sarah Tyacke, 62–80. London: Francis Edwards and Carta Press.

———, ed. 1983. *English Map-Making, 1500–1650: Historical Essays*. London: British Library.

Tyner, Judith. 1994. "Geography through the Needle's Eye: Embroidered Maps and Globes in the Eighteenth and Nineteenth Centuries." *Map Collector* 66:2–7.

———. 2002. "Folk Maps, Cartoon Cartography, and Map Kitsch." *Mercator's World* 7:24–29.

United States Census Office. 1874. *Statistical Atlas of the United States Based on the Results of the Ninth Census 1870 with Contributions from Many Eminent Men of Science and Several Departments of the Government*. Compiled under the authority of Congress by Francis A. Walker. New York: J. Bien.

———. 1898. *Statistical Atlas of the United States Based on the Results of the Eleventh Census*. Compiled by Henry Gannett. Washington, DC: Government Printing Office.

Unno, Kazutaka. 1994. "Cartography in Japan." In Harley and Woodward 1994, 346–477.

Van Ee, Patricia Molen. 2002. "The Coming of the Transcontinental Railroad: The Warren Maps." In *Mapping the West: America's Westward Movement, 1524–1890*, ed. Paul E. Cohen, 172–75. New York: Rizzoli.

———. 2006. Conversation with Susan Schulten, July 25.

Van Loon, Hendrik. 1927. *America*. New York: Boni & Liveright.

Veregin, Howard, ed. 2005. *Goode's World Atlas*. 21st ed. Chicago: Rand McNally.

Verner, Coolie. 1975. "Copperplate Printing." In Woodward 1975, 51–75.

Wainer, Howard. 2005. *Graphic Discovery: A Trout in the Milk and Other Visual Adventures*. Princeton, NJ: Princeton University Press.

Wainer, Howard, and Ian Spence. 2005. Introduction to *The Commercial and Political Atlas and Statistical Breviary / William Playfair*, 1–35. Cambridge: Cambridge University Press.

Wallis, Helen, ed. 1969. *Libro dei Globi.* Facsimile of Vincenzo Coronelli, *Atlante Veneto.* Vol. 1. Amsterdam: Teatrum Orbis Terrarum.

———. 1973. "Maps as a Medium for Scientific Communication." In *Studia z dziejów geografii i kartografii: Etudes d'Histoire de la Géographie et de la Cartography,* ed. Józef Babicz, 251–62. Warsaw: Zaclad Narodowy Imienia.

———. 1997. "Sixteenth-Century Maritime Manuscript Atlases for Special Presentation." In Wolter and Grim 1997, 3–29.

Wallis, Helen M., and Arthur H. Robinson, ed. 1987. *Cartographical Innovations.* London: Map Collector Publications.

Wallis, John. 1790. *Pilgrim's Progress Dissected.* London: John Wallis.

Walters Art Gallery. 1952. *The World Encompassed.* Baltimore: Walters Art Gallery.

Warhus, Mark. 1997. *Another America: Native American Maps and the History of Our Land.* New York: St. Martin's Press.

Waselkov, Gregory A. 1998. "Indian Maps of the Colonial Southeast: Archaeological Implications and Prospects." In *Cartographic Encounters: Perspectives on Native American Mapmaking and Map Use,* ed. Malcolm Lewis, 205–21. Chicago: University of Chicago Press.

Wasinger, Holly. 2005. "From Five Points to Struggle Hill: The Race Line and Segregation in Denver." *Colorado Heritage* (Autumn): 28–39.

Waterschoot, Werner. 1979. "The Title-Page of Ortelius's *Theatrum Orbis Terrarum.*" *Quaerendo* 9 (1): 43–68.

Watson, Helen, and the Yolngu community at Yirrkala. 1993. "Aboriginal-Australian Maps." In Turnbull 1993, 28–36.

Watson, Ruth. 2005. "A Heart-Shaped World: Johannes Stabius, Oronce Fine

and the Meanings of the Cordiform Map." Ph.D. diss., Australian National University.

Welu, James A. 1975. "Vermeer: His Cartographic Sources." *Art Bulletin* 57:529–47.

———. 1977. "Vermeer and Cartography." Ph.D. diss., Boston University.

Wheat, Carl, 1957–63. *Mapping the Trans-Mississippi West, 1540–1861.* 5 vols. San Francisco: Institute of Historical Cartography.

White, C. Albert. 1983. *A History of the Rectangular Survey System.* Washington, DC: U.S. Department of the Interior, Bureau of Land Management.

Whitfield, Peter. 1994. *The Image of the Earth: 20 Centuries of World Maps.* London: British Library.

———. 1996. *The Charting of the Oceans: Ten Centuries of Maritime Maps.* London: British Library.

Whitmore, John K. 1994. "Cartography in Vietnam." In Harley and Woodward 1994, 478–508.

Wilkinson, Leland. 2005. *The Grammar of Graphics.* 2nd ed. New York: Springer.

Willdey, George. [1712]. "The Roads of England according to Mr. Ogilby's Survey." London.

Williams, Raymond. 1983. "Consumer." In *Keywords: A Vocabulary of Culture and Society.* New York: Oxford University Press.

Wilson, Bronwen. 1999. "'The Eye of Italy': The Image of Venice and Venetians in Sixteenth-Century Prints." Ph.D. diss., Northwestern University.

———. 2005. *The World in Venice: Print, the City, and Early Modern Identity.* Toronto: University of Toronto Press.

Winchester, Simon. 2001. *The Map That Changed the World: William Smith and the Birth of Modern Geology.* New York: Harper Collins.

Winichakul, Thongchai. 1994. *Siam Mapped: A History of the Geo-Body of a Nation.* Honolulu: University of Hawai'i Press.

Wolter, John A., and Ronald E. Grim, eds. 1997. *Images of the World: The Atlas Through History.* Washington, DC: Library of Congress.

Wood, Denis. 1993. "The Fine Line between Mapping and Mapmaking." *Cartographica* 30 (4): 50–60.

———. 2006a. "Catalogue of Map Artists." *Cartographic Perspectives: Journal of the American Cartographic Society* 53:61–67.

———. 2006b. "Map Art." *Cartographic Perspectives: Journal of the American Cartographic Society* 53:5–14.

Wood, Denis, and John Fels. 1992. *The Power of Maps.* New York: Guilford Press.

Wooden, Warren W., and John N. Wall Jr. 1985. "Thomas More and the Painter's Eye: Visual Perspective and Artistic Purpose in More's *Utopia.*" *Journal of Medieval and Renaissance Studies* 15:231–63.

Woodward, David, ed. 1975. *Five Centuries of Map Printing.* Chicago: University of Chicago Press.

———. 1977. *The All-American Map: Wax Engraving and Its Influence on Cartography.* Chicago: University of Chicago Press.

———, ed. 1987. *Art and Cartography: Six Historical Essays.* Chicago: University of Chicago Press.

———. 1990. "Roger Bacon's Terrestrial Coordinate System." *Annals, Association of American Geographers* 80:109–22.

———. 1991. "Maps and the Rationalization of Geographic Space." In Levenson 1991, 83–87.

———. 1996. *Maps as Prints in the Italian Renaissance.* London: British Library.

———, ed. 2007. *Cartography in the European Renaissance*. Vol. 3 of *The History of Cartography*. Chicago: University of Chicago Press.

Woodward, David, and G. Malcolm Lewis, eds. 1998. *Cartography in the Traditional African, American, Arctic, Australian, and Pacific Societies*. Vol. 2, book 3 of *The History of Cartography*. Chicago: University of Chicago Press.

WorldofWar.net: The Unofficial WOW Site. http://www.worldofwar.net/cartography/ (accessed July 17, 2006).

Wright, John Kirtland. 1925. *The Geographical Lore of the Time of the Crusades: A Study in the History of Medieval Science and Tradition in Western Europe*. New York: American Geographical Society.

Yale Map Collection. 2006. "Cartographic Curiosities." *The Yale Map Collection Online*. http://www.library.yale.edu/MapColl/curious.html (accessed Apr. 17, 2006).

Yau, John. 1996. *The United States of Jasper Johns*. 1st ed. Cambridge, MA: Zoland Books.

Yee, Cordell D. K. 1994a. "Reinterpreting Traditional Chinese Geographical Maps." In Harley and Woodward 1994, 35–70.

———. 1994b. "Chinese Maps in Political Culture." In Harley and Woodward 1994, 71–96.

———. 1994c. "Traditional Chinese Cartography and the Myth of Westernization." In Harley and Woodward 1994, 170–202.

———. 2001. "The Map Trade in China." In *Approaches and Challenges in a World-Wide History of Cartography*, ed. David Woodward, Catherine Delano Smith, and Cordell D. K. Yee, 111–30. Barcelona: Institut Cartographic de Catalunya.

Yonemoto, Marcia. 2003. *Mapping Early Modern Japan: Space, Place, and Culture in the Tokugawa Period (1603–1868)*. Berkeley and Los Angeles: University of California Press.

Yorke, Douglas A. Jr., John Margolies, and Eric Baker. 1996. *Hitting the Road: The Art of the American Road Map*. San Francisco: Chronicle Books.

Yost, Karl, ed. 1987. *The Ohio and Mississippi Navigator of Zadok Cramer*. Morrison, IL: K. Yost.

Zandvliet, Kees. 1998. *Mapping for Money: Maps, Plans and Topographic Paintings and Their Role in Dutch Overseas Expansion during the 16th and 17th Centuries*. Amsterdam: Batavian Lion.

CONTRIBUTORS

JAMES R. AKERMAN is director of the Hermon Dunlap Smith Center for the History of Cartography at the Newberry Library, Chicago. His publications concern the history of transportation and tourist mapping, popular cartography, atlases, and the use of historical maps in education. He is the editor of *Cartography and Statecraft: Studies in Governmental Mapmaking in Modern Europe and its Colonies* (1998) and of *Cartographies of Travel and Navigation* (2006). He is a co-curator of the Field Museum–Newberry Library Maps exhibition that inspired this book.

DENIS COSGROVE is the Alexander von Humboldt Professor of Geography at the University of California, Los Angeles. He has taught and written extensively on the roles that visual images, including paintings, photographs, and maps, have played in shaping the ways that people perceive, understand, and transform the world's landscapes and places. His map studies have ranged from Renaissance Venetian land surveys to images of the cosmos, the globe, and the whole Earth. His books include *The Palladian Landscape* (1993), *Mappings* (1999), and *Apollo's Eye* (2002).

DIANE DILLON is an art historian and associate director of research and education at the Newberry Library, Chicago. Her research interests include popular cartography, architecture and urban design, world's fairs, and the visual culture of the American West. Dillon's recent essays include "Mapping Enterprise: Cartography and Commodification at the 1893 World's Columbian Exposition," in *Nineteenth-Century Geographies: Anglo-American Tactics of Space*, edited by Helena Michie and Ronald Thomas (2003). She is assistant curator of the Field Museum–Newberry Library Maps exhibition that inspired this book.

MATTHEW H. EDNEY holds the Osher Chair in the History of Cartography at the University of Southern Maine and directs the History of Cartography Project at the University of Wisconsin–Madison. Author of *Mapping an Empire: The Geographical Construction of British India, 1765-1842* (1997) and *The Origins and Development of J. B. Harley's Cartographic Theories* (2005), he is coeditor of *Cartography in the European Enlightenment*, volume 4 of *The History of Cartography*.

MICHAEL FRIENDLY is professor of psychology at York University in Toronto. He is the author of three books on statistical graphics, numerous research papers on data visualization and its history, and the developer of the Milestones Project on the history of thematic cartography and statistical graphics: <http://www.math.yorku.ca/SCS/Gallery/milestone/>.

ROBERT W. KARROW JR. is curator of special collections and maps at the Newberry Library, Chicago. His main research interests are early modern cartography, mapping of the nineteenth-century American West, and cartobibliography. He is the author of *Mapmakers of the Sixteenth Century and Their Maps* (1993) and is a co-curator of the Field Museum–Newberry Library Maps exhibition that inspired this book.

RICARDO PADRÓN is associate professor of Spanish at the University of Virginia. His interest in the literary and cartographic dimensions of the early modern Hispanic imperial imagination culminated in *The Spacious Word: Cartography, Literature and Empire in Early Modern Spain* (2004). He is investigating early modern Spanish interest in Asia and the Pacific, with a special emphasis on emergent globalization.

GILLES PALSKY is assistant professor of geography at the University of Paris 12–Val de Marne and a member of the research unit Epistemology and History of Geography of the National Committee for Scientific Research, Paris. He is the author of *Des chiffres et des cartes: naissance et développement de la cartographie quantitative française au XIXe siècle* (1996). A specialist in the history of contemporary cartography, he also conducts research on the role of images in building geographic knowledge.

SUSAN SCHULTEN is associate professor of history at the University of Denver. She is the author of *The Geographical Imagination in America, 1880–1950* (2001) and, more recently, "How to See Colorado: The Federal Writers' Project, American Regionalism, and the 'Old New Western History,'" *Western Historical Quarterly* (Spring 2005). She is researching a book on the idea of history in American culture.

CONTRIBUTORS

ACKNOWLEDGMENTS

Some of the individual authors have added notes to their chapters with personal acknowledgments, but the editors take this opportunity to thank some of the people who have contributed to the completion of this book and the exhibition that inspired it.

This book began with the exhibition, and our debts, too, begin there. We are extremely grateful first of all to John W. McCarter Jr., President of The Field Museum, and to our good friend Ken Nebenzahl, a trustee of the Newberry Library, for hatching the idea to bring together historically important maps reflecting the diversity of human geographic knowledge and experience. Our collaboration with The Field Museum as co-curators has been long, intense, and rewarding. Although we had mounted exhibitions before at the Newberry Library, our home institution, neither of us had experienced the prolonged planning, letter writing, travel, frustration, and, yes, joy that go with the creation of an exhibition on this scale. That we survived, and that the process was as smooth and pleasant as it was, is due in large part to the wonderful collaboration of our Field Museum colleagues. They began by welcoming us as fellow staff; subsequent weekly meetings and daily, even hourly, e-mails only served to cement further an altogether happy relationship.

There are over one hundred people in the Exhibitions Department at The Field Museum, and a great many of them had a hand in bringing off the exhibition, but we must personally acknowledge these current and former staff members: Laura Sadler, Senior Vice President, Museum Enterprises; Sophia Shaw Siskel, former Vice President, Exhibitions and Education; Robin Groesbeck, Director of Exhibitions; Patrick Ryan Williams, Assistant Curator of Archaeological Science; Laura Biddle Clarke, Director, Sponsorship, Corporate, and Foundation Giving; M.W. Burns, former Exhibition Design Director; Eric Manabat, Exhibition Designer; Lori Walsh, Graphic Designer; Eric Frazer, Production Manager; Pam Gaible, Senior Mount Shop Supervisor; David Mendez, Mount Shop Supervisor; Susan Blecher, Exhibitions Registrar; Angie Morrow, former Exhibitions Registrar; Debbie Linn, Conservator; Tiffany Plate, former Publication Coordinator; and Mark Alvey, former Exhibition Developer for this project.

In particular, we would like to single out three of our Field colleagues for special thanks. Todd Tubutis, the Senior Project Manager for Maps: Finding Our Place in the World and a former Newberry Library colleague, kept a firm hand on the tiller all through the process. Todd kept us on task and on schedule, and did it with a warmth and diplomacy that almost made us forget he was the boss. Both Todd and Matt Matcuk, Exhibition Development Director, accompanied us on several visits to foreign lenders, memorable opportunities to get to know them and to appreciate their manifold talents. In the initial planning of the exhibit concept especially, Matt was extremely helpful to us. And finally, Gretchen Baker, the Exhibition Developer for Maps, got to know our maps intimately and helped us tell their stories in seventy-five words or less. It was a great pleasure working with Gretchen, a fine artist, sometime cartographer, and skillful writer.

At the Newberry Library, we owe thanks to a great many colleagues who encouraged and supported our work in these projects, who gave us material assistance, and who kept things running in our sections while we were running elsewhere. In particular, we are grateful to the past president, Charles Cullen, and our current president, David Spadafora, and to vice presidents James Grossman and Hjordis Halvorson for their wholehearted backing of the joint Field Museum/Newberry Library exhibition and book projects. A number of Newberry trustees and friends deeply committed to the preservation and study of maps encouraged the creation of the exhibition and were most generous in offering advice and, in several instances, loaning maps; these include Roger Baskes, Vin Buonanno, Gerald F. Fitzgerald, Art Holzheimer, Art Kelly, Barry MacLean, Fred Manning, Sandy McNally, Ken Nebenzahl, and Rudy Ruggles. We are indebted to Susan Hanf, Chris Dingwall, JoEllen Dickie, and Patrick Morris, who kept our own sections operating efficiently during our frequent distractions; to Catherine Gass and John Powell, for their Herculean work on a very tight schedule to produce photographs for this book; to Lauren Reno, the Library's de facto registrar, for coordinating what was certainly the largest loan in Newberry history; and to Giselle Simon and her conservation staff, for countless hours of work in repairing, stabilizing, and readying for shipment the many Newberry maps and atlases that figured in the exhibition. Diane Dillon was hired early on in the project to assist us in research, but we came to be so dependent on her ideas, scholarship, energy, and enthusiasm that she has become, in effect, a third curator.

In our planning for the exhibition, we relied heavily on advice and assistance from a panel of expert advisors, consisting of Denis Cosgrove, Matthew Edney, Laura Hostetler, Mark Monmonier, Barbara Mundy, Ricardo Padrón, and Susan Schulten, four of whom (Cosgrove, Edney, Padrón, and Schulten) became even more intimately involved as authors of this book. Other colleagues in the cartographic and library communities were very generous with advice. In particular, we would like to thank Peter Barber, Map Curator at the British Library; Hélène Richard, Director of the Department of Maps and Plans at the Bibliothèque nationale de France; Markus Heinz of the Staatsbibliothek zu Berlin–Preußischer Kulturbesitz, Kartenabteilung; Professor Günter Schilder and Peter van der Krogt of the Faculty of Geographical Sciences at the University of Utrecht; Mary Pedley, Assistant Curator of Maps at the William L. Clements Library; Dr. Tina Cervone, Director of the Italian Cultural Institute of Chicago; Alexei Postnikov, Director of the Institute of the History of Science and Technology, Russian Academy of Sciences; Chris Baruth, Curator, American Geographical Society Library of the Golda Meir Library, University of Wisconsin–Milwaukee; Ewa Barczyk, Director, University of Wisconsin–Milwaukee Libraries; John Hébert, Jim Flatness, and Pam Van Ee of the Geography and Map Division of the Library of Congress, and Ralph Ehrenberg, the retired Chief of the Division; Francis Herbert, Curator of Maps at the Royal Geographical Society; Yuki Ishimatsu, Head Librarian of the Japanese Collection at the East Asian Library at UC Berkeley; Mary Nooter Roberts, Deputy Director and Chief Curator, Fowler Museum at UCLA; Alastair Pearson, Principal Lecturer in Geography at the University of Portsmouth; Susan Gole, editor, *IMCoS Journal*; Mr. Sharat Chawla, New Delhi; Kay Ebel, Assistant Professor of Geography at Ohio Wesleyan University; John Cloud, historian with the NOAA Central Library; Bill Gartner, Associate Professor of Geography and Geology at the University of Wisconsin–Fox Valley; Joseph E. Schwartzberg, Professor Emeritus of Geography, University of Minnesota; and Jude Leimer, Managing Editor of the History of Cartography Project, University of Wisconsin–Madison.

Our colleagues at the Walters Art Museum, where the exhibition will travel for its only other venue, have been very generous as lenders, as hosts, and as supporters. We want to thank particularly the Director, Gary Vikan, and the Curator of Manuscripts and Rare Books, Will Noel.

The entire exhibition would have remained a dream without the active participation of the more than sixty institutional and individual lenders. We are indebted to the many staff members and directors of the lending institutions listed below, who assisted us throughout the process of assembling this exhibition.

Adler Planetarium and Astronomy Museum, Chicago
American Geographical Society Collection, Golda Meir Library,
    University of Wisconsin–Milwaukee

Amsterdams Historisch Museum

Archivo General de las Indias, Seville

Beinecke Library, Yale University, New Haven, Connecticut

Benson Latin American Collection, University of Texas at Austin

Biblioteca Apostolica Vaticana

Bibliothèque nationale de France, Paris

Bodleian Library, Oxford University

Boston Public Library

British Library, London

Centre historique des Archives nationales, Paris

Comune di Roma

East Asian Library, University of California at Berkeley

Ecole Nationale des Ponts et Chaussées, Marne la Vallée, France

Elkhart County Historical Society, Bristol, Indiana

Field Museum, Chicago

Friedrich-Schiller-Universität Jena

Harvard University, Cambridge, Massachusetts

Illinois State Archives, Springfield

James Ford Bell Library, University of Minnesota, Minneapolis

LaSalle Bank N.A., Chicago

Library of Congress Geography and Map Division, Washington, DC

Library of Rush University Medical Center, Chicago

Logan Museum of Anthropology, Beloit College, Wisconsin

London Transport Museum

Missouri Historical Society, St. Louis

Morgan Library & Museum, New York

Musée des Plans-reliefs, Paris

Musée d'Ethnographie, Geneva

Museum für Indische Kunst, Berlin

National Archives (U.K.), Kew

National Archives and Records Administration (U.S.), Washington, DC

National Gallery of Art, Washington, DC

National Maritime Museum, Greenwich, UK

Newberry Library, Chicago

New York Public Library

Peter the Great Museum of Anthropology and Ethnology (Kunstkamera) of the Russian Academy of Sciences, St. Petersburg

Private Collections

Raynor Memorial Library, Marquette University, Milwaukee

Royal Collection, Windsor, UK

Royal Geographical Society, London

Société Asiatique, Paris

Staatsbibliothek Berlin–Preußischer Kulturbesitz

ACKNOWLEDGMENTS

Universitätsbibliothek Basel
Universiteitsbibliotheek Leiden
University of Iowa Museum of Art, Iowa City
Victoria and Albert Museum, London
William Blair and Company, Chicago
William L. Clements Library, University of Michigan, Ann Arbor
Winterthur Library, Winterthur, Delaware
Yale Center for British Art, New Haven, Connecticut
Young Research Library, University of California, Los Angeles

The making of this book has put us in the debt of still more individuals, especially Ariel Orlov, who served as our publication coordinator; Mary Goljenboom, who did yeoman work on a brutal timetable to acquire the images and permissions; the anonymous outside readers who improved the manuscript in so many ways; Peter Beatty and Sandy Hazel at the University of Chicago Press, who brought everything together; and Christie Henry, our very able and cartographically savvy editor at the Press.

Inevitably, with a project of this scope and complexity, we will have omitted the names of people who helped us in one way or another, and for such oversights we can only offer our apologies.

Finally, we wish to acknowledge the extraordinary debt we owe to our good friend, mentor, and colleague, the late David Woodward. This project was supposed to be his, and in almost every respect it is. In the broadest sense, the History of Cartography project he guided from the 1980s to his death in 2004, originally in partnership with Brian Harley, is responsible for the awakening and expansion of the study of the history of cartography we have tried to represent in this book. David was originally hired as curator of the exhibition, and had planned to write a companion volume, which was certain to become a landmark statement of his unrivaled knowledge of map history. He had developed a plan for the exhibition, a preliminary object list, and notes on the project that were graciously lent to us by Rosalind Woodward. Throughout the project, however, as we pondered hard choices or looked for additional inspiration, we asked ourselves what David might have done.

James R. Akerman
Robert W. Karrow Jr.

ILLUSTRATIONS

ILLUSTRATIONS

133 "Charts of the Thermometer, Wind, Rain, and Barometer on the Morning, Afternoon, and Evening on Each Day during December 1861," detail of right half. From Francis Galton, *Meteorographica; or, Methods of Mapping the Weather* (London: Macmillan, 1863) :: 234

134 "Abaque hexagonal donnant sans calcul et sans relèvements la déviation du compass, pour le bateau 'Le Triomphe'" (Hexagonal Abacus Giving the Variation of the Compass without Calculation or Plotting, for the Ship "Le Triomphe"). From Charles Lallemand, *Les abaques hexagonaux* (Paris: Ministère des travaux publics, 1885) :: 236

135 "Mouvement quinquennal de la population par département depuis 1801 jusqu'en 1881" (Quinquennial Change of the Population by Department from 1801 to 1881). From Ministère des Travaux Publics, *Album de Statistique Graphique de 1884* (Album of Statistical Graphics for 1884) (Paris: Imprimerie nationale, 1885) :: 237

136 Francis A. Walker, "Chart Showing the Principal Constituent Elements of the Population of Each State." From United States Census Office, *Statistical Atlas of the United States Based on the Results of the Ninth Census 1870* (New York: J. Bien, 1874) :: 238

137 Henry Gannett, "Rank of States and Territories in Population at Each Census, 1790–1890." From United States Census Office, *Statistical Atlas of the United States Based on the Results of the Eleventh Census* (Washington, DC: Government Printing Office, 1898) :: 239

138 "Carte figurative de l'instruction populaire de la France" (Figurative Map of Popular Education in France). From Charles Dupin, *Forces productives et commerciales de la France* (Productive and Economic Forces of France) (Paris: Bachelier, 1826) :: 241

139 "Crimes contres les personnes" (Crimes against Persons). From André-Michel Guerry, *Statistique morale de l'Angleterre comparée avec la statistique morale de la France* (Moral Statistics of England Compared with the Moral Statistics of France) (Paris: J.-B. Baullière et fils, 1864) :: 243

140 "[Map] Showing the Deaths from Cholera in Broad Street, Golden Square, and the Neighbourhood, from 19th August to 30th September 1854," detail. From John Snow, *On the Mode of Communication of Cholera*, 2nd ed. (London, 1855) :: 245

141 Charles Joseph Minard, "Carte figurative et approximative des quantités de coton en laine importées en Europe en 1858 et 1862" (Figurative and Approximate Map of Quantities of Raw Cotton Imported to Europe in 1858 and 1862). From *Tableaux graphiques et cartes figuratives* (Graphic Tables and Figurative Maps) (Paris: Regnier et Dourdet, 1863) :: 246

142 "Map of the British Isles, Elucidating the Distribution of the Population, Based on the Census of 1841." From Augustus Petermann, *The Atlas of Physical Geography* (London: Wm. S. Orr & Co., 1850) :: 247

143 "Diagram of the Causes of Mortality in the Army in the East" (1857). From Florence Nightingale, *Notes on Matters Affecting the Health, Efficiency, and Hospital Administration of the British Army: Founded Chiefly on the Experience of the Late War* (London: Harrison and Sons, 1858) :: 248

144 "Descriptive Map of London Poverty," northwestern sheet. From Charles Booth, *Labour and Life of the People* (London: Williams and Norgate, 1889) :: 250

145 *Edoras*, by Alan Lee; cover of Karen Wynn Fonstad's *The Atlas of Middle-Earth* (Boston: Houghton Mifflin, 1991) :: 256

146 "The Nine Norse Worlds." From Ingri and Edgar Parin D'Aulaire, *D'Aulaire's Book of Norse Myths* (New York: New York Review Books, 2005) :: 257

147 Everett Henry, "The Voyage of the *Pequod* from the Book, *Moby Dick*, by Herman Melville" (Cleveland: Harris-Seybold Co., 1956) :: 257

148 Ernest Dudley Chase, "The United States as Viewed by California (Very Unofficial)" (Winchester, MA: Ernest Dudley Chase, 1940) :: 258

149 "La Carte du Tendre" (Map of Tenderness). From Madeleine de Scudéry, *Clélie: Histoire romaine dedié à Mademoiselle de Longueville* (Clélie: A Romance Dedicated to Mademoiselle de Longueville) (Paris, 1654) :: 259

150 Sandro Botticelli, chart of Hell (manuscript, ca. 1490) :: 262

151 "Sito et forma della valle inferna" (Site and Form of the Valley of Hell). From Dante Alghieri, *Dante col sito, et forma dell'inferno* (Dante, with the Site and Shape of Hell) (Venice: Aldus Manutius and Andrea Torresano di Asolo, 1515) :: 263

152 "Profilo, pianta, e misure dell' Inferno" (Profile, Plan, and Dimensions of Hell). From Dante Alghieri, *La Divina commedia* (The Divine Comedy) (Florence: Crusca Academy, 1595) :: 264

153 "Treasure Island." From Robert Louis Stevenson, *Treasure Island* (London: Cassell and Company, 1883) :: 266

154 "Utopiae insulae figura" (Map of the Island of Utopia). From Thomas More, *Libellus vere aureus nec minus salutaris quam festivus de optimo reip. statu, deque nova insula Utopia* (A Truly Golden Little Book, No Less Beneficial Than Entertaining, about the Best State of a Commonwealth and the New Island of Utopia) (Louvain: Thierry Martin, 1516) :: 268

155 Ambrosius Holbein, map of Utopia. From Thomas More, *De optimo reip. statu, deque nova insula Utopia* (About the Best

ILLUSTRATION CREDITS

Photos courtesy of The Newberry Library: 1, 6 & 7, Maps © Rand McNally & Company, R.L.06-S-127; 8; 13 [VAULT Case G4042.M5 1866 C6]; 15 [Case oversize G 45005 .638]; 18 [Map6F 5751.P2.2]; 19 [H6083.58]; 20 [Map3C G4101 .P2A 1905 R3 (PrCt)]; 22 [VAULT Ayer Ms Map 3]; 26 [Case fG 13.19 Vol. 3 opp. 2]; 27 [Ayer Art Waldeck E1 #21]; 29, Map © Rand McNally & Company, R.L.06-S-127; 30; 47 [Ayer 6 P9 1482a]; 48, Arthur Holzheimer Collection; 49; 56 [G 007.33, v. 23]; 57 [Case W 1025 .75 v. 2]; 58 [VAULT Ayer oversize G1006 .T515 1580]; 61 [Map 5C 7]; 63 [Novacco 8F7]; 65 [4A 15761]; 67 [Wing +G 32 168]; 74 [Map6F G3201.A 1889 .N3], reproduced with the permission of NG Maps / National Geographic Image Collection; 75 [Ayer 135 07 1573]; 79 [Case +G1007.1, vol. 2]; 80 [map 7C G5830 1619 .L4 (15)]; 85 [Case G1042.367 County DOWN]; 90 [Ayer 150.5 V7 S6 1612]; 91 [Map4F G3700 1718 .L5]; 97; 98 [VAULT drawer map Graff 3360]; 102 [Map4F G4101.P2 1890 R3]; 120 [VAULT Case oversize G1015 .S35 1658 v. 1, pl. 34]; 123 [Novacco 4F 047]; 125, Roger S. Baskes Collection at The Newberry Library; 127 [sc813]; 128 [VAULT Case oversize G1015 .B8 1770]; 131 [H 3145 .5585]; 144 [Map 6F 5754.L7E74]; 153 [Case Y 155 .S851432]; 154 [Case J 205 .582]; 155 [Case J 205 .5862]; 156 [VAULT Ruggles 324]; 167, Kenneth & Jocelyn Nebenzahl Collection; 169 [SC: map 4F 0G4104.C6:2W6A3 1893.R3]; 170 [Case G 1045.78]; 171 [G 89669.05]; 172 [4A 15761]; 174 [G 38 .0609]; 180, Collection of Lori Walsh; 181 [Case+G 117.119 v01.1]; 184A & B, Private Collection; 185

© 2005, Akademische Verlagsanstalt: 31, The Peters Projection World Map was produced with the support of the United Nations Development Programme. For maps and other related teaching materials contact: ODT, Inc., PO Box 134, Amherst MA 01004 USA; (800–736–1293; Fax: 413–549–3503; E-mail: odtstore@odt.org; WEB = www. odtmaps.com

Alinari/The Bridgeman Art Library International: 39, The Mappa Mundi of Fra Mauro, a Camaldolese monk from the monastery of San Michele in Murano, 1459

Copyright © Archaeological Survey of India: 43A, Diagram "al-Biruni's Kitab tahdid nihayat al-amakin" from Ahmed Zeki Velidi Togan, ed., *Biruni's Picture of the World, Memoirs of the Archaeological Survey of India*, no. 53 (Delhi, 1941), page 61.

Authors' Collection: 130; 135

Roger S. Baskes Collection at The Newberry Library: 125, photo courtesy of The Newberry Library

© Biblioteca Apostolica Vaticana (Vatican Library): 41 [URB.GR.82 f.60v–61r]; 150 [Reg. Lat. 1896 f.101]

Biblioteca Nacional, España (National Library of Spain): 161 [GM / Mr 42 nº 497 Mapa Region.1983]

Bibliothèque de l'Ecole des mines de Paris: 134

Bibliothèque nationale de France (National Library of France): 21 [GE B 1118 RES]; 38 [LATIN 8878]; 40 [MSS ESPAGNOL-30]; 59 [GE DD 2987 (63) RES]; 121 [Ge D 10052]; 138 [Ge C 6588]; 149 [RES-Y2–1496]; 165A & B [GE DD 683 RES Folio 3v]

Bildarchiv Preussischer Kulturbesitz/Art Resource, New York: 129

© 2005 Blizzard Entertainment. All rights reserved: 159

The Bodleian Library, University of Oxford: 44 [MS Pococke 375 folios 3v–4r]; 76A & B [MS Arab.c.90, fols. 30b–31a and 29b–30a]; 158 [MS. Tolkien Drawings 97], Reproduced with permission of The J R R Tolkien Estate Ltd. © Christopher Reuel Tolkien 1954

The Bridgeman Art Library International: 162, *Map*, 1963 (oil on canvas), Johns, Jasper (b.1930). Private Collection. Art © Jasper Johns / Licensed by VAGA, New York, NY

Copyright © British Library Board. All Rights Reserved: 9 [Roy. 14. C. VII, fol. 2]; 23 [Maps C.22.d.18]; 45A [MSS. Mar.G.28]; 55 [Maps K.Top.4.36.1.11

TAB]; 119 [Maps K. Top.5.84]; 139 [Maps 32.e.34]; 142 [Maps * 1125.(1.)]

Centre historique des Archives nationales (CHAN), Paris (History Center of the French National Archives): 78, photo courtesy of Gordon M. Sayre, University of Oregon

Chicago Daily Tribune, Feb. 25, 1894: 173

Chicago History Museum: 178 [ICHi-31126]

Governing Body of Christ Church, Oxford: 72 [Maps Woodnorton 1]

From *D'Aulaires' Book of Norse Myths* published by New York Review Books. Copyright © 1967 by Ingri and Edgar Parin d'Aulaire. Copyright © Renewed 1995 by Nils Daulaire and Per Ola d'Aulaire. Used by Permission: 146

Courtesy of the Deakin University Art Collection, Australia: 66

Courtesy of the East Asian Library, University of California, Berkeley: 35, Hotan, "Nansenbushu bankoku shoka no zu," 1710. 1 map ; 115 × 145 cm, folded to 36 × 20 cm; Woodblock print; in Japanese.

Bibliothèque, Ecole nationale des Ponts et Chaussées: 132; 141

Ralph E. Ehrenberg Collection: 14

Evanston Northwestern Healthcare: 5

The Field Museum: 10 [A109831, Photo by Ron Testa]; 25 [A587T, Photo by Ron Testa]

Collection of Gerald F. Fitzgerald: 9, photo courtesy of The Newberry Library

Fonstad, Karen Wynn. *The Atlas of Middle-earth*. Boston: Houghton Mifflin, 1991: 145

Freer Gallery of Art, Smithsonian Institution, Washington, DC: 12, Gift of Charles Lang Freer, F1911.168 (detail)

© 2004 M. T. Gastner, C. R. Shalizi, and M. E. J. Newman: 115; 116

Harvard Map Collection, Harvard College Library: 16 [MA 5315.590 pf*]; 111, 112, 113A & B [LC G1019 .R19 1944 f]

Maps © Erwin Raisz. Reprinted with permission by Raisz Landform Maps, Brookline, MA.; 148 [MAP-LC G3701. A67 1940.C4]

Arthur Holzheimer Collection: 48, photo courtesy of The Newberry Library

Collection of Horyuji Temple. Photo: Askaen Co., Ltd: 36

By permission of the Houghton Library, Harvard University: 17 [*51–2477 PF (horz)]

Humboldt, A. von. "Carte des lignes isothermes." *Annales de Chimie et de Physique*, 2nd ser., vol. 5 (1817): 124

Art © Jasper Johns / Licensed by VAGA, New York, NY: 162, *Map*, 1963 (oil on canvas), Johns, Jasper (b.1930). Private Collection. Photo: The Bridgeman Art Library International

Collection of Kobe City Museum: 77

Collection of Modupe G. Labode: 109A & B

© Adrian Leskiw: 163

Library of Congress, Geography and Map Division: 11 [G9930 1908 .B8]; 37 [g9930 cto01173]; 83 [Hauslab folio 30, 11 & 12]; 89 [g3200 cto00725C]; 93 [g3300 aro03900]; 94 [g3881s ar300600]; 95 [g3700 cto01175]; 96 [g4050 cto00654]; 100 [g3701e cto00604]; 101 [g3861e cw0013200]; 103 [g3701gm gcto0008 ca000063]; 104 [g3701gm gcto0008 ca000083]; 105 [g3701gm gcto0008 ca000069]; 106 [g3701gm gcto0008 ca000044]; 107 [g3701e cto00802]; 122 [g9112g cto00753]; 136 [g3701gm/gcto0008/ ca000070]; 137 [g3701gm gcto0010 ca000010]; 147 [G3201.E65 1956 .H4 MLC]; 160 [g9930 cto01818]; 182 [G1796.A6 H3 1868 Vault]; 183 [G3802. H9A9 1776 .P6 Vault Cabinet]; 186 [G7821.F7 1890. S5 vault]

London Transport Museum, Copyright Transport for London: 24

MapQuest, Inc.: 2, The MapQuest logo is a registered trademark of Map-

INDEX

Note: Italicized page numbers indicate illustrations.

INDEX